D1174507

HANDBOOK OF ENVIRONMENTAL ENGINEERING

Volume 2
Solid Waste Processing and Resource Recovery

HANDBOOK OF ENVIRONMENTAL ENGINEERING

Volume 1: Air and Noise Pollution Control, 1979

Volume 2: Solid Waste Processing and Resource Recovery, 1980

Volume 3: Biological and Natural Control Processes

Volume 4: Solids Separation and Treatment

Volume 5: Physicochemical Technologies for Water and Wastewater Treatment

HANDBOOK OF ENVIRONMENTAL ENGINEERING

Volume 2

Solid Waste Processing and Resource Recovery

Edited by

Lawrence K. Wang

Department of Mechanical Engineering
Stevens Institute of Technology
Hoboken, New Jersey

and

Norman C. Pereira

Monsanto Company
St. Louis, Missouri

The HUMANA Press • Clifton, New Jersey

Library of Congress Cataloging in Publication Data

Main entry under title:
Solid waste processing and resource recovery.
(Handbook of environmental engineering; v. 2)
Includes bibliographical references and index.
1. Refuse and refuse disposal. 2. Recycling (Waste, etc.) I. Wang, Lawrence K. II. Pereira, Norman C.
TD170.H37 vol. 2 [TD791] 628.5′08s [628′.445]

ISBN 0-89603-008-3 79-91087

Printed in the United States of America

Preface

The past few years have seen the emergence of a growing, widespread desire in this country, and indeed everywhere, that positive actions be taken to restore the quality of our environment, and to protect it from the degrading effects of all forms of pollution—air, noise, solid waste, and water. Since pollution is a direct or indirect consequence of waste, if there is no waste, there can be no pollution, and the seemingly idealistic demand for "zero discharge" can be construed as a demand for zero waste. However, as long as there is waste, we can only attempt to abate the consequent pollution by converting it to a less noxious form. In those instances in which a particular type of pollution has been recognized, three major questions usually arise: 1, How serious is the pollution? 2, Is the technology to abate it available? and 3, Do the costs of abatement justify the degree of abatement achieved? The principal intention of this series of books is to help the reader to formulate answers to the last two of the above three questions.

The traditional approach of applying tried-and-true solutions to specific pollution problems has been a major factor contributing to the success of environmental engineering, and in large measure has accounted for the establishing of a "methodology of pollution control." However, realization of the complexity of current environmental problems, and understanding that, as time goes on, these issues will become even more complex and interrelated, renders it imperative that intelligent planning of pollution abatement systems be undertaken. Prerequisite to such planning is an understanding of the performance, potential, and limitations of the various methods of pollution abatement available for environmental engineering. In this series of books, we will review at a tutorial level a broad spectrum of engineering systems (processes, operations, and methods) currently being utilized, or of potential utility, for pollution abatement. We believe that the unification to be presented in these books is a logical step in the evolution of environmental engineering.

The treatment of the various engineering systems presented will show how an engineering formulation of the subject flows naturally from the fundamental principles and theory of chemistry, physics, and mathematics. This emphasis on fundamental science is based on the recognition that engineering practice has of necessity in recent years become more firmly based on scientific principles rather than depending so heavily on empirical accumulation of facts, as was earlier the case. It was not intended, though, to neglect empiricism where such data lead quickly to the most economic design; certain engineering systems are not readily amenable to fundamental scientific analysis, and in these instances we have resorted to less science in favor of more art and empiricism.

Since an engineer must understand science within a context of application, we first present the development of the scientific basis of a particular subject, followed by exposition of the pertinent design concepts and operations, and detailed explanations of their applications to environmental quality control or improvement. Throughout, methods of practical design calculation are illustrated by numerical examples. These examples clearly demonstrate how organized, analytical reasoning leads to the most direct and clear solutions. Wherever possible, pertinent cost data have been provided.

Our treatment of pollution-abatement engineering is offered in the belief that the trained engineer should more firmly understand fundamental principles, be more aware of the similarities and/or differences among many of the engineering systems, and exhibit greater flexibility and originality in the definition and innovative solution of environmental pollution problems. In short, the environmental engineer ought by conviction and practice be more readily adaptable to change and progress.

Coverage of the unusually broad field of environmental engineering has demanded an expertise that could only be provided through multiple authorship. Each author (or group of authors) was permitted to employ, within reasonable limits, the customary personal style in organizing and presenting a particular subject area, and consequently it has been difficult to treat all subject material in a homogeneous manner. Moreover, owing to limitations of space, some of the authors' favored topics could not be treated in great detail, and many less important topics had to be merely mentioned or commented on briefly. In addition, treatment of some well established operations, such as distillation and solvent extraction, has been totally omitted. All of the authors have provided an excellent list of references at the end of each chapter for the benefit of the interested reader. Each of the chapters is meant to be self-contained and consequently some mild repetition among the various texts was unavoidable. In each case, all errors of omission or repetition are the responsibility of the editors and not the individual authors. With the current trend toward metrication, the question of using a consistent system of units has been a problem. Wherever possible the authors have used the British system (fps) along with the metric equivalent (mks, cgs, or SIU) or vice versa. The authors sincerely hope that this inconsistency of units usage does not prove to be disruptive to the reader.

The series has been organized in five volumes:

I. Air and Noise Pollution Control
II. Solid Waste Processing and Resource Recovery
III. Biological and Natural Control Processes
IV. Solids Separation and Treatment
V. Physicochemical Technologies for Water and Wastewater Treatment

As can be seen from the above titles, no consideration is given to pollution by type of industry, or to the abatement of specific pollutants. Rather, the above categorization has been based on the three basic forms in which pollutants and waste are manifested: gas, solid, and liquid. In addition, noise pollution control is included in Volume I.

This Engineering Handbook is designed to serve as a basic text as well as a comprehensive reference book. We hope and expect it will prove of equal high value to advanced undergraduate or graduate students, to designers of pollution abatement systems, and to research workers. The editors welcome comments from readers in all these categories. It is our hope that these volumes will not only provide information on the various pollution abatement technologies, but will also serve as a basis for advanced study or specialized investigation of the theory and practice of the individual engineering systems covered.

The editors are pleased to acknowledge the encouragement and support received from their colleagues at the Environmental and Energy Systems Department of Calspan Corporation during the conceptual stages of this endeavor. We wish to thank the contributing authors for their time and effort, and for having borne patiently our numerous queries and comments. Finally, we are grateful to our respective families for their patience and understanding during some rather trying times.

LAWRENCE K. WANG
Hoboken, New Jersey
NORMAN C. PEREIRA
Charleston, Tennessee

Contributors

RAUL R. CARDENAS, JR. • *Department of Civil Engineering, Polytechnic Institute of New York, Brooklyn, New York*

JARIR S. DAJANI • *Department of Civil Engineering, Stanford University, Stanford, California*

EUGENE A. GLYSSON • *Department of Civil Engineering, University of Michigan, Ann Arbor, Michigan*

WALTER R. NIESSEN • *Camp Dresser and McKee, Boston, Massachusetts*

NORMAN C. PEREIRA • *Monsanto Company, St. Louis, Missouri*

P. MICHAEL TERLECKY, JR. • *Frontier Technical Associates, Inc., Buffalo, New York*

P. AARNE VESILIND • *Department of Civil Engineering, Duke University, Durham, North Carolina*

LAWRENCE K. WANG • *Department of Mechanical Engineering, Stevens Institute of Technology, Hoboken, New Jersey*

DENNIS WARNER • *Department of Civil Engineering, Duke University, Durham, North Carolina*

Contents

1

Introduction to Solid Waste Management

P. Aarne Vesilind

Department of Civil Engineering, Duke University, Durham, North Carolina

Norman C. Pereira

Monsanto Company, St. Louis, Missouri

I. INTRODUCTION

Much of our knowledge of ancient civilizations, as well as more contemporary times, has been derived from the solid waste discarded by the people of those times. In ancient cities, waste was seldom a problem—it simply accumulated until the city was leveled in a war and a new community was built on top of the rubble [1]. The debris of war, as well as the discards of bygone eras and civilizations, provide a rich history of lifestyle and cultural development [2]. For example, archaeologists are currently digging in the town dump of Williamsburg, Virginia, in order to obtain a more complete picture of life in colonial times.

During the dawn of civilization, solid waste was not a major concern since very little was discarded, and what trash there was could be easily disposed of by throwing it over the city wall. Solid waste became a problem only when populations increased and homes were built outside the city walls. Urban refuse could no longer be so easily disposed of.

In the middle ages, improper disposal of urban waste in overcrowded cities contributed significantly to the transmission of disease, and the solid waste problem can be indirectly blamed for the plagues that swept over Europe. In England, a street which was named Shambles because of its unsanitary condition, gave us a common descriptive term for cluttered and dirty areas.

The first real progress in urban sanitation, known as the "sanitary awakening," occurred around the turn of the century. In 1888, the British Parliament outlawed the throwing of refuse into roadside ditches and into the streets. At about the same time the germ theory was gaining general public acceptance, and sanitarians, epidemiologists, and public works engineers demonstrated how public health could be affected by the pollution and contamination of air and water. Disease transmission caused by improper solid waste disposal was also becoming well documented.

In the New World, the early settlers led mostly frugal rural lives and had very little to throw away. Moreover, the country was big, the population pressure small, and natural resources were seemingly without limit.

The modern United States, however, aided by the technological developments of the 20th century, is now in an era of affluence and we throw so much away that the people living "near the city walls" are complaining. The environmental impact of disposing of our affluence is creating increasingly serious problems. We discard about 3.5 lb/day (1.6 kg/day) of solid waste material per person. Furthermore, since today over 70% of the population is counted as urban, our per capita solid waste discards amount to staggering quantities in urban areas. In a typical year, Americans discard about 45 million tons (41 million metric tons) of paper, 5 million tons (4.5 million metric tons) of plastics, 50 billion metal cans, and 25 billion bottles. In addition to being voluminous, this "urban ore" also contains a great deal of nonrenewable natural resources, many of which are becoming increasingly scarce.

Consequently, solid waste management has emerged as a national problem of considerable magnitude. The push to decrease the waste quantities as well as to recover scarce resources from these wastes has in the last few years significantly broadened the scope of solid waste management.

In this chapter, the solid waste problem is first discussed with respect to its magnitude and importance, followed by a brief review of collection, disposal, and recovery practices. This is followed by a discussion on some special types of solid wastes, and legislative activities pertaining to solid waste management. The subjects of collection, disposal, and recovery are examined in greater detail in subsequent chapters of this volume.

II. SOLID WASTE GENERATION

Before embarking on a discussion of the sources and quantities of solid waste, it is necessary to first define some terms used for describing various types of solid waste.

Refuse is considered to be the total solid waste collected by the municipality, including commercial, institutional, and residential waste. *Garbage* is defined as strictly food waste, that is, organic material resulting from food preparation. *Rubbish* is a term given to almost everything else that is thrown out as domestic solid waste, such as tin cans, glass, paper, wood, cloth, grass and shrub clippings, etc. The third type of waste under refuse is *ashes*. There are cities and towns where a considerable number of people still use coal for heating homes and these ashes must be disposed of. Refuse is thus composed of many different components, and it is obvious that the terms "garbage truck" or "garbage man" are really misnomers.

The term *trash* has been introduced recently, and is defined as those large bulky items not usually placed in refuse cans, such as tree branches, big boxes, etc. Many towns have in fact a separate trash collection crew that specializes in such material.

A. Quantities of Municipal Refuse

Not many years ago, the quantities of solid waste produced were of little interest to anyone except the haulers, and truck and landfill equipment manufacturers. It was in their interest to promote greater quantities of waste.

In the late 1960s, a few cities suddenly found themselves in a crisis situation, with no landfill volume left, and nowhere else to dump the refuse. Concern for solid waste quantities began to be directed toward reducing the quantity of waste produced. On a national scale, little could be accomplished until some estimates were obtained on just how much waste there really was. The federal government, through the Environmental Protection Agency, began a study to develop the necessary data. A much-quoted study, *National Survey of Community Solid Waste Practices*, was published in 1968 [3]. Since that time, other data have been obtained and analyzed [4–9].

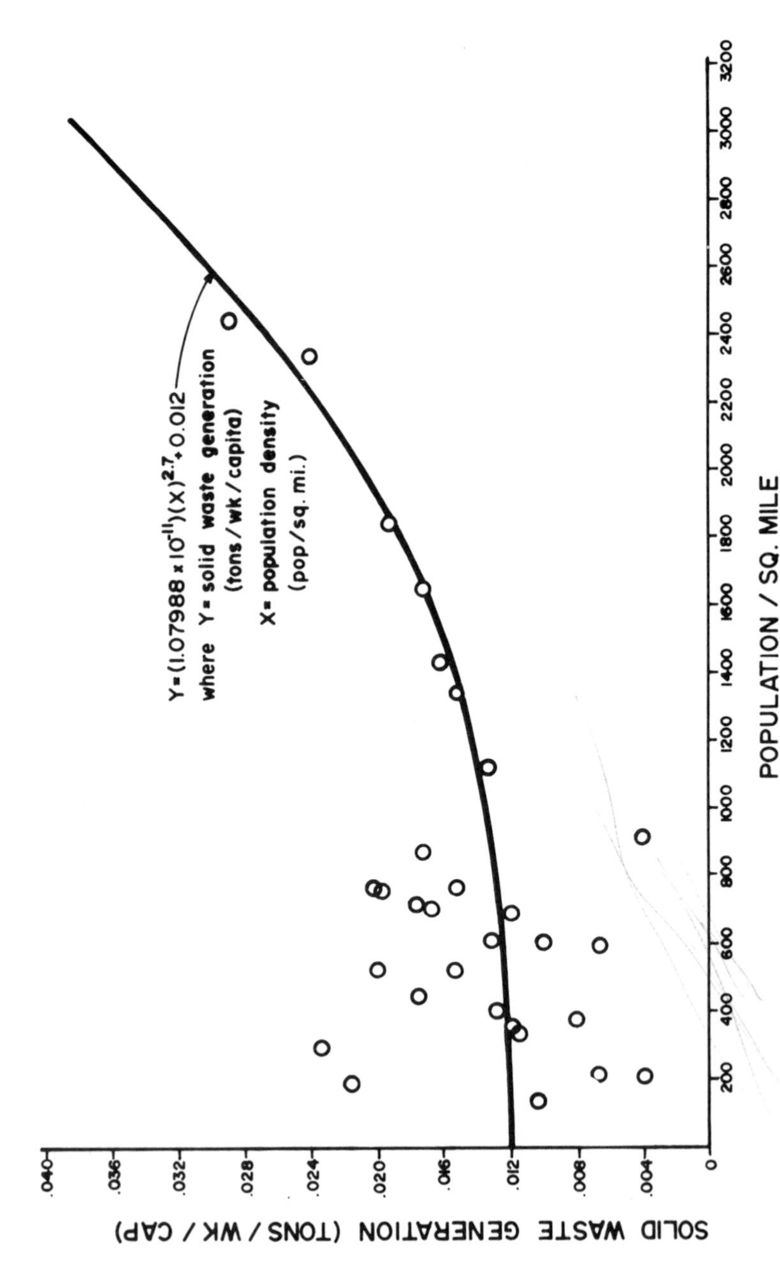

FIG. 1 Solid waste generation as a function of the degree of urbanization.

Table 1
Estimated per Capita Municipal Solid Waste Production[a] in the United States

Component	lb per person per day	kg per person per day
Household and commercial refuse	3.1	1.41
Litter and public refuse cans	0.05	0.02
Street sweepings	0.1	0.05
Trash	0.1	0.05
Other	0.15	0.07
Total	3.5	1.60

[a] These figures were derived from a number of sources and are considered to be a reasonable estimate for the year 1976.

Table 1 is a summary of the approximate quantities of municipal refuse generated annually in the United States. The figure of 3.5 lb per person per day (1.6 kg per person per day) has been disputed by several sources. Some original EPA documents placed the figure at closer to 5 lb per person per day (2.3 kg per person per day); other studies indicated a level less than 3 lb per person per day (1.3 kg per person per day).

In reality, these differences (or for that matter, Table 1) are of little importance, since every section of the country, every town or city, and indeed every neighborhood has its own waste generation rate. It is foolhardy to use national estimates to calculate the expected refuse in any given area. Rural people, for example, tend to use products longer before discarding them, and find ingenious ways of reusing waste products that urban dwellers might have thrown away. Lifestyle also clearly affects production of municipal refuse. That solid waste production is a function of population density (a gross measure of urbanization) can be seen from Fig. 1 [10].

Overall national solid waste figures therefore are of considerable importance when national policy is of concern, but of little use when local waste disposal systems are designed or evaluated. The only means to measure such quantities is still the reliable and irreplaceable refuse truck scale.

B. Composition of Municipal Refuse

Ordinarily, a community is not much interested in the components of its refuse. The weight and volume are sufficient for most collection

Table 2

Material Flow Estimates[a] of Residential and Commercial Postconsumer Net Solid Waste[b] Disposed of, by Material and Product Categories, 1973

Material	Product category, in millions of tons, as generated[c]							Totals			
								As-generated weight[c]		As-disposed weight[d]	
	News-papers, books, maga-zines	Con-tainers, packag-ing	Major house-hold appli-ances	Furni-ture, furnish-ings	Cloth-ing, foot-wear	Food prod-ucts	Other prod-ucts	Million tons	Percent	Million tons	Percent
Paper	11.3	23.3	—	tr.	tr.	—	9.6	44.2	32.8	53.4	39.6
Glass	—	12.1	tr.	tr.	—	—	1.1	13.2	9.9	13.4	10.3
Metals	—	6.5	1.9	0.1	tr.	—	4.0	12.5	9.3	12.7	9.9
Ferrous	—	5.6	1.7	tr.	—	—	3.7	11.0	8.2		
Aluminum	—	0.8	0.1	tr.	—	—	0.1	1.0	0.7		
Other nonferrous	—	0.1	0.1	tr.	—	—	0.2	0.4	0.3		
Plastics	tr.	3.1		0.1	0.2	—	1.6	5.0	3.7	5.6	4.1
			0.1								
Rubber and leather	—	tr.		tr.	0.5	—	3.0	3.6	2.7	3.7	2.7
Textiles	tr.	tr.	—	0.6	0.6	—	0.7	1.9	1.4	2.1	1.6
Wood	—	1.9	—	2.5	tr.	—	0.5	4.9	3.6	4.9	3.6
Total nonfood product waste	11.3	46.9	2.1	3.4	1.3		20.5	85.4	63.5	96.0	71.1
Food waste	—	—	—	—	—	22.4	—	22.4	16.6	18.0	13.3
Total product waste	11.3	46.9	2.1	3.4	1.3	22.4	20.5	107.8	80.1	114.0	84.4
Yard waste								25.0	18.5	19.0	14.1
Misc. inorganics								1.9	1.4	2.0	1.5
Total								134.8	100.0	134.8	100.0

[a] From Ref. 11.

[b] Net solid waste disposal defined as net residual material after accounting for recycled materials diverted from waste stream.

[c] "As-generated" weight basis refers to an assumed normal moisture content of material in its final use prior to discard, for example: paper at an "air-dry" 7 percent moisture, glass and metals at zero percent.

[d] "As-disposed" basis assumes moisture transfer among materials in collection and storage, but no net addition or loss of moisture for the agregate of materials.

and disposal analyses. Only when the recovery of resources from the solid waste is considered is its composition of concern.

There are basically two methods of determining the composition of refuse, that is, input and output analysis.

Input analysis (also known as materials flow analysis) requires consideration of the total production of each raw material (paper, aluminum, glass, metals, rubber, plastics, etc.), and then systematic tracing of each material through the production system to its final use in one or more of the product categories that appear in municipal (residential and commercial) solid waste. The use of this methodology [7] presupposes the availability of extensive data from either governmental or industrial sources. Furthermore, familiarity with the reporting conventions and production stages of the particular industry are mandatory if realistic estimates are to be expected.

Consider, for example, an input analysis on figures such as those provided by the American Paper Institute. Of the 61.2 million tons (55.5 million metric tons) of paper consumed in 1973, 6.15 million tons (5.6 million metric tons) ended up as industrial scrap, and 2.2 million tons (2.0 million metric tons) were diverted to the liquid waste stream or dissipated in use. The remaining 52.85 million tons (47.94 million metric tons) of paper were discarded as waste, of which 8.65 million (7.85 million metric tons) were recovered for reuse, resulting in a net tonnage of 44.2 million tons (40.10 million metric tons) entering the municipal solid waste stream.

The as-generated column of Table 2 summarizes the results of such input analysis calculations and provides a breakdown by material and product end use in 1973.

The major drawback to input analysis is that not all waste materials in municipal refuse pass through the production sector. Examples of these materials are food waste, yard waste, and miscellaneous inorganic materials constituting important waste fractions. This waste is not readily estimated by an input analysis, and this is where the second method of estimation, i.e., output analysis, becomes useful.

Output analysis is performed by examining or measuring the solid waste (discarded stream) by actually weighing and separating collected solid waste samples. Such analysis is usually subject to much variability as well as interpretation of the results, depending on the region and season in which the sampling was made. A major problem to be recognized in output analysis involves the moisture content of the various waste stream components. For example, paper is discarded on an air-dry basis of about 7% moisture, but absorbs moisture from other components such as garbage. The water thus absorbed yields a mistakenly high paper fraction unless the moisture content is stated as a separate component. Typical moisture levels in municipal refuse components are shown in Table 3. The tabulations report "as discarded" (before moisture transfer) and "as disposed" (after moisture transfer) percentages to reflect the moisture transfer in mixed municipal refuse.

The "as-generated" statistics for food waste, yard waste, and miscellaneous inorganics categories in Table 2 were obtained from an output analysis of data for 1973. The "as-disposed" column of Table 2 was calculated using the figures of Table 3. Adjusting for moisture does not materially change the total moisture content of the refuse mix, only the distribution of moisture among the refuse categories.

It should be clear that the input analysis method provides refuse quantity and composition estimates that are more descriptive of the overall or nationwide waste generation picture. However, such macroscale analysis is useless in gauging regional or local refuse generation patterns. These microscale data are better obtained via the output analysis method which more properly depicts the seasonal and compositional variation in municipal refuse. An examination of refuse compositional data from across the United States shows great variability as shown in Table 4. Some cities whose data are listed in Table 4 have been analyzed by different surveys, and the variation in results among these surveys for the same city is unsettling and sobering. This points to the need to exercise great caution and judgment in accepting

Table 3
Assumed Percentage Moisture Content of Materials in Municipal Solid Waste[a]

	As discarded	As disposed
Paper	7.0	23.1
Glass	0	3.0
Metals	0	5.5
Plastics	2.0	13.0
Rubber and leather	2.0	13.0
Textiles	7.0	20.0
Wood	15.0	15.0
Food	70.0	63.0
Yard	50.0	34.0
Miscellaneous	2.0	4.0
Weighted average	0.27	0.27

[a] From ref. 5.

such data and utilizing them in any decision making regarding disposal or resource recovery.

The solid waste compositions referred to thus far have been weight fractions of various waste categories. A further compositional analysis of solid waste provides an elemental breakdown of the various solid waste components. Table 5 provides such a breakdown and gives an interesting insight into the elemental makeup of municipal refuse components. The organic, polymeric wastes such as paper, plastics, rubber, and others have consistent values for carbon, hydrogen, and oxygen. Note the large amount of aluminum and iron available in the metal group. There is, of course, a wide range of heating values for some of the groups, with glass and metals having negligible heat content. It would seem profitable to separate and segregate these materials for their utility rather than to simply burn them. (See also the Tables in the Appendix of Chapter 3.)

C. Industrial Solid Waste [13]

The category of industrial solid waste is so broad and varied that nothing general can be said about it except that all industries generate residues they do not want and the sum total of these residue quantities is overwhelming. There is no generally agreed-upon convention for categorizing industry in order to identify the various types of industrial solid waste. It is, however, convenient to arrange industry into three broad categories, each having specific types of activity which, in turn, have specific waste generating characteristics. These three categories are (a) extractive industries, (b) basic industries, and (c) conversion and fabricating industries.

1. *Extractive Industries*

These are industries in which raw materials are taken from the earth and marketed in essentially their original state. Four extractive industries, namely mining, quarrying, agriculture, and logging, are of particular significance as generators of solid wastes. The solid wastes generated by such industries are simply components or products of the earth. These industries concentrate wastes at specific locations; the wastes are normal products of the earth and its living things, but differ in nature, some being inert materials and others are biodegradable organic matter.

a. *Mining.* The mining industry produces vast amounts of solid waste, more than one billion tons per year. The main reason for this is

Table 4
Composition of Municipal Solid Waste in the United States,[a] %

Study	Food	Yard	Misc.	Glass	Metal	Est. ferrous	Ferrous	Aluminum	Nonferrous	Paper	Plastics	Textiles	Wood	Leather, rubber
Oceanside, N.Y.	9.6	33.3		9.7	8.0	7.0				32.8		3.0	1.2	
	10.2	19.0		9.5	8.2	7.2				39.8		3.3	6.6	
Cincinnati, Ohio	28.9	6.4		7.5	8.7	7.7				42.0	1.6	1.4	2.7	
Oceanside, N.Y.	16.7	0.3		11.9	10.6	9.6				53.3		2.2	1.5	
Flint, Mich.	29.1	26.7		12.7	14.5	13.5				13.0		0.3	1.0	
Johnson City, Tenn.	21.1	0.9	0.6	7.0	7.5	6.5				59.8	0.9		0.3	0.6
San Diego, Cal.	0.8	21.1		8.3	7.7	6.7				46.1	0.3	3.5	6.4	4.7
Berkeley, Cal.	12.5	12.5	7.1	11.3	8.7	7.7				44.6	1.9	1.1		0.3
Raleigh, N.C.	31.8	8.4		11.9	9.2	8.2				38.9				
Santa Clara Co., Cal.	2.1	34.5	0.5	10.9	7.4	6.4				36.2	1.5	1.3		1.1
Flint, Mich.	36.0	0.3	0.7	23.2	14.5	13.5				21.1		0.8	0.8	

Weber Co., Utah	8.5	4.2	5.9	4.6	8.4	7.4				61.8		2.0	2.2	
Johnson City, Tenn.	34.6	2.3	0.2	9.0	10.4	9.4				34.9	3.4	2.0	0.8	2.4
New Orleans, La.	18.9	9.2		16.2	12.2	11.2				39.4		2.6		
Alexandria, Va.	7.5	9.5	3.4	7.5	8.2	7.2				55.3		3.7	1.7	
Atlanta, Ga.	12.3	1.6	3.4	10.3	8.6	7.6				58.6		1.8	0.4	
	17.5	2.8	3.4	6.5	8.8	7.8				53.2		2.0	3.2	
New Orleans, La.	18.9	9.2	1.5	16.2	12.2	11.2				39.4		2.6		
				13.0			8.8	1.0			3.8			
Tampa, Fla.	9.1	41.5	6.1	6.0			4.8	0.8	0.3	24.1	2.4	2.8	1.5	0.6
Wilmington, Del.	16.5	9.0	2.9	14.7			5.7	0.6	0.3	33.7	3.3	8.9	2.5	1.9
San Diego, Cal.	0.8	21.1		8.3	7.6	6.6				46.2	0.3	3.5	7.5	4.7
Madison, Wis.	15.3	13.8	9.0	10.1	6.7	5.7				42.4		1.6	1.1	
Purdue University	12.0	12.0	15.4	6.0	8.0	7.0				42.0	0.7	0.6	2.4	0.9
Kaiser, E. R.	8.4	6.9	12.0	7.7	6.9	5.9				53.5	0.8	0.8	2.3	0.8
Little, A. D.	16.6	12.6	1.7	8.5	8.7	7.7				44.2	1.2	2.3	2.5	1.7
Battelle Inst.	14.0	5.0	3.0	9.0			7.5	1.0	0.5	55.0	1.0		4.0	
Averages	14.6	12.5	4.5	10.3	9.2	8.2	6.7	0.9	0.4	52.7	1.7	2.4	2.5	1.8

[a] From Ref. 12.

Table 5
Estimated Elemental Analysis of Municipal Refuse Components,[a] %

Category	C	H	O	N	Ash	S	Fe	Al	Cu	Zn	Pb	Sn	P[b]	Cl	Se	Fixed carbon[c]	Heating value[c] Btu/lb	Heating value[c] kJ/kg	Expected range of composition[c] wt.%
Metal	4.5	0.6	4.3	0.05	90.5	0.01	77.3	20.1	2.0	—	0.02	0.6	0.03	—	—	0.5	740	1,720	7–10
Paper	45.4	6.1	42.1	0.3	6.0	0.12	—	—	—	—	—	—	—	—	Trace	11.3	7,930	18,400	37–60
Plastics	59.8	8.3	19.0	1.0	11.6	0.3	—	—	—	—	—	—	0.01	6.0	—	5.1	11,500	27,700	1–3
Leather and rubber							—	—	—	2.0	—	—	—	—	—	6.4	10,175	22,700	0.5–5
Textiles	46.2	6.4	41.8	2.2	3.2	0.2	—	—	—	—	—	—	0.03	—	—	3.9	8,030	18,700	1–5
Wood	48.3	6.0	42.4	0.3	2.9	0.11	—	—	—	—	—	—	0.05	—	—	14.1	8,400	19,500	1–4
Food wastes	41.7	5.8	27.6	2.8	21.9	0.25	—	—	—	—	—	—	0.24	—	—	5.3	8,540	19,900	12–18
Yard wastes	49.2	6.5	36.1	2.9	5.0	0.35	—	—	—	—	—	—	0.04	—	—	19.3	7,300	17,000	4–10
Glass	0.52	0.07	0.36	0.03	99.02	—	—	—	—	—	—	—	—	—	—	0.4	65	151	6–12
Miscellaneous	13.0[d]	2.0[d]	12.0[d]	3.0[d]	70.0	—	—	—	—	—	—	—	—	—	—	7.5	3,500	8,140	—

[a] From Ref. 5.
[b] Excludes phosphorus in $CaPO_4$.
[c] Dry basis.
[d] Estimated (varies widely).

that the ores of most desirable metals are very lean in the metals. For example, copper ores contain less than 1% copper, and even the richest iron ore yields only 30% iron. Although some 80 mineral industries generate quantities of solid waste, only eight major industries are responsible for 80% of the total. These are copper, which contributes almost 41%, followed by iron and steel, bituminous coal, phosphate rock, lead, zinc, alumina, and anthracite industries. Table 6 shows approximately the tons of solid waste that are produced merely during the mining operation. The major waste problem is the local creation of large piles of inert material that is unsightly and destructive of the land resource it occupies, and that may contribute to water pollution by leaching over a period of time. Mining solid waste amounts to over a billion tons each year, and in addition there are some 23 billion tons of accumulated waste of the past 30 years spread across the nation.

b. *Quarrying.* Open pit, or strip mining and the quarrying of glass sand, stone, and sand and gravel are typical of this type of extractive industry. The solid waste problem is similar to that of mining except that the quantities involved are much smaller (approximately 0.5–5% of mining wastes).

Table 6
Waste Produced by Processing Minerals and Fuels,[a] 1000 tons

Industry	Mine waste[b]	Mill tailings	Washing-plant rejects	Slag	Processing-plant wastes	Total
Alumina					5,350	5,350
Anthracite			2,000[c]			2,000
Bituminous coal	12,800		86,800			99,600
Copper	286,600	170,500		5,200		462,300
Iron and steel	117,599	100,589		14,689	1,000	233,877
Lead and zinc	2,500	17,811	970			21,281
Phosphate rock	72		54,823	4,030	9,383	68,308
Other	n.a.[d]	n.a.	n.a.	n.a.		229,284[e]
Total	419,571	288,900	144,593	23,919	40,233	1,122,000

[a] From Ref. 14.
[b] Other than surface mine overburden.
[c] Includes negligible quantities of mine waste.
[d] Not available.
[e] Wastes of remaining mineral mining and processing industries, 20% of total wastes generated.

c. *Agriculture.* The solid waste generating aspects of this category include both plant and animal wastes. Plant residues consist of crop and orchard residues, and forest trash. The tonnage of plant residues left on the farms exceeds by far the tonnage of crops taken to market. These wastes consist of straw, stubble, leaves, hulls, vines, tree limbs, and similar trash. Most of these wastes are burned to eliminate plant diseases and pests; a small part is used for mulch, ensilage, bedding for animals, etc.

Livestock and poultry wastes (mostly manure) are of considerable concern because of the huge amounts produced and the concentration of animals at central production points. A feedlot handling 10,000 cattle accumulates about 300 tons (270 metric tons) of solid waste per day; similarly, a poultry operation with 270,000 hens creates about 40 tons (36 metric tons) of manure daily.

Table 7 gives estimates of agricultural solid waste production rates. Animal wastes are beneficial to the soil if dispersed or allowed to decompose under controlled conditions. However, when the wastes are

Table 7
Estimated Agricultural Solid Waste Production Rates[a]

Category	Annual Waste Production Rate
Manure	
Chickens (fryers)	6.4 tons/1000 birds
Hens	55.0 tons/1000 birds
Sheep	2.0 tons/head
Hogs	3.2 tons/head
Horses	12.0 tons/head
Beef cattle (feedlot)	10.9 tons/head
Dairy cattle	14.6 tons/head
Fruits and nuts	
Grapes, peaches	2.4 tons/acre[b]
Apples, pears	2.25 tons/acre
Plums, prunes	1.5 tons/acre
Field and row crops	
Corn	4.5 tons/acre
Cauliflower, lettuce, broccoli	4.0 tons/acre
Tomatoes, beets, cabbage, potatoes, peanuts	3.0 tons/acre
Cotton, beans, peas, soybeans	2.0 tons/acre
Oats, wheat, barley, rye	1.5 tons/acre
Flax, alfalfa	0.8 tons/acre
Tobacco, sugarcane	0.5 tons/acre

[a] Partly adapted from Ref. 15.
[b] 1 ton/acre = 2.72 metric tons/hectare.

produced in one place in large quantities they are likely to decompose anaerobically and become breeding grounds for insects and vermin and contribute high BOD (biochemical oxygen demand) leachates to surface and underground waters.

Agricultural solid waste production is estimated at approximately 2 billion tons per year.

d. *Logging.* In an average year approximately 25 million tons of logging debris are left in the forest; that is, about one ton of debris for every 1000 board feet of logs harvested. This debris serves as a harbor for insects and tree diseases, and is also a severe fire hazard.

2. *Basic Industries*

Basic industries utilize the products of the extractive industries as raw materials and convert them into refined materials from which other industries produce consumer goods. Examples of products from this industrial category are: metal sheets, tubes, and wires, coke, industrial chemicals, paper, lumber and plywood, plastics, glass, and synthetic fabrics. The solid wastes generated by these industries are more diverse in composition. Only a fraction of the refined materials appears as solid waste, and much of the solid waste can be recycled directly within this industrial category. The eight basic industries most prominent as solid waste generators are discussed below.

a. *Metals.* Mined ores are shipped directly to a plant where the metal is extracted and refined. This process results in significant solid wastes, e.g., 20% of steel ingot production results in slag. Similarly, aluminum and copper each produce about 5 million tons (4.5 million metric tons) of inert waste annually. A subsequent stage at which ingots are formed into shapes, yields smaller amounts of solid wastes. These are mostly trimmings from the product itself and residues from other refined products associated with the process.

b. *Chemicals.* This industry produces the greatest variety of solid wastes, ranging from slurries to dry solid cake, from combustible organic tars to inert inorganic salts, from toxic materials such as chromates to common salt. However, they all stem from three sources: unreacted raw materials, contaminants in the raw materials, or by-products of chemical reactions. The solid wastes of organic chemical production are primarily tars formed as undesired by-products of chemical reactions. Inorganic solid wastes result mostly from unreacted raw materials or contaminants in the raw material or ore. A recent study has estimated that 33 representative chemicals (16 inorganic, 17 organic) produced about 57 million tons (52 million metric tons) of solid waste in 1972 [16].

c. *Paper.* The production of paper and paperboard results in solid wastes of two types: residues of materials used in the process, and residues of the product itself. The first type consists of items such as treebark, wood fiber, paper pulp, and inert filler; the second type of trimmings and wastage. Pulp and paper mills produce about 45 million tons (41 million metric tons) per year of solid waste.

d. *Plastics.* Basic industries that convert basic chemicals into plastic sheets or other forms used by fabricators, generate plastic wastes that are mostly trimmings or off-specification materials. Estimates of such wastes are quite difficult to obtain.

e. *Glass.* The major portion of solid waste generated in the basic glass industry is recycled within the industry. This waste usually consists of cullet (glass fragments) from breakage and trimming, off-grade material, and slag from the purification of glass sand.

f. *Textiles.* The three major basic textile industries are cotton, linen, and wool. Cotton textile mills generate wastes such as strapping and burlap used in baling, plus comber wastes and fibers damaged in storage and shipment, which are recycled. Linen textile mills have similar wastes related to flax. Preparation of wool results in wastes such as fiber, twine, wool fat, and dirt. In addition, residues from spinning, weaving, and trimming operations also add to the total solid waste picture.

g. *Wood Products.* Tree bark, sawdust, shavings, splintered wood, and trimmings (about 10% of the original tree) constitute the major solid waste items in the lumber industry. In addition, plywood trimmings, knots, and ashes may be added to the overall solid waste. This amounts to about 55 million tons (50 million metric tons) a year.

h. *Power.* Solid wastes commonly associated with coal-burning power stations are fly ash, bottom ash, and boiler slag. Large volumes of coal ash are produced at generating stations. For example, a 1000-MW station can generate on the order of 288,000 dry tons/year (261,000 metric tons/year) of coal ash, which in 20 years would occupy 2530 acre-ft (312 hectare-m). Statistics indicate that in 1969 the U.S. power industry produced 21, 7.6, and 2.9 million tons (19, 6.9, and 2.6 metric tons) of fly ash, bottom ash, and slag, respectively.

In the case of oil-burning stations, an ash is produced as a result of sediment and residue. This ash is much less in quantity than that produced by a coal-burning station, e.g., a 1000-MW station would produce on the order of 21,900 tons/year (19,900 metric tons/year) of ash, or about 9 acre-ft/yr (1.11 hectare-m).

3. *Manufacturing Industries*

This industrial sector consists of the conversion and fabricating industries. They convert the products of the basic industries into the goods that are consumed by the general populace and that characterize our economy and standard of living. The greatest fraction of solid waste generated by these industries consists of residues of the basic materials they utilize. Moreover, unlike basic industries that recycle much of their own trimmings and rejects, the manufacturing industries can seldom utilize such residual wastes and must instead depend on some secondary industry to reclaim these wastes. Furthermore, the measures necessary to move products from one sector of the industry to another impose a secondary solid waste burden (in the form of packaging and containers) on the receiver.

The range of manufacturing industries, and the consequent range of solid wastes produced, is so wide that only a few representative categories are discussed here.

a. *Packaging.* Aluminum, steel, glass, plastics, cardboard, corrugated paperboard, and plastic and paper laminates are among the materials used by the packaging industry. The solid waste stream from this industry depends on the type of material and range of activities at the plant, and constitutes a fraction of the material utilized.

b. *Automotive.* There is probably no other sector of industry that compares with automobile assembly in the amount of solid wastes it inherits from its relationship with other industrial activities. By far the greatest component of automobile assembly plant waste is the discarded packaging and shipping materials associated with the delivery from other industrial suppliers of such components as tires, batteries, generators, carburetors, wheels, bumpers, hub caps, and dozens of other items that make up an automobile. Painting and upholstering also adds container and material residues to the solid waste stream.

c. *Electronics.* Solid waste associated with the production of electronic components is relatively small when compared with the waste from other manufacturing industries, and consists mostly of plastics, glass, and wire and sheet metal scrap. Packaging materials utilized in shipping electronic components is the major waste generating problem.

d. *Paper Products.* Conversion of paper into various products ranging from facial and toilet tissue, paper towels and napkins, to books, magazines, and newspapers, results in high-quality solid waste in the form of paper trimmings and filled paper residues that can be recycled through a secondary industry.

e. *Hardware.* This is the metals industry, which produces machines, tools, utensils, and gadgets used by all classes of industry and by the public. The solid waste from such an industry mostly consists of residues from trimming and sizing of tubes, plates, and structural shapes, from boring and machining of metals, and miscellaneous residues from casting and forging processes.

f. *Soft Goods.* This industry produces commercial commodities from leather, textiles, and plastics. The residues from the material processed are the major item of solid waste generated.

g. *Food Processing.* Solid wastes from the food processing industry are predominantly fractions of the material processed. Fruit and vegetable processing industries produce moderate amounts of solid waste, which are, however, difficult to handle. Peels, skins, pulp, and seeds may be suspended in billions of gallons of saline, alkaline, or acidic water. Of the 14 million tons (12.7 million metric tons) produced in 1970 only about 4.5 million tons (4.1 million metric tons) required disposal, the rest being used for animal feed. Solid wastes from the dairy industry are quite low (220,000 tons, 198,000 metric tons in 1970) but require costly handling. Milk solids have high pollution potential. They are valuable as animal feed, feed supplements, and as basic ingredients in the growth medium for microorganisms used for pharmaceuticals. The meat processing industry produces few solid wastes requiring disposal. In 1970, over 19 million tons (17 million metric tons) of solid waste was produced, of which almost all was utilized in the manufacture of soap, leather, glue, gelatin, and animal feed. Other animal parts yield fatty acids, oils, greases, and glycerine. Bones are processed to proteins, fats, and bone meal.

h. *Construction.* Typical residues are obtained from lumber, plasterboard, wire, paper, cement bags, sheet metal scrap, etc. However, the most significant solid wastes result from peripheral activities such as demolition debris, breaking up of pavement, and site preparation, which produce large volumes of rock, tile, broken concrete, brick, tree stumps, piling, and miscellaneous rubble.

4. *Discussion*

Although the above list of industries discusses the extractive, basic, and manufacturing industries as discrete entities, it should be borne in mind that these categories may be only sectors of a single large industry which owns and operates every aspect from raw material to the finished consumer product. It is furthermore important to understand the relationship of industrial practice to society's total solid waste problem.

Each sector of the industrial system discards what it cannot pass to the next in line. Thus the product of the extractive industry is the raw material of the basic industry and the product of the basic industry becomes raw material for the manufacturing industry. Finally the entire output of the manufacturing industry becomes the solid waste of society (this is an assumption implicit in the input analysis method described earlier). Since the overall waste generation of the commercial and municipal sectors of society is a function of what flows from industry into these sectors, and since the rate at which the industrial system operates determines the nation's economy, it is fair to conclude that the entire industrial effort is inadvertently dedicated to the production of solid waste.

It is logical to expect industry to play a significant role in the control of its wastes and, consequently, of the total waste load on society. Unfortunately, the free market nature of our economic system makes it necessary that in the manufacture of consumer goods, such factors as consumer acceptance, shelf-life, the ability to resist damage, ease of manufacture, novelty and obsolescence, and other concerns be of importance, and that absolutely no thought be given to the final disposal of the product.

Responsibility for solid waste control and for resource conservation requires that during the design of consumer products serious consideration be given to specifications such as degradability of synthetic materials, ease of dismantling for segregating component materials, minimum number of types of materials, and other materials recovery or disposal objectives. Furthermore, it is important that society understand the interrelationships within industry and the dependence of our level of affluence upon industrial activity. An informed public plays an especially significant role in controlling industrial solid wastes because public opinion is respected by industrialists and public attitude forms the basis for public policy.

D. Solid Waste from Air and Water Pollution Controls [17]

The Federal Clean Air Act of 1970 and the Federal Water Pollution Control Amendments of 1972 require reduced discharges of pollutants to air and water. Although increased pollution abatement is beneficial, pollution controls generate solid residues once the pollutants to be controlled are trapped.

The major air pollutants capable of producing solid waste residues are primarily sulfur oxides and particulates. Other air pollutants are inconsequential in the production of solid waste residues. Major water pollutants capable of producing solid residues are more difficult to determine because of the great diversity of substances released to receiving waters. Consequently, suspended solids and BOD_5 are used as indicators of solid residue quantities resulting from water pollution control activities.

Table 8 lists the total solid wastes from air and water pollution control for 1971 and 1985 (projected) broken down by the pollutant from which they were derived. Particulates, sulfur oxides, and miscellaneous air pollutants are identified, whereas all water pollutants are grouped together. For the years 1971 through 1985, these solid wastes are expected to increase from about 60 million metric tons in 1971 to over 230 million metric tons in 1985. Mine tailing wastes are excluded from the totals since their quantities are very large. They are disposed of in the general locale of the mines and are normally not a part of the urban disposal system.

Pollution control of power plants plays a major role in contributing solid waste. These solid wastes are projected to increase from 25.6 million metric tons in 1971 to over 154 million metric tons in 1985. Power plants contributed 42% of all solid wastes from pollution control activities in 1971, and this should increase to 67% by 1985. The major factor in this increase is the use of limestone scrubbers to control sulfur oxides.

Aside from power plants, other major industrial sectors generating solid wastes from air and water pollution control are nonferrous smelting and refining, chemicals and allied products, paper and allied products, cement and dry products, steel and hazardous wastes. Together with power plants, these sectors contributed 90% of all solid wastes from air and water pollution control in 1971 and are expected to contribute up to 93% in 1985. Most of the 29.8 million metric tons of solid hazardous wastes generated in 1985 will be from uranium mine tailings; these wastes may increase rapidly because of the increasing role of nuclear power plants.

Tables 9 and 10 list the total solid wastes from air and water pollution control from 1971 and 1985 (projected), respectively, identified by the air and water treatments that generate the residues.

Table 8

Air and Water Pollutants whose Control Generates Solid Waste Residues,[a] 1000 metric tons[b]

Industry	1971 Residues				1985 Residues			
	Particulates	Sulfur oxides	Other air	Water	Particulates	Sulfur oxides	Other air	Water
Feedlots	0	0	0	1,020	0	0	0	1,150
Mining	460	0	0	8×10^5	1,410	0	0	1.3×10^6
Meat and dairy	0	0	0	60	0	0	0	220
Fruits and vegetables	0	0	0	50	0	0	0	100
Grain mills	500	0	0	0	2,700	0	0	0
Paper and allied products	2,440	0	0	4,470	5,800	0	0	9,540
Chemicals and allied	3,140	470	20	6,050	5,860	1,300	90	10,150
Petroleum refining	200	0	0	570	420	0	0	1,070
Cement and clay	3,690	0	0	90	2,550	0	0	90
Blast furnaces and steel	300	0	170	2,550	320	0	310	3,970
Iron foundries	130	0	0	0	650	0	0	0
Nonferrous metals	800	0	0	5,190	2,350	9,070	0	16,100
Power plant	25,600	0	0	0	62,530	91,530	0	20
Sewerage systems	0	0	0	1,540	0	0	0	2,830
Other sources	370	0	0	350	1,770	0	0	260
Totals	37,170	470	190	21,940	84,950	101,900	400	45,500
Hazardous waste streams	1,030				29,770			
Percent of total	62	0.7	0.3	37	36	44	0.2	20

[a] From Ref. 17.
[b] Dry weight.
[c] Excluding mining.

Table 9
Pollution Treatment Processes Contributing to Solid Waste Generation,[a,b] 1971, 1000 metric tons

Industry	Air treatment				Water treatment				Industry, total
			Wet			Secondary			
	Mechanical	Electrostatic	Water	Chemical	Primary	Chemical	Biological	Advanced	
Feedlots	0	0	0	0	250	0	770	0	1,020
Mining	0	0	0	0	800	1,260	0	0	800
Meat and dairy	0	0	0	0	20	20	20	0	60
Fruits and vegetables	0	0	0	0	40	10	1	2	50
Grain mills	500	0	0	0	0	0	0	0	500
Paper and allied products	0	1,450	990[c]	0	320	370	1,920	1,860	6,910
Chemicals and allied products	890	730	1,540	470	4,480	1,570	0	0	9,680
Petroleum refinery	0	200	0	0	550	0	0	20	770
Cement and clay	1,120	460	2,100	0	0	80	20	0	3,780
Blast furnaces and steel	30	80	200	170	0	2,540	0	0	3,020
Iron foundries	80	0	50	0	0	0	0	0	130
Nonferrous metals	800	0	0	0	5,190	100	0	0	5,990
Power plant	0	0	25,600	0	0	0	0	0	25,600
Sewerage systems	0	0	0	0	370	580	590	0	1,540
Other sources	110	200	60	0	90	50	210	0	720
Total[d]	3,500	3,080	29,400	1,780	12,470	5,920	2,950	1,800	59,770
Percent of total	65	5	49	3	20	10	5	3	

[a] From Ref. 17.
[b] Does not include hazardous wastes.
[c] Unless otherwise shown, effluent from wet air pollution controls is assumed to be handled by sedimentation.
[d] Excluding mining.

Table 10
Pollution Treatment Processes Contributing to Solid Waste Generation,[a,b] 1985 projected, 1000 metric tons

Industry	Air treatment				Water treatment				
			Wet			Secondary			
	Mechanical	Electrostatic	Water	Chemical	Primary	Chemical	Biological	Advanced	Industry, total
Feedlots	0	0	0	0	310	0	840	0	1,150
Mining	0	0	0	0	1,300	(32,690)	0	0	1,300
Meat and dairy	0	0	0	0	10	70	20	120	220
Fruits and vegetables	0	0	0	0	90	20	0	30	140
Grain mills	2,700	0	0	0	0	0	0	0	2,700
Paper and allied products	0	4,900	900[c]	0	310	5,120	1,710	2,400	15,340
Chemicals and allied products	1,760	1,010	3,180	1,300	7,850	2,300	0	0	17,400
Petroleum refinery	0	420	0	0	1,000	0	0	70	1,490
Cement and clay	1,770	570	210	0	0	90	0	0	2,540
Blast furnaces and steel	90	400	240	310	0	3,560	0	0	4,600
Iron foundries	350	0	300	0	0	0	0	0	650
Nonferrous metals	1,130	0	150	10,590	15,610	40	0	0	27,520
Power plant	3	0	62,530	91,530	0	0	0	20	154,080
Sewerage systems	0	0	0	0	510	11,600	1,160	0	2,830
Other sources	470	1,080	270		70	70	340	0	2,300
Total[d]	8,270	8,380	67,780	103,730	25,760	12,430	4,070	2,640	233,060
Percent of total	4	4			11	5	2	1	

[a] From Ref. 17.
[b] Does not include hazardous wastes.
[c] Unless otherwise shown, effluent from wet air pollution controls is assumed to be handled by sedimentation.
[d] Excluding mining.

The air pollution treatments are categorized into mechanical, electrostatic, water scrubbing, and wet chemical treatment. Mechanical air treatment includes those treatments that use dry physical removal mechanisms such as cyclones and baghouses, whereas electrostatic treatment includes mainly precipitators. Water treatment scrubbers use pure water to scrub solids from flue gases, and wet chemical treatments include methods that use chemicals in the scrubber water in order to capture gaseous pollutants. The most significant wet chemical method is limestone scrubbing, which reacts limestone with sulfur oxides in order to form calcium sulfate and sulfite, which may be precipitated from water. Although these treatments require further water treatment to remove the solids created, the residues produced are categorized with the air treatment residues.

Water pollution treatments are categorized into primary, chemical secondary, biological secondary, and advanced. Primary treatments include physical systems, such as screening, flotation, and sedimentation. Chemical secondary treatments react chemicals with the pollutants in order to cause them to precipitate from solution, whereas biological treatments utilize bacteria in order to decompose organic compounds in water. Advanced treatments include methods that are efficient in the removal of dissolved solids, such as ion exchange, reverse osmosis, etc.

In 1971, 49% of all solid residues were produced by water scrubbing systems, 20% by primary water treatment, and 10% by chemical water treatments. By 1985, chemical scrubbing will have increased to 44% of all solid wastes from air and water pollution control reducing water scrubbing and primary water treatment to 29 and 11%, respectively. The main source of this increase in the contribution of chemical scrubbing is the projected application of limestone scrubbing of electric power plant sulfur oxide emissions.

E. The Total Solid Waste Picture

Table 11 is based on estimates of solid waste generation from the various sources discussed in the previous subsections for 1978. These calculations are at best approximate, but they do point to the fact that the annual solid waste load generated in the United States is of the order of 4 billion tons (about a 100 lb/person/day). The bulk of the solid waste, namely agricultural and mining wastes, is generated in areas far removed from urban centers and is never moved very far. The remaining 500 million tons or so are the main concern from the

Table 11
Annual Solid Wastes Generated in the United States[a]

Origin	Waste per year: Million tons	Waste per year: Million metric tons
Residential, commercial, institutional	140	125
Basic and manufacturing industries	380	345
Agriculture	2000	1815
Mining	1200	1090
Pollution control (air and water)	125	115
Total	3845	3490

[a] Estimated for 1978.

point of view of collection, storage, transportation, and disposal. At a modest overall cost of $10.00 per ton, the annual bill for disposal could reach around $5 billion. To make matters worse, solid waste production seems to be on an exponential curve. The EPA has predicted that by 1980 the total municipal solid waste alone will approach 175 million tons, and that by 1990 this figure will be closer to 225 million tons [12]. This is a gloomy but reasonable projection if the current trend is maintained, and there does not seem to be much room for optimism.

Two kinds of restrictions are expected to help decelerate the exponential waste generation rates.

The first kind is restrictive legislation. The most famous example of this type of legislation is the so-called Oregon Bottle Law (now adopted by at least four other States) which prohibits the use of pop-top cans and discourages the use of nonreturnable beer and soft drink containers. Since the beverage containers under such a system would be used many times over, the total amount of solid waste produced is reduced, but only slightly. As seen from Table 4, glass solid waste averages to about 10% by weight, so that even a 50% reduction will yield a reduction of only about 5% in the total solid waste. Such a law, however, has been successful in reducing the amount of beverage container litter.

Other types of restrictive legislation such as tax on packaging, or a penny-a-pound disposal tax, or a tax prorated according to the ease of disposal, are still under consideration.

The second restriction that will affect waste generation rates is the shortage of raw materials. For example, if the bauxite-producing nations form a cartel and drive the price of bauxite up, the price of aluminum will go up to perhaps where it will become a fairly scarce metal relative to today's supply. This would mean that an aluminum beer can could become very expensive; too expensive to throw away, and hence we will have an economic restriction on the kind of waste that we will be producing. Given our finite natural resources, it seems quite obvious that such economic restrictions (which may result in legislative restrictions) are in the offing and that the total amount of solid waste produced will show a leveling off and perhaps even a drop in the future.

In the next several decades recycling will become a necessity as reprocessing becomes less costly than obtaining the raw materials from increasingly distant and inaccessible places and from high labor cost sources.

III. SOLID WASTE COLLECTION AND TRANSPORTATION

It is somewhat paradoxical that, whereas the prime concern with solid waste is usually its final disposition, the major costs associated with solid waste result from its collection and transportation. In most communities about 80% of the cost of overall solid waste management is in collection. A community of 500,000 people, for example, might spend 4 million dollars each year for solid waste management, about 3.5 million of which is for collection alone. On the national scale, the collection of residential and commercial solid wastes costs over 4 billion dollars, which represents the fifth largest public expense after education, highways, welfare, and public safety.

This section is a discussion of solid waste collection with regard to collection agencies, residential and commercial collection practices, and finally some recent innovations in solid waste collection.

A. Collection Agencies

There are many different organizational arrangements for solid waste collection, which can be classified according to recipient, provider, arranger, and type. No less than 16 different arrangements can be found, although only four are important in terms of use: municipal, contract, private, and franchise.

Under municipal collection, city employees collect the refuse with city equipment under the supervision of a governmental department such as the department of public works. In contract collection, local government hires a private firm to collect refuse in a given area and pays the firm directly. Under this approach the contractor owns the equipment, but must meet all performance criteria established by the contract. Under private collection, a private firm makes arrangements with the customer for pickup, and does not have an exclusive territory assigned to it by the local government. In a franchise arrangement, local government awards to a private firm the exclusive right to collect refuse in a specified area, and the firm is paid directly by its customers.

It has been argued that municipal systems should cost less since they do not earn a profit or pay taxes, and pay lower interest rates when they borrow. Furthermore, the official in charge of the collection agency has as his basic motivation the maintenance of sanitary conditions in the area where refuse is collected since public scrutiny is always directed at his work. The disadvantages of public ownership and operation of a collection system include the monopolistic nature of such operations, which can result in a lack of stimulus toward efficiency. Labor pressures for higher wages, less work, and greater job security limit the flexibility of many public systems to implement labor saving techniques. Furthermore, since the major issue in relations between governmental officials and the voting public often appears to be taxes and monies spent, the emphasis in operating a municipal collection system may be placed upon saving money and reducing costs to the detriment of the collection operation.

The potential advantage of contracting with private companies for refuse collection services is that the competition between various firms should keep the cost down. Where contracts are awarded under a competitive bidding system, the community can retain control of collection policies and derive the benefit of a competitive, profit-motivated collection system. Furthermore, the burden of expenditures for equipment and other capital outlay is placed on the private company. The main disadvantage to this setup centers upon the need for active regulation by a public agency.

The case of private collectors contracting with customers is most prevalent in suburban and rural areas. The obvious advantage to this arrangement is that individuals in these areas can avail themselves of a service that may not be available from nearby municipalities. The disadvantages arise from the fact that high competition and severe price cutting can lead to a high rate of business failure and interruptions in service. There is also the danger that the collectors will informally agree to honor each other's territories; a monopoly on services in a given area can lead to unjustified price schedules.

Table 12
Potential Advantages and Disadvantages of Types of Public and Private Ownership and Operation of Collection Services, and the Conditions Favored[a]

Alternative	Potential advantages	Potential disadvantages	Conditions that favor alternative
Public			
Municipal	Tax-free Nonprofit Economies of scale City has administrative control Option of separate collection for recycling Option of mandatory collection Management and policies are continuous over time, resulting in experienced personnel and permitting long-range planning Records can be kept over a long time	Monopolistic Lack of incentive to maximize efficiency Financing and operations often influenced by political constraints Frequently financed from general tax fund and subject to 1-year budgeting process Solid waste management often low-priority item in budget Labor pressures may result in inefficient labor practices and strikes Restrictive budget policies may affect equipment replacement and maintenance Policies of job support inflate labor costs	Past history of unsatisfactory contractual operations for public services Public predisposition toward government operation of public services Quality of service provided more important criterion than economics

Private firms			
With contract from governmental unit	Competitive bidding for contract(s) helps keep prices down City retains administrative control Option of separate collection for recycling Option of mandatory collection	Danger of collusion in bidding Public agency must regulate contractors	Flexibility is needed to make changes in operations that would result in labor savings and other cost reductions Existence of qualified private contractors Public predisposition toward private sector involvement in public services Newly incorporated communities, or where population growth is outpacing ability of community to provide public services
In open competition	Competition may reduce costs Self-financing	City has no administrative control Danger of collusion among haulers to reduce competition and keep prices high Cutthroat competition can result in business failures and service interruptions Overlapping routes, waste of fuel No option for citywide separate collection for recycling Difficult to enforce mandatory collection ordinances	Unacceptable alternative

(*continued*)

Table 12 *continued*

Alternative	Potential advantages	Potential disadvantages	Conditions that favor alternative
With exclusive franchises	Self-financing	City has no administrative control Monopolistic, can lead to high prices No option of separate collection for recycling Difficult to enforce mandatory collection ordinances	Unacceptable alternative
Combination			
Municipal system and private firms under contract	Competition helps keep price down Alternative available if either sector cannot deliver service City has administrative control Option of separate collection for recycling Option of mandatory collection		Municipality is expanding through annexation or merger with other jurisdictions Changing from separate garbage and trash collection to combined collection
Competition between municipal system and private firms	Competition helps keep prices down	Overlapping routes, waste of fuel No option of citywide separate collection for recycling Lack of mandatory collection	Unacceptable alternative

[a] From Ref. 18.

The reason behind franchised collection is to counteract the negative aspects of excessive competition (business failures, discontinuities in collection service). However, the exclusive franchise does create a monopolistic situation.

Table 12 lists the advantages and disadvantages of public and private ownership of collection services.

A recent survey [19] covering 2060 cities with populations under 700,000 inhabitants and a combined population of over 52 million people has shown some striking statistics regarding collection arrangements. Private firms collect commercial, institutional, and industrial refuse in about three times as many cities as municipal agencies, and collect residential refuse in almost twice as many cities (66.7%) as municipal agencies (37.4%). Furthermore, substantially more communities rely entirely on private firms (45.2% of all cities have only contract, or franchise, or private, collection) than on municipal agencies (32.5% of all cities) for residential refuse collection.

However, a rather different picture emerges with regard to the number of people rather than the number of cities. Because larger cities are more likely to have municipal collection, 61.3% of the population is serviced by municipal agencies despite the fact that only 37.4% of the cities have any municipal collection.

No specific guidelines can be given for selecting between public or private collection agencies. The success of any arrangement depends on good personnel, sound management, and minimum political interference. In examining whether public or private personnel, equipment, and facilities should be used for solid waste collection, the following issues must be considered: (a) the relative efficiencies and economics of public or private ownership and operation; (b) the ability of the governmental agency to manage a public system and/or contracts; (c) possible legal constraints on the powers of the governmental unit to enter into contracts for services; and (d) the public attitude. From the broader viewpoint of social costs and benefits, it should be remembered that a change from municipal to contract collection may reduce the cost to the household for refuse collection service, but if this is achieved by increasing the number of unemployed residents—whether or not they receive unemployment or welfare benefits—the net costs to society and to taxpayers across the nation may outweigh the direct savings to local households.

B. Residential Collection

The high cost of residential refuse collection arises from the large labor force required. Furthermore, in an average community many different varieties of waste may require separate crews with separate equipment dealing with curbside and alley pickup, street sweeping and litter cleanup.

The first step in the collection of residential solid waste is its deposit in containers. This should be carried out in a manner that does not constitute a fire, health, or safety hazard, or provide food or breeding ground for insects or rodents. For multiunit or apartment collection, the use of bulk containers designed for mechanized collection is recommended. These come in a great variety of shapes and sizes and are used with specialized collection vehicles. They can be emptied mechanically by a vehicle with an arm controlled by the driver who never has to leave the cab. In addition, on-site compactors with detachable containers for transfer pickup are coming into use in some apartment developments. Here too special collection vehicles are required. Individual household units should use properly maintained, lightweight metal or plastic cans with tight-fitting lids and no more than 35 gal (0.15 m^3) capacity. Paper or plastic sacks are also acceptable for residential solid waste. Apart from the obvious advantage of cleanliness, sacks do not have to be returned to the roadside and later to the residence.

In the event that the householder is required to separate the refuse, extra containers are necessary for each category of refuse such as food wastes or reusable materials (glass, metals). Newsprint can be bundled and stored without containers.

The next step in residential collection is the point of collection, that is, where the refuse is picked up. Two basic options are available: (a) Residents may be required by city ordinance to place their containers at the curbside or alley where the collectors will pick them up. The resident then retrieves the empty container. (b) In backyard collection, the collectors empty the containers and either return them to their place, or leave them at the curb and the resident retrieves the containers. Approximately 60% of the collection systems in the United States are curbside and alley, and 40% are backyard. Curbside/alley collection is more economical, and on the basis of increased efficiency and productivity, fuel conservation, and reduced injuries to collectors, collection from curbsides and alleys rather than backyards is recommended.

The frequency of collection is another important factor which varies significantly from one location to another. There seems to be a shift from a twice-a-week collection schedule to a once-a-week collection, which could in part owe to the fact that the fraction of garbage (food wastes and other putrescible matter) in municipal refuse is decreasing considerably, and thus the reason for collecting twice a week is becoming less important.

Collection frequency should be the minimum consistent with public health and safety in order to keep collection costs and fuel consumption down. When citizens are required to separate waste, a frequency should be set for each waste category. For instance, newsprint may be collected once a month, other household refuse once a week.

The choice of collection vehicle can have a significant impact on the overall efficiency of the collection effort. Many types of collection vehicles are available and the most common are the side and rear loaders. Side loader capacities range between 16 and 37 yd^3 (12 and 28 m^3) and have hopper loading on either side of the vehicle body. In low-density single-family curbside collection a one-man side-loading vehicle should suffice, whereas in higher population density areas and alley collection two-person crews may be needed. Rear loaders range in size from 16 to 25 cubic yards (12–19 m^3). Backyard service or alley collection in high population density areas may be achieved with rear loaders and three-man crews. Both side and rear loaders utilize continuous-action compaction mechanisms that use a moving blade or ram to push the refuse against itself. Such compaction can result in bulk refuse density of about 500 lb/yd^3 (295 kg/m^3). Average density of residential refuse is about 150 lb/yd^3 (90 kg/m^3).

In addition to side and rear loaders, two types of vehicles generally employed for collecting commercial wastes are sometimes used in residential collection. The front loader, which ranges in capacity from

Table 13
Cost of Curbside Collection by Frequency of Collection and Crew Size, in Four Cities, 1973[a]

City	Crew size	Frequency of collection	Cost[b] per ton, $	Percent difference
1	1	1	8.29	39
2	1	2	13.48	
3	3	1	12.82	13
4	3	2	14.67	

[a] From Ref. 20.

[b] Labor rates for the cities have been normalized to permit intersystem comparisons; therefore, these figures do not reflect actual collection costs.

24 to 41 yd^3 (18 to 31 m^3) collects from containers varying in size from 2 to 10 yd^3 (1.5 to 7.5 m^3). Tilt-frame trucks with roll-off containers are also used for residential collection in rural areas. Vehicle selection is affected by such local factors as haul time to the disposal site, street or alley width, and types and amount of waste.

Table 14
Recommended Crew Size and Vehicle Type for Residential Solid Waste Collection by Point of Collection and Housing Density[a]

Point of collection	Housing density: Suburban	Housing density: Inner city high-density area
Curbside/alley	One man using a side-loading right-hand-drive vehicle with a low step-in cab	Two men using a side-loader with low step-in cab For very heavy waste loads, three men using a rear-loading vehicle with a low step-in cab
Backyard	Two men with tote barrels, using a vehicle with a low step-in cab Satellite vehicles	Three men with tote barrels, using a rear-loading vehicle with a low step-in cab

[a] From Ref. 18.

Table 15
Typical Ranges in Packer Truck Prices (Including Chassis), 1975[a]

Standard sizes, yd^3	Price, $
Rear loaders,	
16–25	20,000–45,000
20[b]	25,000–32,000
Side loaders, 16–37	20,000–38,000
Front loaders, 24–41	35,000–50,000
Roll-off, 35,000–75,000 lb	30,000–45,000

[a] From Ref. 18.
[b] The prices for the 20 yd^3 rear loader are shown separately because of the prevalence of this vehicle size.

Tables 13–15 show some typical costs related to crew size, frequency of collection, point of collection, and collection vehicle. Table 16 shows yearly costs for typical residential collection.

The determination of optimum combination of vehicle, crew size, and frequency of collection is in reality very complex, and local influencing factors must be continually reviewed in order to assure optimal selection.

Table 16
Typical Yearly Collection Costs for 2- and 3-Person Crews Including Vehicle, 1975[a]

Cost components	2-Person crew, $	3-Person crew, $
Depreciated vehicle procurement cost[b]	6,900	6,900
Maintenance cost	4,000	4,000
Consumable items		
Fuel (6,065 gal at $0.50)	3,025	3,025
Oil	480	480
Tires	1,680	1,680
Miscellaneous (insurance, fees)	3,000	3,000
Labor, including 20% fringe benefits		
Driver ($5/h)	12,480	12,480
Helper(s) ($4.50/h)	11,232	22,464
Management and administrative overhead		
(30% of direct labor)	7,114	10,483
Total annual costs, $	49,911	64,512
Cost per ton[c], $	15.36/ton	19.85/ton

[a] From Ref. 18.

[b] Straight-line depreciation over 5 years at 6% interest. Vehicle is 20 yd^3 rear loader.

[c] Average compacted waste density is 500 lb/yd^3 and 2.5 loads are carried each day; 20 yd^3 body × 500 lb = 10,000 lb or 5 tons; 5 tons × 2.5 trips/day = 12.5 tons/day; 12.5 tons/day × 260 days = 3,250 tons per year.

C. Commercial Collection

Commercial solid waste is defined as those wastes generated by business or industrial establishments and by residential apartment buildings with more than four to six units. Most commercial collection is made in containers between 1 and 20 yd^3 (0.75 and 15 m^3) in capacity. The containers are adapted so that they may be picked up by special vehicles, usually with only the driver being needed to effect the transfer. The frequency of collection required depends on the amount of food and wet wastes deposited. In addition, the containers may be equipped with compaction devices to reduce the size or number of containers that have to be maintained and the frequency of collection. The most commonly used collection vehicles are the front-loading and the tilt-frame trucks. The front loader is not designed for hand loading and the container is designed so that the truck can be driven to it; the truck lifts the container over the front of the truck cab, inverts it into a hopper, and then lowers the container back to its previous position. The body capacity of the front loader ranges from 20 to 40 yd^3 (15–31 m^3). The tilt-frame truck has a pair of guide rails that tilt to facilitate the loading and unloading of a boxlike roll-off container. In normal operation, the truck arrives with an empty container, which it rolls off, and then loads a full container in its place. The roll-off containers vary in size from 10 to 55 yd^3 (7.5–42 m^3).

The above two systems are so efficient in their operation that they are being adopted in some cities and towns for residential pickup; if not for all categories of waste then at least for bulk waste such as furniture and large appliances.

In communities where the commercial solid waste quantities are insufficient to justify the additional cost of a commercial system, it is more cost-effective to collect the commercial waste with residential collection equipment such as side and rear loaders. Figure 2 illustrates various vehicles utilized in solid waste collection.

D. Recent Concepts in Collection

Present collection systems and techniques are for the most part antiquated and have scarcely changed since the days of the horse and cart. (The carts have been covered and fitted with doors and load-compaction devices). Sophistication and modernization in collection systems has been slow in coming and in acceptance. This is a task for industry in cooperation with government and is being done at a rate commensurate with the economic returns in this field.

Some recently introduced concepts and equipment in solid waste collection are discussed below.

1. *Transfer Stations*

The concept of transfer stations is not new, but has gained popular acceptance only in the last few years. The exhaustion of close-in land disposal sites and the failure to acquire sites before either rising land costs or resistance of neighbors to solid waste disposal in their territory, has lengthened the haul from collection to disposal. Rising labor costs

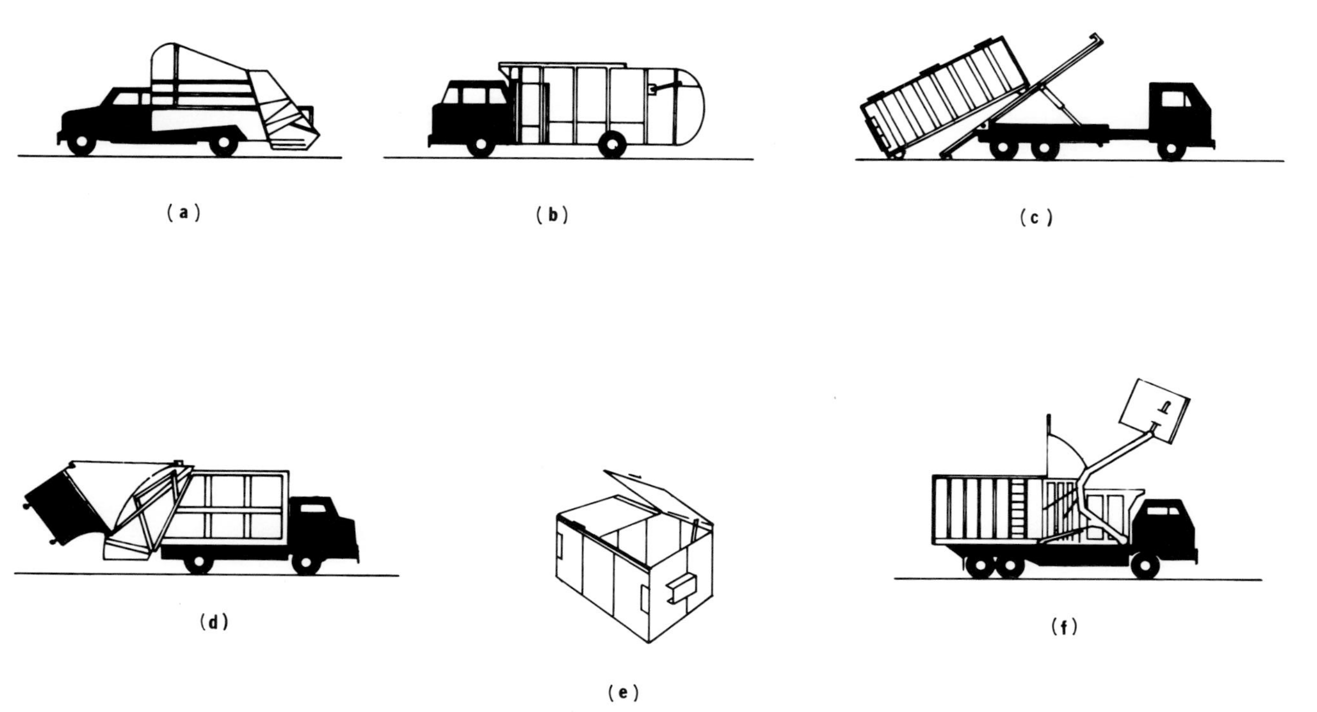

FIG. 2 Collection vehicles: (a) rear loader; (b) side loader; (c) tilt-frame truck with roll-off container; (d) bulk container being emptied into rear loader; (e) detached bulk container; (f) bulk container being emptied into front loader.

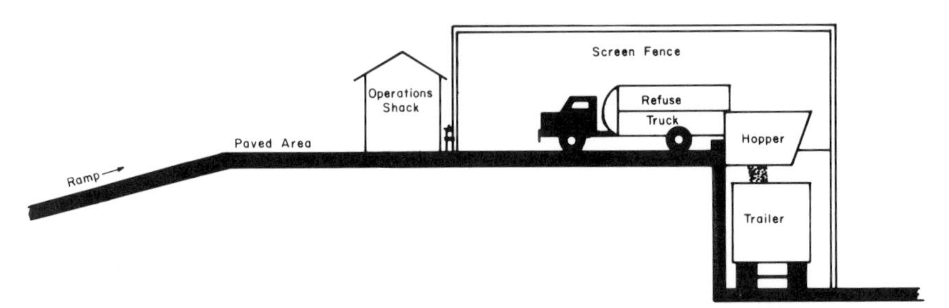

FIG. 3 Refuse transfer station.

and capital costs of collection vehicles have caused a search for an alternative to the special-purpose vehicle and a two- or three-man crew enroute to and from the disposal site for an hour or more per trip. The solution is the transfer station—a facility where the solid waste from several relatively small collection vehicles is placed into one large vehicle (60 yd^3, 45 m^3, or more) before being hauled to the distant disposal site (see Fig. 3). Compaction of the refuse at the transfer station can attain 300–600 lb/yd^3 (180–360 kg/m^3) density. Further shredding and baling can achieve densities of around 1500 lb/yd^3 (890 kg/m^3) that is, almost ten times the density of refuse in garbage cans.

Various means of long-haul transport, from transfer station to disposal, have been developed. The most common among these are trailer trucks with built-in compaction systems operating at just under the legal highway load limits of 15–25 tons (13.6–22.7 metric tons).

One alternative to trailer-truck transport is rail haul. However, the political problems associated with this concept are yet to be solved. When rail haul does become a widespread reality, it will allow a wide flexibility in locating disposal sites (especially strip mines, mine pits, and quarries) and result in considerable savings for communities where long-haul distances are necessary.

Another alternative is barge transport, especially suitable for ocean dumping. However, with ocean dumping being increasingly restricted the utility of this mode of transport is severely limited.

Transfer stations offer potential savings even though they require an extra material handling step. For a transfer station to function adequately, the system, including the station, must in essence be equal to or better than the system without it. Thus, operations with a transfer system should prove more reliable, flexible, and adequate, than a system where the collection trucks haul the refuse directly to the disposal site.

The construction of a transfer facility requires large capital expenditures for land, facilities, and stationary and vehicular equipment. Furthermore, a high quality of supervision is required to get the maximum return on capital expenditure, and the crew manning the station. Good maintenance of equipment and cleanliness of the site is mandatory in order to gain acceptance on the part of the community providing the site.

The factors that must be considered in deciding the suitability of a transfer station for a particular area, include haul distance to the ultimate disposal site, the time of travel to this site, and the overall efficiency of the particular transfer operations.

The characteristics of 12 transfer operations are shown in Table 17. The capital and operating cost breakdowns are given in Tables 18 and 19 respectively.

2. *Pipeline Transportation* [22]

Although the concept of pumping bulk solids through pipelines is not new, its application to the field of solid waste collection is relatively new. It is a well-known fact that liquids or gases traveling at high velocity are capable of entraining solid matter. This characteristic of flowing fluids can be utilized for pipelining solid wastes, with either air or water as the transport fluid, in a number of modes of flow. The three most promising are pneumatic suspension, water slurry flow, and waterborne slug flow.

Pneumatic suspension systems use air as the carrier fluid. Depending on the solid waste dimensions prior shredding may or may not be necessary. A vacuum-sealed pipeline system collects the refuse fed through vertical refuse chutes. The air velocity that sweeps the solid waste is typically 90 ft/s (27 m/s). Underground pipes range in diameter from 20 to 24 in. (0.5–0.6 m) and sufficient vacuum is produced by turboextractors. Power requirements are high, as much as 1000 hp (746 kW) per 1000 ton-miles/day (1460 metric ton-km/day). Therefore this method is best suited for short distance hauling (1 or 2 miles) into inaccessible areas, or for feeder systems within a city.†

When the haulage is longer than a few miles and/or the total volume of waste in the pipe at a single moment is excessive, the pneumatic system becomes impractical or uneconomical. Slurry systems are most economical for intermediate distances and quantities. The refuse is shredded, mixed with water, and fed into the pipeline under pressure.

† Pneumatic transport systems are at present being used in the Walt Disney World amusement park at Orlando, Florida, and Sundeborg, a suburb of Stockholm, Sweden. The Sundeborg system has been in use since late 1966 in an apartment complex where 5000 units are being served.

Table 17
Characteristics and Costs of 12 Transfer Systems in 1975[a]

Characteristic	Site No.[b]												
	1	2	3	4	5	6	7	8	9	10	11	12	Average
Average tons/day	387	120	289	670	112	190	600	228	155	220	200	880	
Tons/yr (1,000)	110	30	33	174	35	49	180	83	49	57	60	229	
Days/yr of operation	284	250	130	260	312	260	300	364	318	260	302	260	
Site acreage	0.5		3	2	1.75	2	2	1,500	1	5	1	4.5	
Round trip distance, miles	40	63	22	(2)	20	23	35	45	27	25	20	53	
Round trip time, hr	1.5	2.5	1	(2)	1	1	1.3	1.5	1.5	1.5	1.16	1.8	
Number of shifts	2	1	1	1	1	1	1	1	1	1	1	1.5	
Number of supervisors	1	1	1	1	0.5	1	1	2	0.5	1	4	4.5	
Number of station employees	11	3	4	7	1	2	5	7	3	4	10	44	
Number of drivers	15	4	3	0	4	4	7	8	3	4	5	45	
Total operating costs													
Total/yr, $1,000	723	236	94	1,293	93	163	330	321	127	129	176	2,337	—
Cost/ton, $	6.58	7.82	2.49	7.43	2.78	3.29	1.84	3.87	2.57	2.26	2.92	10.72	4.55
% of total cost	98	74	58	87	80	83	80	80	91	80	86	88	82
Total annualized capital costs													
Total/yr, $1,000	17	81	68	198	24	33	84	79	13	33	29	319	—
Cost/ton ($)	0.15	2.70	1.82	1.14	0.68	0.67	0.47	0.94	0.26	0.58	0.48	1.46	0.94
% of total cost	2	26	42	13	20	17	20	20	9	20	14	12	18
Total cost													
Total/yr, $1,000	740	316	162	1,492	121	196	415	400	140	162	205	2,656	—
Cost/ton, $	6.73	10.52	4.31	8.57	3.46	3.96	2.31	4.81	2.83	2.84	3.40	12.18	5.49

[a] From Ref. 21.
[b] Average of two transfer stations in same community.

Table 18
Capital Costs for 12 Transfer Systems in 1975[a] (1,000 $)

Cost item	Site No. 1	2	3	4	5	6	7	8	9	10	11	12	Average
Land													
Total cost	—	—	—	—	—	—	—	19	9	20	—	—	—
Site construction													
Total cost	118	270	530	2,815	118	0	272	184	9	26	11	125	
Annualized cost	5.9	27.0	26.5	140.8	5.9	—	13.6	9.2	0.4	1.5	0.5	12.5	
Cost/ton	0.06	0.90	0.71	0.81	0.17	—	0.08	0.11	0.01	0.03	0.01	0.05	0.24
Stationary equipment													
Total cost	6	37	45	865	42	91	189	180	68	22	7	149	
Annualized cost	0.4	3.7	3.0	56.7	3.5	6.1	18.9	9.0	3.4	2.2	0.5	14.9	
Cost/ton	0.004	0.12	0.08	0.33	0.10	0.12	0.10	0.11	0.07	0.04	0.01	0.07	0.10
Vehicular equipment													
Total cost	52	251	194	0	114	200	337	477	45	176	162	1,165	
Annualized cost	10.4	50.2	38.9	—	14.3	26.9	51.8	60.6	9.0	28.9	27.9	291.4	
Cost/ton	0.09	1.68	1.03	—	0.41	0.55	0.29	0.72	0.18	0.51	0.46	1.27	0.60
Total													
Total cost	176	558	769	3,680	274	291	798	860	131	244	180	1,439	
Annualized cost	16.7	80.9	68.4	198.4	23.7	33.0	84.3	78.8	12.8	32.6	28.9	818.8	
Cost/ton	0.15	2.70	1.82	1.14	0.68	0.67	0.47	0.94	0.26	0.58	0.48	1.39	0.94

[a] From Ref. 21.

Table 19
Annual Operating Costs for 12 Transfer Systems in 1975[a] ($1,000)

Cost item	Site No. 1	2	3	4	5	6	7	8	9	10	11	12
Labor												
Salaries and wages	385.8	127.6	36.6	239.9	56.5	109.6	160.0	174.0	73.4	31.5	89.6	1,042.0
Fringe benefits	85.9	31.9	14.6	95.5	9.4	22.1	21.8	22.4	16.9	4.4	21.5	1.0
Total	471.7	159.5	51.2	335.4	65.9	131.7	181.8	196.4	90.3	35.8	111.1	1,133.0
Stationary equipment												
Repair and maintenance	10.6	0	2.1	24.2[b]	1.4	0.6[b]	15.3	10.0[b]	0.4[b]	1.6[b]	4.5[b]	147.0
Parts and supplies	23.5	0	0.5	—	6.0	—	9.0	—	—	—	—	50.0
Other equipment expense	6.4	0	0	—	0.3	0	5.5	0	0	0	0.8	8.8
Total	40.5	0	2.6	24.2	7.7	0.6	29.8	10.0	0.4	1.6	5.3	205.8
Vehicular equipment												
Repair and maintenance	0	46.8	18.5[b]	0	8.6	17.8[b]	13.2	49.1[b]	17.4	26.7	30.9[b]	327.3
Parts and supplies	0	12.8	—	0	3.6	—	37.8	—	4.1	12.3	—	387.8
Rental	168.1	0	0	882.3[c]	0	0	2.0	0	1.8	24.7	0	0
Fuel and oil	21.8	—	4.9	0	2.3	10.0	20.8	26.0	8.7	17.5	6.7	90.0
Total	189.9	59.6	23.4	882.3	14.5	27.8	73.8	75.1	32.0	81.2	37.6	806.0
Facilities and site maintenance	2.0	0	2.1	0	6.0	1.0	5.6	1.6	0.6	0.2	0	34.3
Utilities	2.1	15.6	7.0	2.9	2.5	1.0	9.0	5.4	1.9	2.9	2.4	77.5
Administrative expense	9.8	0	0	17.1	0.5	0.5	19.4	10.8	1.6	1.8	16.1	15.6
Other expenses	7.5	0	7.2	31.4	0.4	0	11.0	21.5	0	5.6	3.8	66.7
Total	723.5	234.7	93.5	1,293.3	97.5	162.6	330.4	320.8	126.8	129.1	176.3	2,337.1

[a] From Ref. 21.
[b] Includes parts and supplies.
[c] Includes all parts and supplies, repair and maintenance, and fuel and oil.

Typical power requirements for slurries are 50 hp (37 kW) per 1000 ton-miles/day (1460 metric ton-km/day). The shredding and presizing of refuse at each point of generation can be expensive. In addition, water must be plentiful and wet disposal must be possible, otherwise further costs can be incurred in separation, drying, and water purification. Zandi [23] has developed an economic analysis to show a favorable benefit/cost ratio for the installation of pneumoslurry systems in existing communities. In such systems pneumatic pipelines would collect the refuse from several generating points to a central presizing facility where the wastes would be crushed and mixed with sewage to produce a slurry to be pumped over long distances for disposal. This system eliminates the need for expensive presizing units at each point of refuse generation.

Slugs are cylinders of shredded refuse, slightly smaller than the pipeline, and are propelled by water that can be recirculated and reused. This mode of pipeline transport has the greatest potential for becoming an efficient method for pipelining large quantities of refuse over long distances greater than 30 miles. Its major advantage is its low power requirement of approximately 1 hp (746 W) per 1000 ton-miles/day (1460 metric ton-km/day). The high density of compressed slugs (50–70 lb/ft^3, 800–1120 kg/m^3) also simplifies landfill disposal and extends the landfill capacity.

The economics of pipeline systems are not very clear at the present time, but it does appear that economies of scale will render them potential candidates for solid waste collection systems. These economies of scale could be realized in the case of large population centers where large quantities of solid waste must be transported over long distances for disposal or resource recovery processing.

3. *Improved Equipment and Practices*

A recent innovation in solid waste collection equipment has been the introduction of a German compactor truck (the KUKA Shark), which can continue the compaction cycle while being loaded or moved. Being comparatively noise-free it can be used in nighttime operation. Maintenance costs are also claimed to be low because of the absence of hydraulically operated gear. Capital costs are competitive with conventional compactor trucks.

Three innovations aimed at lowering costs and getting the most ton-miles from high-cost packer trucks are collection trains, scooter pickups, and packers manned by a single person who drives and collects. The first two cover assigned areas and meet with a packer manned solely by a driver. Packer trucks designed for one-man handling include dual drives, left and right, or front and rear controls, operation from standing position and low side-feeder points. Experiences with one-man crew operations are reported by Ralph Stone Co. [24]. In short, there are many alternatives to the "three guys on a truck" system, but it takes imagination and a certain degree of initiative by local officials to implement them.

E. Planning a Collection System

As with any good planning effort, the key in planning a solid waste collection system is the ability to ask the right questions. What to collect? Who will collect it? How often to collect it? Where to collect from? What vehicles to use? What routing, crew size, and schedule to use? How many transfer stations are needed and where should these be located? How many incinerators, landfills, or other types of disposal sites are needed and where should they be located?

The question of what geographic area to cover is usually determined on the basis of existing political jurisdictions. There might, of course, be a lot of merit in combining a number of these jurisdictions into a single solid waste district. Such an arrangement may capitalize on the economies of scale for processing and disposal facilities. Intergovernmental cooperation between adjacent jurisdictions should be sought whenever the savings resulting from such large-scale transfer and disposal facilities outweigh the additional costs incurred as a result of collecting the waste generated from a large geographical area.

A community can operate its own collection system, or it can leave that responsibility to private enterprises operating under contracts with the community. A third form of collection alternative is for communities to require home owners to individually engage private entrepreneurs.

The choice of collection point (curbside, alley, or backyard), frequency of collection, and collection vehicle, all have significant effect on the overall efficiency of the collection effort.

A major management decision facing solid waste collection agencies is that of allocating routes to collection crews. One widely used approach is to allocate a definite daily task to each crew, thus dividing the community into areas that provide about one day's work for each crew.

The economics of overall solid waste collection, processing and disposal are a function of all of the above considerations, as well as others such as the nature of the community, and its selected method of disposal. Population density is a significant determinant because a community having a density of less than 10 dwelling units per acre could expect up to 50% more in collection costs per ton than a community which is developed at twice that density. Land use and zoning regulations have a definite effect on the distribution of the activities in the community and thus the distribution of waste generation and the distances to be traveled by collection vehicles. They also play a significant role in determining the location of transfer stations or other processing and disposal facilities, and these locations in turn determine haul distances. Other local factors, such as topography, climatic conditions, and the quality of the road system, can also have a significant influence on the cost.

It certainly is no easy task to determine the most economical combination of equipment types, management strategies, and location decisions. This is further complicated by the variety of possible institutional arrangements, some of which have been described above. Tradeoffs exist between the different elements of the overall system. For example, it is known that larger landfills, transfer, or processing systems cost less per ton than smaller units. It is also clear that the larger a unit gets, the larger collection area it will need to serve and thus higher the collection costs per ton. In order to capture the complexities of the larger numbers of such sets of tradeoffs, often working in opposite directions, a solid waste system planner needs a great deal of experience and all the help available from sophisticated mathematical and computerized techniques [25–27]. Some of these are discussed in detail in Chapter 7.

IV. SOLID WASTE DISPOSAL

Solid waste disposal is an extension of the process that starts with collection. It involves the processing and ultimate disposal of solid waste by methods that meet acceptable standards of public health and community environments.

In this section a brief overview of several disposal methods will be presented. Those methods considered more suitable for municipal or general solid waste disposal will be dealt with in detail in subsequent chapters.

There are usually three alternative methods of solid waste disposal: (a) direct disposal of waste, (b) processing of waste to facilitate the subsequent disposal, and (c) processing of waste to recover materials and/or energy and subsequent disposal of the residues. The third alternative properly falls into the subject of "resource recovery" and will therefore be discussed in the next section which deals with that subject.

A. Direct Disposal

Direct disposal of solid wastes involves only the final step of waste disposal with very little or no preprocessing of the waste.

1. *Open Dumps*

This is the cheapest method of disposing of solid wastes. It involves literally dumping the waste in a convenient out-of-the-way spot. In England this process is known by the appropriately descriptive term of "tipping".

Dumps are associated with all of the health hazards and all of the affronts to our senses that one expects might be associated with solid waste. They are breeding places for rats and flies, produce odors, frequently catch fire, produce air pollution from smoke and particulate matter, and constitute eyesores.

An adjunct to the open dump is a procedure known as "open burning" wherein the waste is combusted openly. This practice is especially prevalent in rural areas. Some towns use a method of disposal on a steep slope and a controlled rate of burning. Any form of burning is a source of smoke and odor, contributes to air pollution, encourages fly and rodent breeding, and invites nuisance conditions.

It should be amply clear that open dumps or burning have no place in solid waste disposal technology and will not be discussed further.

2. *Marine Disposal*

Another alternative is the dumping of waste in the marine environment. This form of disposal continues to be a controversial subject, with some engineers expressing the view that such disposal of solid wastes would be a highly effective utilization of the ocean's vast reservoir [28]. However, the concept that the ocean is a gigantic sink with unlimited capacity has, in recent times, given way to the concept of the ocean as a limited and valuable resource.

Dumping wastes so that they become irretrievable, should retrieval ever be necessary, or placing them in an ecologically fragile environment, such as a marine desert or the deep oceans, is contrary to sound environmental engineering.

Some marine disposal under planned procedures can, however, be beneficial. The reef of automobile bodies dumped off Long Island at Sheepshead Bay has provided much needed protection for fish to spawn and grow. With proper research, compatible wastes may be matched with selected oceanic areas, resulting in economic disposal.

Nevertheless, any venture into this means of disposal has to be carried out with caution.

3. *Grinding of Garbage*

The use of domestic garbage grinders is steadily increasing. These units grind garbage and dispose of it into the sewer. The advantage of such grinders lies in their convenience to the householder. However, their impact on the total solid waste disposal picture is quite minimal since domestic garbage constitutes only about 10% of the total municipal refuse.

Concern is sometimes expressed that the use of grinders will impose an additional load on sewage treatment plants (increased oxygen demand and solids by about 20–50%). This is especially problematic in small community sewage treatment systems.

4. *Hog Feeding*

Feeding garbage to hogs is practiced in only a few locations in the United States. The bulk of the garbage disposed in this manner comes from hotels, restaurants, institutions, and industry. Nevertheless, the low fraction of garbage in municipal refuse, as well as the potential problem of disease transmission (mainly trichinosis) have eliminated this option as a major alternative in the selection of a disposal method.

5. *Sanitary Landfill*

The various disposal methods noted above represent at best a partial solution to the solid waste disposal problem. The sanitary landfill offers a proven method of direct disposal that satisfies present day environmental standards.

The sanitary landfill is defined by the American Society of Civil Engineers [29] as "a method of disposing on land without creating nuisance or hazard to public or safety, by utilizing the principles of engineering to confine the refuse to the smallest practical volume and to cover it with a layer of earth at the completion of each day's operation, or at more frequent intervals as may be necessary." Above all else the landfill is an engineering operation; it is not a dump; it does not simply happen.

There are four basic steps in the planning and construction of a landfill:

1. The deposition of solid waste in a prepared section of a site in such a way that a minimum working area is presented at any one time.

2. The spreading and compaction of the waste in fairly thin layers in order to get it to the maximum density.
3. The covering of the waste, with a layer of compacted cover. This should be done daily.
4. The final covering of the entire construction should be with approximately 2–3 ft of earth. This prevents rodents from burrowing down into the refuse.

The cost of landfills was estimated by the EPA [30] and the resulting curve showing cost per ton versus daily capacity has been published. Figure 4 shows this curve, and includes some actual data from smaller communities in eastern North Carolina. The fit seems quite good. For the smaller landfills, the fixed costs such as the operator's salary, the equipment, etc., require a larger cost per ton. Larger landfills can experience economies of scale except when land costs are prohibitive (as tends to be the case in most large metropolitan areas today).

Chapter 4 will discuss landfills in detail.

B. Processing Prior to Disposal

The primary objective of processing solid waste prior to its disposal is to reduce the volume of wastes to be disposed of. Volume reduction has definite advantages since it reduces hauling costs and ultimate disposal costs in landfilling. Although volume reduction of solid waste offers advantages, the capital and operating costs for achieving volume reduction are significant and must be balanced against the benefits achieved.

3. *Incineration*

Incineration is a controlled combustion process for burning combustible wastes to gases and to a residue containing little or no combustible material. Modern incinerators will accommodate all ordinary municipal refuse. The waste that must be excluded from the incinerator is inert material in large amounts and large or bulky objects whose size or shape prevents their admission into the furnace.

Typically 80–90% of the total volume of refuse can be put through an incinerator. After incineration there will still be a residue amounting to 10–15% of the volume of the total refuse. Consequently, some 20–35% of the total volume of refuse remains to be disposed of even after incineration. Incinerator residue consists of noncombustibles such as metals, glass, and ashes, as well as combustible materials not completely consumed in the burning process. Such residue is quite stable, and easy to compact and dispose of by landfilling.

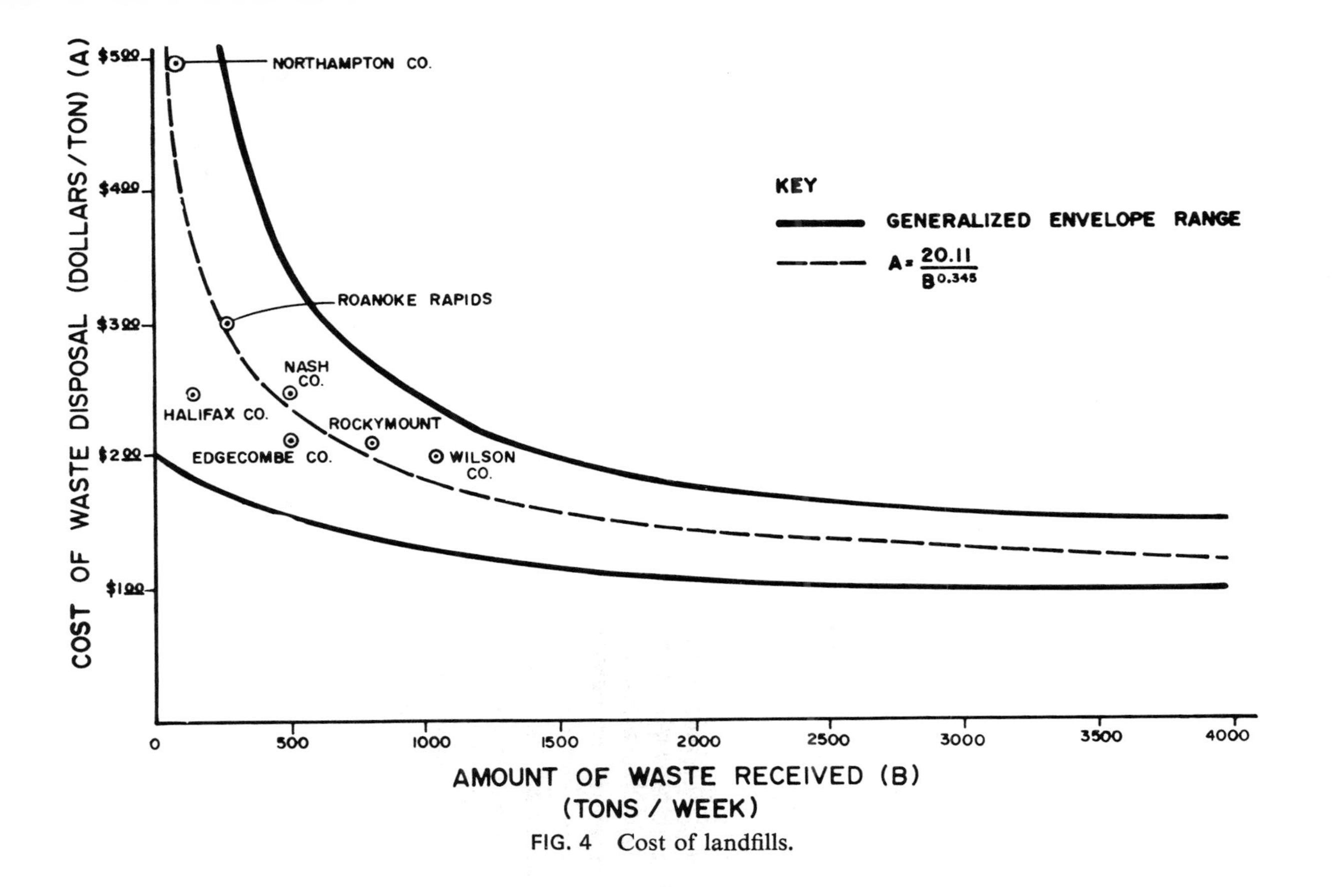

FIG. 4 Cost of landfills.

Stringent air pollution restrictions require the installation and maintenance of air pollution control devices on incinerators. These add considerably to the high capital and operating costs of incinerators.

Incineration is the subject of Chapter 3.

2. *Pyrolysis*

One of the major problems with incineration has long been the air pollution created by inefficient combustion. The large amount of excess air required in order to achieve reasonable temperatures and reduction in volatiles dictates intricate and expensive air pollution control equipment. Pyrolysis has been gaining ground as an alternative to incineration because of the relative ease of air pollution control.† Pyrolysis, in a simple sense, is the process of combustion in an air-deficient atmosphere resulting in lower volumes of off-gas than incineration, a benefit for air pollution control.

Pyrolysis of municipal refuse has three end products: a mixture of combustible gases; a liquid that is usually a tar-like substance; and a solid residue that resembles charcoal.

Approximately 80% volume reduction (comparable to incineration) can be expected from pyrolysis; this, plus the fact that the noncombustible residue is fairly inert, can substantially aid in extending the life of landfills.

Although volume and weight reduction of wastes can be accomplished by pyrolysis, it has greater potential for converting solid wastes into gaseous, solid, and liquid fuels. Pyrolysis will be further studied in Chapters 3 and 6.

3. *Shredding*

The third method of volume reduction is shredding. The objective of shredding, sometimes called pulverizing, is to reduce the size of the material to the point where it no longer looks or behaves like refuse. The systems in use at the present time all use the principle of brute force, devoid of finesse or design imagination.

The most common type of shredder in use today is the hammermill. The hammermill works much like a kitchen garbage grinder, and has a series of swinging hammers rotating on a horizontal shaft which crash against the refuse and break the material caught between the hammers and the grate below. The shredding operation reduces hetero-

† There is one notable exception. The large system, constructed in Baltimore, was found to have significantly higher particulates. See Section IV in Chapter 6.

geneous and difficult-to-handle material to homogeneous and relatively easy-to-handle material.

Shredding in itself (without compaction) is a useful preprocessing step prior to landfilling, pyrolysis, composting, mixing shredded refuse with other solid fuels prior to incineration, and materials separation in resource recovery.

One of the great advantages of a shredder is that it is perfect for the disposal of "white goods" (stoves, refrigerators, and other large kitchen appliances). These are made mostly of sheet steel, and a shredder is an ideal method of reducing their volume. In fact, some shredder operators put one or two stoves or refrigerators through the shredder at the end of each day as a cleaning process, like a metallic toothpick.

Shredding is the subject of Chapter 2.

4. *Compaction or Baling*

The objective of compaction is to increase the density of the refuse and concurrently to make it easier to handle and transport. The process is one of mashing the materials together with brute force to create a dense bale. A typical bale of refuse is about 36 × 36 × 70 in. (0.9 × 0.9 × 1.78 m) and the density can range anywhere from 1600 to 1800 lb/yd^3 (950–1070 kg/m^3) which is approximately twice the density usually achieved in a landfill. These bales are transported to the field and set out without any earth cover. There do not seem to be problems with rodents, flies, or settlement. Baling is economical in situations where the material is to be transported for a long distance before disposal. Accordingly, this method has come to be used especially in larger towns and cities where the landfill sites are far away.

Compaction will be studied in conjunction with shredding, in Chapter 2.

V. RECOVERY OF RESOURCES

A. Reasons for Recovery and Recycling

The first question that should come to mind when resource recovery is suggested is, "Why should we try to recover resources in the waste?" The answer is two-fold: (a) The conservation of natural resources, and (b) reduction of waste to be disposed of and the resulting reduction in detrimental effects on the environment.

Many of our raw materials are nonreplenishable, and are getting to be in shorter and shorter supply [31]. For example, copper is already

15–30% imported; tin, chromium, manganese, and other metals are almost 90% imported into the United States because we have nearly depleted our own resources. As the population increases and the standard of life improves, there is an increasing use of natural resources. We will simply run out of the sources of these materials and concurrently, perhaps, bury ourselves in the trash.

The second reason for recovery and recycling is to reduce the volume of refuse created; that is, to reduce the amount of material that we have to dispose of and forget. Any detrimental environmental effects resulting from solid waste disposal will be significantly reduced if more of the waste can be recycled.

The problems with recovery and recycling involve both the collection of the material as well as the question of purity. The secondary materials industry is probably the only industry that has a reversed relative position of market and source of raw materials. Most industries have a central source of raw material (such as, for example, coal or oil) and a scattered market. These industries take their raw material from one source and then distribute the product in a diffuse market. The secondary materials industry is just the opposite, having a source of raw material that is scattered and producing a product that is then delivered to a central point. The problems with collecting a raw material from a diffuse source presents unique and severe challenges to these industries.

Since the secondary materials industry has little control of the

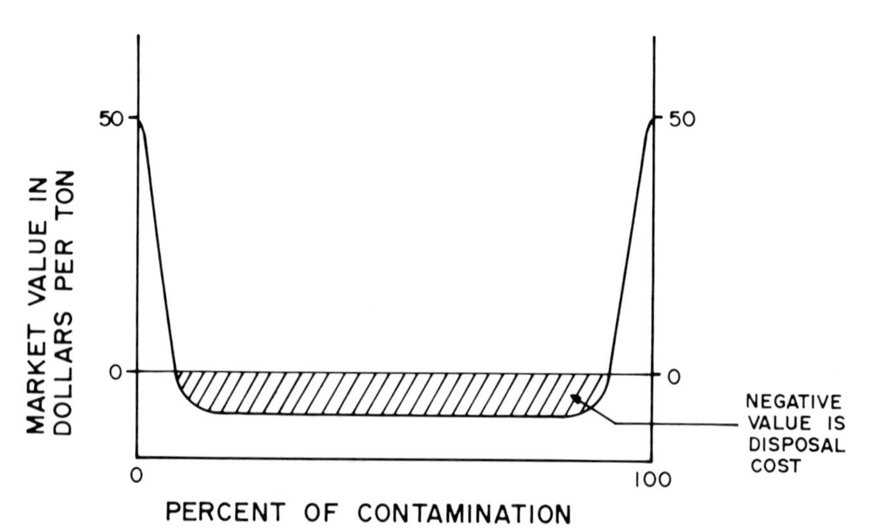

FIG. 5 Ideal value of two materials as determined by their purity.

raw material received, it is concerned with the question of purity. The effect of purity on the market value of a product is shown ideally in Fig. 5. Note that this is a graph representing the percent of "contamination" from 0 to 100% versus the market value in dollars per ton. At zero contamination, the product has a high value, which drops rapidly to below zero dollars (where the fraction of the contaminant is high and one has to pay to dispose of it). As the percentage of contamination increases and approaches 100% (when it becomes a pure contaminant) the product again has a high market value. There are only two small peaks in this graph where a material is sellable. Purity, therefore, is an important consideration in the recovery of waste and all of the resource recovery programs and processes are aimed at achieving it.

A classic example of purity problems is in the recovery of glass from beverage bottles with metal twist-off caps. Removing the cap leaves a metal ring around the neck, which substantially reduces the market value of the glass since the presence of the metal, even in small quantities, creates serious problems in the glass manufacturing process. Another example is paper reprocessing, where the paper mills refuse to accept paper that has been plastic coated since a very small amount of plastic will cause the machinery to produce an unacceptable paper product. The science of resource recovery is thus in essence a quest to make pure materials out of impure (and often unpredictable) raw materials.

Materials and energy recovery from solid wastes is the subject of Chapter 6.

B. Methods of Obtaining Pure Materials from Mixed Solid Waste

The obvious solution to getting pure materials is to avoid contamination in the first place; in effect to separate out the material at the source. For example, machinists in a metal fabricating shop will put shavings and waste pieces of a certain kind of steel in one barrel and another kind of steel in another, and these wastes become valuable raw materials because they are pure. In residential areas though, it is more difficult to get homeowners or apartment dwellers to separate their solid waste into specific buckets and barrels. In some communities such programs have existed for many years, but this was more to increase the efficiency of collection and not to ease the work of materials recovery. Only partial cooperation has been gained in most communities attempting source separation. With financial and moral support from the U.S.

Environmental Protection Agency (EPA), it may be possible sometime in the future to obtain sufficient cooperation to make such a plan successful, but this has not yet been the case.

The alternative to obtaining pure material at the source is, of course, to have a central facility that processes the waste into various components. The remainder of this discussion is devoted to such waste processing methods.

The most primitive (low technology) method for separating out various components in waste is the use of pickers; that is, people who stand alongside a conveyor belt and manually pick out the material. This operation is efficient and thorough because people make the decision as to what a component is. It is also inefficient since it is difficult to find people who want to devote their careers to picking refuse, and thus the wage scales must be high in order to attract the needed labor. Furthermore, many of the wastes cannot be separated by pickers (e.g., aluminum tops on steel beverage cans).

Another popular method of separation is air classification. This method depends on the differences in density and aerodynamic drag in order to achieve separation. The heavy fraction (inorganics) of solid wastes drops to the bottom and the light fraction (organics) is carried off with the vertical air stream. Air classification requires that the refuse first be shredded into uniform size particles.

Another binary separation device is the magnet, which can be constructed in a number of configurations. This is a widely used device especially in the cleaning of steel and iron from shredded automobile bodies, and the separation of steel cans from refuse.

Screening is a method of separating material by size. Shredded refuse dumped onto a series of various mesh size screens can be separated in different size fractions by a "sieving" effect of the screens. This is useful in the classification of metals from glass, with the glass being ground up finely and metal coming out as the large particles.

A popular method for separating organic material from inorganic material is called the ballistic separator. This device uses a conveyor on which the shredded material is brought up and thrown into a tank. The heavy resilient particles carry further and drop into the inorganic bin, and the lighter, sloppier organic materials do not carry very far and drop into the nearer bins. This device is commonly used in Europe for the separation of inorganic materials from compost.

There are, of course, many other possibilities for mechanical classification. Much of our present technology is either very old (from the mining industry) or very new and untested. There is a tremendous future in the classification industry, of taking raw waste material and treating it so the end product is a pure material useful for further beneficial application.

A more comprehensive review of separation techniques for resource recovery purposes is presented in Chapter 6.

The organic fraction of solid waste can be processed physically and decomposed by microorganisms to produce a nuisance-free humus-like substance that may be used as a soil conditioner. This operation is known as composting and its application to mixed municipal refuse has not displayed widespread success in the United States. A basic problem area in composting is development of a market for the product. Compost is marketed primarily to some segment of agriculture. If sold in small bagged lots to the nursery industry or to the homeowner, municipal compost must compete with animal manures, peat moss, and other similar products. Furthermore, the quality of the compost has to be stringently controlled if it is to compete with these organic garden products.

The composting of leaves and municipal sludge may well have a future, since the raw material is much more homogeneous and predictable than refuse, and quality control over the product will be better. Composting is the subject of Chapter 5.

C. Recovery of Energy

Municipal refuse has enough organic matter to make it a potentially valuable source of energy. About 70–80% of solid waste is combustible, having an average heat content of 9 million BTU/ton. If all of the municipal waste could be converted to energy, it could result in an "oil equivalent" of over 500,000 barrels per day [11]. Obviously, not all of this energy can be obtained without expending energy, and furthermore, paper (a significant constituent of municipal waste) is not "equivalent" to oil, since the methods and means of combustion differ significantly between oil and waste paper. Nevertheless, if all of the combustible waste from metropolitan areas was used for energy production, it would equal about 10% of the BTUs derived from the combustion of coal by all the utility companies, or about 1% of all of the energy expended in the United States [11]. This is significant and much work has recently been directed toward this end.

Two basic energy recovery systems are incineration and pyrolysis. Water-wall incinerator technology, much of it imported from Europe, has been successfully used in a number of cities for burning untreated refuse and producing steam. Other municipalities process the refuse and feed only the organic fraction as a supplemental fuel in coal fired boilers.

Pyrolysis systems are used for either producing steam (as in the case of incineration) or for producing liquid or gaseous fuels (pyrolysis products).

Both incineration and pyrolysis will be further discussed in detail in Chapters 3 and 6.

VI. SPECIAL TYPES OF SOLID WASTE

In Section 2 of this chapter it was shown that the major types of solid waste are municipal, industrial (including mining and agricultural), and those resulting from air and water pollution control. Within these three categories occur certain types of solid waste that merit special attention. What makes these wastes "special" is not their quantities (in fact in comparison with the total solid waste burden as shown in Table 11, these waste quantities would appear to be miniscule), but rather their persistence or conspicuousness in the urban environment, and/or their potential for causing severe public health and environmental hazards. These wastes have existed for many years but have come into prominence only recently.

In the following sections, eight such types of waste will be discussed, each with regard to its sources, quantities, and current disposal and resource (material/energy) recovery practices.

A. Nonradioactive Hazardous Waste

Every community produces certain hazardous waste materials that cannot or should not be handled and disposed of in the same manner as are ordinary solid wastes. The reason for this is that a significant potential is created for causing adverse public health or environmental impacts if such waste is handled, stored, transported, treated, or disposed of in the manner generally accepted for ordinary solid wastes.

Hazardous waste includes acids, toxic chemicals, explosives, biological toxins, flammable chemicals, and other harmful or potentially harmful chemicals. The EPA has identified about 160 wastes and waste streams as being hazardous [32], and the appearance of such wastes warrants special attention.

EPA estimated that approximately 10% of the annual production of about 350 million metric tons of industrial waste is hazardous, and these wastes are expected to increase by 3% annually. Major generators of hazardous waste among the 17 industries EPA studied in detail [33] are listed below (1977 estimates, in million metric tons).

Organic chemicals	11.7	Textiles	1.9
Primary metals	9.0	Petroleum refining	1.8
Electroplating	4.1	Rubber and plastics	1.0
Inorganic chemicals	4.0	Miscellaneous (7 sectors)	1.0
		Total	34.5

It is also estimated that about 40% of this waste is truly solid waste with the rest of it being mostly in the form of liquid or sludge.

The distribution of this waste generation is generally concentrated in the industrialized areas. Ten states generate about 65–70% of the nation's industrial hazardous waste: Texas, Ohio, Pennsylvania, Louisiana, Michigan, Indiana, Illinois, Tennessee, West Virginia, and California.

Environmentally acceptable technology for the treatment and disposal of hazardous waste is generally available (see Table 20). Such treatment processes perform the following functions: volume reduction, component separation, detoxification, and materials recovery. Obviously, no single process can perform all these functions and a series of processes are required to provide adequate treatment. Often such treatment can be significantly more expensive than conventional treatment or disposal, and in the absence of strong legislation or regulations, is not widely practiced.

The Resource Conservation and Recovery Act of 1976 (RCRA), Public Law 94-580, requires the EPA to establish a major new regulatory program for the control of hazardous waste. Both EPA and State governments will share the responsibilities for implementing and monitoring the program. However, unlike the RCRA program for managing nonhazardous waste—a program that is controlled entirely by the States subject to Federal guidelines—the hazardous waste management program will be a strictly regulated system. Federal controls were deemed necessary by Congress because of the special dangers these wastes pose to public health and the environment.

RCRA's sections 3001 to 3006 and 3010 are designed to regulate hazardous waste from "cradle to grave," that is, from the point of generation to its ultimate disposal. The Act requires EPA to publish standards for generators, i.e., one whose activity or process produces hazardous waste; transporters; and owners or operators of facilities

Table 20

Functions, Applicability, and Resource Recovery Capability of Currently Available Hazardous Waste Treatment and Disposal Processes[a]

Process	Functions performed[b]	Types of waste[c]	Forms of waste[d]	Resource recovery capability
Physical treatment:				
Carbon sorption	VR,Se	1,3,4,5	L,G	Yes
Dialysis	VR,Se	1,2,3,4	L	Yes
Electrodialysis	VR,Se	1,2,3,4,6	L	Yes
Evaporation	VR,Se	1,2,5	L	Yes
Filtration	VR,Se	1,2,3,4,5	L,G	Yes
Flocculation/settling	VR,Se	1,2,3,4,5	L	Yes
Reverse osmosis	VR,Se	1,2,4,6	L	Yes
Ammonia stripping	VR,Se	1,2,3,4	L	Yes
Chemical treatment:				
Calcination	VR	1,2,5	L	
Ion exchange	VR,Se,De	1,2,3,4,5	L	Yes
Neutralization	De	1,2,3,4	L	Yes
Oxidation	De	1,2,3,4	L	
Precipitation	VR,Se	1,2,3,4,5	L	Yes
Reduction	De	1,2	L	
Thermal treatment:				
Pyrolysis	VR,De	3,4,6	S,L,G	Yes
Incineration	De,Di	3,5,6,7,8	S,L,G	Yes
Biological treatment:				
Activated sludges	De	3	L	No
Aerated lagoons	De	3	L	No
Waste stabilization ponds	De	3	L	No
Trickling filters	De	3	L	No
Disposal/storage:				
Deep-well injection	Di	1,2,3,4,6,7	L	No
Detonation	Di	6,8	S,L,G,	No
Engineered storage	St	1,2,3,4,5,6,7,8	S,L,G	No
Land burial	Di	1,2,3,4,5,6,7,8	S,L	No
Ocean dumping	Di	1,2,3,4,7,8	S,L,G	No

[a] From Ref. 18.

[b] Functions: VR, volume reduction; Se, separation; De, detoxification; Di, disposal; and St, storage.

[c] Waste types: 1, inorganic chemical without heavy metals; 2, inorganic chemical with heavy metals; 3, organic chemical without heavy metals; 4, organic chemical with heavy metals; 5, radiological; 6, biological; 7, flammable; and 8, explosive.

[d] Waste forms: S, solid; L, liquid; and G, gas.

that treat, store, or dispose of hazardous waste. The mechanism that will connect these separate standards into a continuous system of hazardous waste control is the *manifest system*. The manifest is a document used to record the movement of hazardous waste from the generator's premises to an authorized off-site treatment, storage, or disposal facility. The names, addresses, and signatures of the generator, transporter, and facility operator will appear on the manifest so that an individual wasteload may be tracked from the generator to disposal.

Seven regulations are proposed or are being developed pursuant to the Act to carry out the hazardous waste program. These regulations will be designated as Part 250 of Title 40 of the Code of Federal regulations (CFR). Their major aspects follow:

Hazardous Waste Identification (Subpart A). This regulation defines a hazardous waste. EPA has proposed two methods for determining whether a waste is hazardous. A waste is considered hazardous if it is listed in the regulation or if it has one of the following characteristics: ignitability, corrosivity, reactivity, or toxicity. The regulation lists about 160 wastes and waste streams. EPA may add or delete materials from the list and may add or delete characteristics that identify hazardous waste not on the proposed list. Any such change will be made by revising Subpart A through the rule-making process, giving the public an opportunity to comment on EPA's reasons for making the change.

Hazardous Waste Generators (Subpart B). This regulation defines the responsibilities of hazardous waste manufacturers. They have to determine if any of their wastes are hazardous; containers used for storage, transport, or disposal of hazardous waste have to be labeled properly; records should identify quantities, constituents, and disposition of waste deemed hazardous; a system of documents (manifest system) has to be used to assure that the waste is delivered to a licensed treatment, storage, or disposal facility; and reports have to be submitted to EPA or an authorized state agency on the quantity of hazardous waste generated and their disposal.

This Subpart exempts those who generate less than 100 kg of hazardous waste a month, provided the waste is disposed of in an approved facility.

Hazardous Waste Transporters (Subpart C). This regulation establishes a system which controls the transportation of hazardous waste. It calls for proper labeling, compliance with the manifest system, and keeping records on the transportation of hazardous waste, that identify the source and delivery point of the waste. It also specifies that hazardous waste may be transported only to an authorized treatment, storage, or disposal facility that the generator/shipper designates on the manifest form.

Hazardous Waste Treatment, Storage, and Disposal (Subpart D). This regulation establishes minimum standards for owners and operators of facilities that treat, store, and dispose of hazardous waste. These standards are designed to assure such facilities are safe. They include design and engineering standards to contain, neutralize, or destroy the wastes so they cannot contaminate groundwater, surface water, or the air. Furthermore, each site must be prepared to pay damages of up to $5 million for each case of damage while the site is operating.

Hazardous Waste Facilities Permits (Subpart E). This regulation requires the establishment of a permit program for owners and operators of facilities that treat, store, or dispose of hazardous waste. To obtain a permit the applicant must submit information on the site and on the composition, quantities, and concentrations of hazardous waste proposed to be handled at the facility. Also, information must be provided on the volume and frequency of such activities over a given period of time.

State Hazardous Waste Programs (Subpart F). This regulation sets forth guidelines to assist states in the development of hazardous waste management programs and the procedures for states to follow in seeking authorization from EPA for their programs. States seeking authorization must demonstrate that their programs are equivalent to or stricter than the Federal program. Authorization by EPA may be withdrawn if EPA later determines that the program is not being properly administered or enforced.

Notification Requirements (Subpart G). This regulation requires all generators, transporters, and facility operators to notify EPA (or an authorized State) of their hazardous waste activities. The notice must state the location and general description of the activity and the hazardous waste handled by the person submitting the notice.

The above set of regulations lay the framework for a rather comprehensive regulatory program for managing this nation's hazardous wastes. EPA is committed to issuing all final regulations by December 1979, and is authorized, under Section 3008 of RCRA, to prosecute any person who fails to comply with any requirements under the regulations. Violations can result in civil or criminal penalties or both.

B. Radioactive Solid Waste

The total amount of radioactive material in the world has been significantly increased by man caused by the testing of nuclear weapons and the use of radioactive materials in research laboratories, industrial operations, and medical research and treatment. From the point of view of solid waste management, the most significant source of possible radioactive contamination (not considering atmospheric testing of nuclear weapons) is the commercial nuclear electric power industry. Total installed nuclear electrical generating capacity in the United States in 1974 was 25,000 MW and is projected to increase to about 1 million MW by the year 2000. The problems of waste handling and disposal will undoubtedly increase as the nuclear energy program is extended and diversified over the next several decades.

The technology for handling the growing waste quantities does exist [34]; the real problems lie in making the critical decisions related to environmental standards and criteria, that will affect the environmental policies of the US toward future radioactive waste management. Some of the present directions will be described in the following paragraphs, but first the origins of solid wastes from nuclear power plants must be understood.

1. *Sources and Quantities*

A nuclear power reactor is an installation in which nuclear fission can be maintained as a self-sustaining yet controlled chain reaction. It is a kind of furnace in which uranium is the fuel burned, and heat, neutrons, and radioisotopes are produced. A reactor is often designated according to the moderator or coolant used within it. The function of the coolant is to carry off the heat of the nuclear reaction from the reactor and to transfer this heat to the power generating system. Various coolants used are liquids and gases, e.g., light water, heavy water, organic liquids, molten metals, and helium. Light water reactors (LWR) are the predominant type of nuclear power plant currently being installed and most of what follows will pertain to this type of reactor.

To understand the various types of solid wastes resulting from nuclear power production, one must first understand the various phases of the nuclear fuel cycle [35]. The nuclear fuel most used today is uranium, and thus the nuclear fuel cycle starts with the mining of uranium ore, and processing the ore in a mill to yield a high-grade U_3O_8 concentrate (yellow cake). This concentrate is sent to a refinery where the uranium is extracted with an organic solvent and converted, by heating, to almost pure UO_3. This refined material contains the natural content of 0.7% U-235 (which fissions in a reactor, yielding heat) and 99.3% U-238 (which does not fission, but can be converted to the fissionable Pu-239). Since U-235 concentration must be around 3% in order for it to be used as nuclear fuel, the next step involves enrichment. First the uranium is converted to gaseous UF_6 and then a gaseous diffusion process is used to separate (enrich) the faster diffusing U-235 hexafluoride isotope from U-238 hexafluoride. The enriched UF_6 (2–3.5% U-235) is converted to a different chemical form, e.g., UO_2, and sent to a fuel-fabricating plant where pellets are formed and loaded into steel or zircalloy tubes (cladding) called elements. The elements are mounted into tube bundles called assemblies. In the reactor, the fuel undergoes fission and produces energy and also an excess of neutrons which in turn form new radioactive materials. At some stage the fission products interfere with the efficiency of the fission chain reaction to the point where it becomes necessary to remove the fuel elements, which still contain unreacted fuel, and to send them to a fuel reprocessing facility. Here the uranium and plutonium remaining in the fuel elements is recovered. The recovered uranium, depending on the U-235 content, would be sent either for enrichment or blended with other fuel materials. The plutonium would be stored for future use in breeder reactors. This completes the nuclear fuel cycle.

The fuel cycle described above is currently not fully operational in the United States because the back end of the cycle, namely the reprocessing of spent fuel and recovery of uranium and plutonium, has not been fully implemented†; however, the cycle as described above is often assumed for future operations.

All phases of the fuel cycle from mining to fuel fabrication involve solid wastes having relatively low radiation intensity equivalent to natural or slightly enriched uranium. Residual solids and tailings from milling operations contain a large fraction of the radium that was originally present in the ore. The blowing of fine radium particles can create an inhalation problem, whereas leaching of radium into nearby streams contaminates water that might be used for drinking or irriga-

† The only commercial reprocessing facility in the United States is the Nuclear Fuel Services plant at West Valley, N.Y. However, this company has withdrawn from the reprocessing business because of economics and regulatory requirements. The Allied-General Nuclear Services plant in Barnwell, S.C. could process 1500 metric tons per year if it were operational. But the company stopped construction when it became obvious that commercial, near-term operation was not going to be licensed [36].

tion. The blowing problem is avoided by chemical stabilization of the residual piles, and the leaching is contained by proper retaining walls. It is estimated that the annual solid waste resulting from mining and milling operations associated with a 1000 MW(e) nuclear power plant is about 2.62×10^6 tons (2.38×10^6 metric tons) with a total radioactive level of about 1125 curies† [37]. This works out to a specific radioactivity level of 4.73×10^{-10} Ci/g, which can be classified as low-level waste.

A wide assortment of solid wastes slightly contaminated with radioactivity is routinely generated in nuclear power plants. These wastes include paper, plastics, wood, rags, glassware, protective clothing, small tools, concentrates from the treatment of liquids containing low level radioactivity, contaminated resins from ion exchange demineralizers, high efficiency particulate air filters, and some liquid wastes mixed with adsorbent materials. These wastes, categorized as low level wastes, will typically amount to something of the order of 2500–5000 Ci/yr of mixed radioactive species from a 1000 MW(e) power plant, and will occupy a volume of approximately 270–540 55-gall drums/year (average density of about 1100 kg/m^3) [38]. The total volume of these solid wastes generated in 1975 was estimated to be about 3 million ft^3 (85,000 m^3) and is projected to double in volume by 1980.

The really significant generation of radioactive solid waste begins with the use of the fuel in the reactor and the subsequent reprocessing of spent fuel.

Nuclear plants use fuel rods with a life span of approximately three years. Each year, roughly one-third of the spent fuel rods are removed and stored in cooling basins, either at the reactor or elsewhere. Depending on the reactor type, fuel characteristics, and plant capacity factor, a 1000-MW plant discharges between 23 and 31 tons of spent fuel per year. At present (early 1979), over 4000 tons of spent fuel are being held in temporary storage basins.

For more than 20 years, the assumption was that the spent fuel would be reprocessed. In 1977 a new national policy was revealed to preclude reprocessing for the foreseeable future.

Reprocessing of spent fuel rods from nuclear power plants entails

† A sample of radioactive material is said to have an activity level of 1 Ci when it disintegrates at the rate of 3.7×10^{10} disintegrations per second. The mass of a substance having an activity of 1 Ci is $(At_{1/2})/1.12 \times 10^{13}$ g, where A is the atomic mass and $t_{1/2}$ is the radioactive half-life in seconds. Thus for a given atomic mass, the mass of a radioisotope having an activity of 1 Ci depends on its radioactive half-life.

recovering uranium and plutonium from spent fuel for reuse in light water reactors. Up to the time of reprocessing, the fission products along with the unused fissionable materials have been well contained in the nuclear fuel during its life in the reactor core. Now it is necessary to dissolve these materials into solution, separate the fissionable material, and return it to fabrication plants to be made into new fuel elements again. Fuel reprocessing is the step where the tremendous inventory of fission products, activation products, and fissionable materials is opened up, and consequently great care must be exercised in order to prevent contaminating the environment with deadly radioactive substances.

Table 21 shows the estimated annual generation of solid wastes resulting from a proposed 750 metric tons/year spent fuel reprocessing facility [39]. The activity level of this solid waste varies over seven orders of magnitude from 10^5 to 10^6 Ci/ft^3 (3.5×10^6 to 3.5×10^7 Ci/m^3) down to 10^{-2} or less Ci/ft^3. The most notorious solid waste is the solidified high-level liquid waste (row 1, Table 21). In the first cycle of the solvent extraction process used in fuel reprocessing, most of the fission products remain in the aqueous phase and are separated from the uranium and plutonium that enter an organic (solvent) phase [40, 41]. It is this fission product aqueous waste stream that makes up the high level liquid waste. This waste contains over 99.9% of the nonvolatile fission products and from 0.1 to 0.2% of the uranium and plutonium originally present in the spent fuel. It is highly corrosive and contains sufficient decay heat to cause boiling.

The leached hulls referred to in row 2 of Table 21, result from a preliminary step during which the spent fuel elements are mechanically disassembled into their components (claddings, end pieces, other hardware) and are then treated with strong, hot nitric acid in order to leach out the metallic oxides of the uranium, plutonium, and other transuranic compounds and fission products in the fuel rod pieces. The result is the leached hulls to be disposed of separately, and a solution of nitrates that is passed on to a solvent extraction unit.

Intermediate level wastes (row 3 of Table 21) consist mostly of sodium nitrate from solvent cleanup (67%), sand and filter aids from fuel pool cleanup (23%), and spent resins from liquid treatment (10%).

Failed equipment refers to plant hardware such as dissolvers, tube bundles in evaporators, condensers, fractionators, pumps, valves, etc., that wear out in service and have to be replaced.

Alpha particle contaminated combustible wastes consist of paper, cloth, gloves, plastic, and miscellaneous contaminated trash, much of which originates in the plutonium oxide process areas. This type of

Table 21

Annual Solid Waste Generation from Fuel Reprocessing Plant (750 metric tons/year)[a]

Type of waste	Original Volume ft³	(m³)	Percent of total	Radioactivity level Ci/ft³	(Ci/m³)	Shipping container	Final Volume ft³	(m³)	Percent of total	Comments
High-level waste	1,400	(40)	1.3	10^5–10^6	(3.5×10^6 to 3.5×10^7)	1½ by 15 ft canisters	2,800	(80)	3.9	High-level liquid waste solidified by calcination and dilution prior to disposal
Leached hulls	9,200	(260)	8.3	500	(1.8×10^4)	Steel canisters	9,200	(260)	12.8	Decontaminated by washing prior to disposal
Intermediate-level wastes	7,120	(200)	6.5	1–100	(35–3500)	55-gall drums	7,230	(205)	10.1	Resins and filter aids are mixed with concrete; sodium nitrate salts are evaporated and dried prior to disposal
Failed equipment	10,360	(290)	9.3	0.1–5	(3.5–180)	Special containers	15,500	(440)	21.6	Decontaminated as necessary by washing
Alpha waste (combustible)	17,000	(480)	15.3	0.05–0.5	(1.8–18)	55-gall drums	3,700	(105)	5.2	Volume reduction by combustion or acid digestion
High-efficiency particulate air filters	2,000	(60)	1.8	0.05–1.0	(1.8–35)	Special containers	500	(14)	0.7	Compaction prior to packaging for disposal
Low-level wastes	63,800	(1810)	57.5	0.005–0.05	(0.18–1.8)	55-gall drums	32,700	(930)	45.7	Combustibles are incinerated or compacted, sludges are dried, prior to disposal
Total volume	111,000	(3140)	100.0				71,600	(2030)	100.0	

[a] From Ref. 37.

combustible waste is to be distinguished from the combustible waste mentioned in the next paragraph.

The low level wastes account for about half the total solid waste from the reprocessing plant. These wastes contain no significant plutonium and comprise mostly combustible trash (55%), slurries (37%), and salts (8%) from various process stages dealing with low-level contaminated liquids.

As can be seen from Table 21, solid wastes from a reprocessing facility are quite varied in quantity and character and pose a diversity of disposal problems. It must be reemphasized that fuel reprocessing is as yet a rather limited activity and the figures presented in Table 21 only represent potential, not existing, solid wastes. Until such time that commercial reprocessing becomes common practice, all spent fuel rods must be placed in interim storage facilities provided with adequate cooling.

2. *Handling and Disposal*

The handling and disposal of radioactive solid wastes is largely dictated by the characteristics of the wastes; i.e., whether they are low-level or high-level wastes.

a. *Low-level Wastes.* Low-level solid wastes are those which do not generate sufficient heat or radiation to require special cooling or shielding and can be disposed of under controlled conditions. These wastes are generally at an activation level of about 1 Ci/ft^3 (35 Ci/m^3) or less, and make up the bulk of all solid radioactive wastes generated. Typical examples are resins or filters that trap activation or fission products, alpha and other radioactivity contaminated trash such as paper, rags, plastics, etc., solids resulting from treatment of liquid wastes, certain salts or sludges that are dried to a solid, and low-level liquid wastes that are dried or evaporated, or absorbed on absorbent materials such as vermiculite.

Traditionally, near-surface land burial (similar to sanitary landfill methods) has been the method for disposal of solid low-level radioactive wastes in the United States, and it still remains the principal approach. Burial operations consist of temporary storage, burial in trenches of packaged radioactive wastes, and continual monitoring of the radioactive characteristics of the surrounding grounds, air, and water to ensure the detection of any significant migration of radioactivity. The packages (cartons, crates, or drums) of radioactive wastes received at burial facilities must meet the Department of Transportation requirements [42] providing protection against high-speed impact, puncture, fire, and water immersion. The burial process may vary from simple burial in trenches and covering with earth, to the placement of containers in excavations that have been carefully surfaced to prevent access to groundwater [43]. The selection of the burial site should be based on an analysis of site topography, geographic location, geology, meteorology, hydrology, and the normal usage of ground- and surface-waters in the general area. The migration of radionuclides from burial trenches can be substantially reduced if water-to-waste contact is eliminated and trenches are kept dry.

Low-level wastes are buried at six federal or state-owned sites in Kentucky, South Carolina, Illinois, Washington, Nevada, and New York. These sites are estimated to have the capacity to handle low level wastes until the years 1995–2000. The projection of waste generation rates from the commercial nuclear fuel cycle indicate future needs to dispose of very large volumes of radioactive wastes (approximately 8–24 million m^3 accumulated by the year 2000), and consequently it has been deemed desirable to reduce the volume of such wastes as much as possible prior to disposal. The prime method of volume reduction currently being used is compaction. A number of different incineration concepts (acid digestion, controlled air incineration, cyclone incineration, pyrolysis) are being developed for treatment of the combustible fraction (especially alpha-contaminated combustibles) of low-level wastes [44]. Low-level waste is a significant fraction of the total solid wastes generated throughout the nuclear fuel cycle and reducing its volume not only aids in reducing final disposal requirements, but also reduces the shipping and storage costs. Furthermore, incineration results in an immobilized ash or residue which stabilizes the waste for terminal isolation.

The category of low-level radioactive wastes receiving the most emphasis from regulatory agencies is the highly toxic transuranic (TRU) waste; these are wastes contaminated by elements heavier than uranium such as neptunium, plutonium, americium, curium, etc., with their relatively mobile alpha contaminants. It has been indicated that special disposal techniques other than simple near-surface land burial will be necessary for TRU wastes. These disposal techniques have not yet been defined, but in the interim shipments of low-level wastes to disposal sites will be limited to 10 nCi (10^{-9} Ci) per gram of transuranic elements. This is an arbitrary "safe level" and not so easily ascertained because of the difficulty in obtaining representative samples.

Currently, all TRU low level wastes buried at federal sites are being segregated, packaged, and stored in such a fashion as to be readily retrievable as contamination-free packages for an interim period of 20 years [45]. Future decisions regarding TRU wastes will have a very significant effect on the commercial nuclear power industry.

b. *High-level Wastes.* High-level wastes include spent fuel rods and some of the wastes generated during the reprocessing of these spent fuel rods. During the reprocessing step, chemical dissolution and solvent extraction results in an aqueous nitric acid solution that contains better than 99.9% of the nonvolatile fission products, the transuranium actinides formed in the reactor, and about 0.5% of the fuel not recovered during reprocessing. Because of the high levels of radioactivity involved (typically 10^5–10^6 Ci/ft^3), high-decay heat rates (about 5000 Btu/h-ft^3), the long half-lives of many of the radionuclides (some transuranics have half-lives ranging between 10^3 to 10^6 years or more) and their biological hazards, these wastes must be perpetually isolated from man's environment. There are currently about 600,000 gallons (2.27×10^6 L) of this waste to be disposed of, resulting from reprocessing nuclear power plants and more than 100 times that quantity from government operations. At present such wastes are kept in solution and stored in large tanks equipped with cooling systems. Under current regulations in the United States applicable to commercial fuel reprocessors, these wastes may be stored in liquid form for up to 5 years and then must be converted to a suitable solid form and delivered to a government repository within 10 years. The first deliveries may be expected in the late 1980s.

From the standpoint of solid waste management, two aspects of high-level waste management are worthy of discussion. The solidification of high-level liquid wastes and their confinement or isolation (in a government repository) from our environment.

The general criteria which a successful solidification process should fulfill are economics, simplicity, and capability of operation by remote control. Furthermore, the solid product should be a dense, compact, and stable solid mass which will not revert to liquid or gas in the presence of high radioactivity or temperatures, and should withstand leaching by either rainwater or natural ground water.

Four solidification processes have been developed in the United States to the point of demonstration on an engineering scale: fluidized bed calcination, spray calcination, pot calcination, and phosphate glass solidification. In all four processes heat is applied to drive off volatile constituents, primarily water and nitrates, resulting in either a calcined solid or a melt that will cool to a monolithic solid. The latter generally requires dilution of the waste with nonradioactive materials (20–40% of the total solid) to incorporate the waste into materials (glass or ceramics) that have low solubility in water and are fusible at reasonably low temperatures (below about 2200 °F or 1200 °C). Characteristics of typical final solid waste forms from the four processes are shown in Table 22. The processes are briefly described below.

In *fluidized bed calcination* liquid waste is sprayed into a heated fluidized bed through pneumatic atomizing nozzles located on the wall of the calciner vessel. The primary solidification processes occurring during calcination are evaporation of water and the thermal decomposition of metallic nitrates to oxides. The resulting calcine is formed in layers on the fluidized bed particles and the particles are intermittently withdrawn from the bed. The resulting mixture of powdery solids and granules is usually in the size range of 0.002–0.02 in. (0.05–0.5 mm). The temperature within the fluidized bed should not be too high (normally around 600 °C) otherwise the particle size becomes too great for efficient fluidization. Sufficient attrition normally occurs to counteract particle growth to result in a steady-state particle size and bed weight. The resulting calcine may be the final waste form or may be coupled with an in-can melter where glass-forming additives are added for production of borosilicate glass. The melter usually operates at 1000–1100 °C and the glass is cooled and stored in canisters [47, 48].

The *spray calciner* consists of a vertical heated-wall spray drier with an atomizing nozzle located at the top center of the vessel. Liquid waste is pumped to the nozzle where it is atomized by pressurized air, producing small droplets which are dried and calcined in flight in the 1300 °F (700 °C) temperature maintained at the vessel walls. Calcination time is extremely short. The calcine solids drop vertically and at the bottom of the calciner are mixed with glass-forming additives. The mixture then flows through a diversion valve to batch-operated in-can melting canisters where the calcine reacts with the glass formers. After each canister is filled and contents are completely melted, it is cooled, capped, seal welded, and stored [49].

In *pot calcination*, liquid wastes are preevaporated to a concentration factor of about 1.4, then fed to a stainless steel cylindrical pot which also serves as the storage or disposal container. The liquid is then continuously boiled away, and the vapor is fractionated to recover nitric acid and recycled to the pot in a closed loop. When the pot is full of solids, the addition of aqueous wastes is stopped and the temperature of the solids is raised and held at about 1650 °F (900 °C) to complete denitration and dehydration. Feed additives can be added to result in a glass rather than a calcine cake.

Phosphate glass solidification is a continuous process that converts waste directly to a glass. The process consists of mixing phosphoric

Table 22
Characteristics of Solidified High-Level Waste[a]

	Pot calcine	Spray phosphate ceramic	Phosphate glass	Borosilicate glass[b]	Fluidized bed calcine
Form	Scale	Monolithic	Monolithic	Monolithic	Granular
Description	Calcine cake, friable	Ceramic hard, tough	Glass hard, brittle	Glass hard, brittle	Calcine, mean particle diam. 100–500 μm
Bulk density, g/cm^3	1.2–1.4	2.7–3.3	2.7–3.0	3.0–3.5	1.0–1.7
Wt% fission product oxides (max)	90	30	25	50	50
Thermal conductivity, W/(m^2)(°C/m)	0.3–0.4	1.0–1.4	0.8–1.2	1.0–1.4	0.2–0.4
Leachability in cold water, g/cm^2-day	1.0–10^{-1}	10^{-3}–10^{-5}	10^{-4}–10^{-6}[c]	10^{-5}–10^{-7}	1.0–10^{-1}

[a] From Ref. 46.
[b] Produced by either spray or fluidized bed calcining followed by melting, or by in-canister vitrification processing.
[c] Devitrified phosphate glass exhibits increased leachability (leach rates = 10^{-2} to 10^{-3} g/cm^2-day).

acid with highly concentrated waste in a 1:5 ratio. The mixture is then concentrated to a thick slurry by heating to about 275 °F (135 °C) to drive off the water and nitric acid to an off-gas handling system where it is condensed. The slurry is routed to a melter where dehydration and denitration are completed and the material is melted. The molten phosphate glass is then dropped into a canister where it cools and solidifies.

The most advanced solidification process is a French design which comprises an inclined rotary calciner, coupled to a melting furnace [50]. The process is designed to operate continuously, and an industrial scale plant has been operational since mid-1977. This plant is capable of producing 100 tons/yr of glass, the equivalent of 600 tons/yr of reprocessed fuel.

Some of the more dangerous isotopes contained in solidified radioactive wastes are fission products such as strontium-90 (half-life 28 yr), cesium-137 (half-life 30 yr), and iodine-129 (half-life 1.6×10^7 yr), and transuranics such as plutonium-239 (half-life 24,000 yr). The long-term potential risks posed by such materials dictate that any disposal scheme should provide for the permanent storage of the wastes in such a manner that isolation from man's environment is provided for the periods of time required for effective removal of the hazard by decay; this means, from a few hundred years to thousands of years. Furthermore, such disposal should require minimal reliance on human surveillance and intervention.

Current expectations are that high-level solidified wastes will be disposed of in Federal repositories. The wastes will reside from 1000 to 3000 ft (300–900 m) below the ground in stable geologic formations. Either salt, crystalline, or agrillaceous rock formations could be used as storage area. However, salt beds have received the most attention to date because in many areas salt deposits have remained virtually undisturbed since the time they were formed hundreds of millions of years ago. Salt dissipates heat better than other types of rock, thus serving as an excellent sink for radioactive decay heat. Furthermore, the very presence of salt in these formations attests to the fact that salt has, in general, been isolated from circulating ground water. For any form of geologic storage, it is only through transport in groundwater and through aquifers that buried radioactive wastes could come in contact with man's environment.

In assessing possible repository sites, area tectonics, seismicity, erosion, structure, hydrogeology, and mineral resources have to be analyzed. Intensive drilling and core sampling aids in these analyses. In addition, tests have to be performed for rock stress, heat transfer, radiological effects, and soil movement.

To develop the repositories, large vaults will be excavated in the chosen formation. High level solidified wastes will be set in metal-lined holes carved in these vaults. Waste will then be capped with a removable shield plug. The metal shield around the waste will allow it to be retrieved if necessary. The site will be continuously monitored for heat and radioactive releases into the salt formations. All waste will be retrievable.

At present (1979), it is uncertain what the final high-level waste form will be. If the Department of Energy (DOE) allows the use of mixed oxide fuel (uranium and plutonium oxides) in light water reactors, then reprocessing of spent fuel will expand to commercial levels, and the high-level waste will be glass or calcined solid cast in stainless steel canisters. The highest projected inventory of high-level solidified wastes for the year 2000 is about 80,000 canisters 1 ft. in diameter and 10 ft long. Typically, ten to twelve such canisters will contain the high-level wastes produced each year by a 1000-MW nuclear power plant.

If DOE decides against fuel reprocessing, then the spent fuel rods themselves (with proper encapsulation) will be placed in repositories. Prior to repository storage, interim storage at water pool facilities will be provided. In this case the volume of waste to be stored will be quite high (about seven times the volume of solidified liquid waste from reprocessing) and will probably increase the size or number of required repositories over current projections [36, 51].

The requirement that high-level radioactive solid wastes be isolated for hundreds of years seems alarming to some because few things in our environment last that long. The major questions regarding the repository storage proposition are: (a) will an adequate timetable be provided to cope with the backlog of spent fuel, (b) will the storage solution be effective and economic and, (c) will this be the final safe and permanent place for our nuclear wastes [52]. In the meantime, potentially attractive disposal concepts are also being studied. These include seabed disposal, which entails controlled emplacement of solidified waste canisters in stable geologic formations in the deep ocean; ice sheet burial in Antarctica or Greenland, either free flow or anchored; extraterrestrial disposal by shooting the waste into solar orbit or even out of the solar system; and transmutation which would convert long-lived transuranics and fission products to shorter-lived elements (with a lesser risk potential) by bombarding the nuclei with either photons or subatomic particles [34, 46].

c. *Intermediate-level Wastes.* In addition to the high- and low-level solid wastes, there is the category of intermediate-level solid wastes. These wastes are not self-heating like the high-level wastes, but are contaminated to a level of anywhere from 1 to 1000 Ci/ft^3 (35–34,000 Ci/m^3) and consequently must be disposed of in a safe manner. Intermediate-level solid wastes such as sludges or salt slurry are either dried to a solid or mixed with concrete to form a solid. One source of intermediate-level solid wastes that has received much attention is cladding hulls (see Table 21). These are the metal hardware pieces that are mechanically separated from fuel assemblies during reprocessing. These hulls are contaminated with fission products and transuranics from the incomplete dissolution of the fuel and with induced activity, including plutonium, from the irradiation of uranium impurities in the zirconium metal.

The former practice of disposing of intermediate level wastes has been through cribs or large boxes open at the bottom only and buried with sufficient cover to preclude surface radiation problems. This process takes advantage of the long soil exchange column between the crib and the groundwater to fractionate the radioactive cations. Such practice is acceptable only when the radioactive levels are low enough and where adequate decontamination and decay within the soil is assured, somewhat as in the case of low-level wastes. However in the event that alpha-emitting transuranics are present, a shielded storage facility for short-term storage is necessary along with the provision for final shipment to a federal repository. At present there is a need for specifications on the container dimensions and activity levels that DOE will accept at a repository.

Much research is currently being conducted toward the treatment of cladding hulls. The most promising approach appears to be decontamination of the hulls to a low alpha-activity level, followed by chopping up and melting the metal for reuse. If the zirconium cladding is separated from the other materials and decontaminated before melting, it could be cast into ingots of sufficiently high quality to be suitable for reuse as cladding when the induced activity had decayed sufficiently to permit fabrication.

C. Hospital Waste

The total amount of solid waste produced in hospitals is impressive. Hospitals were among the first places to be hit by the "disposable revolution," and consequently disposable syringes, bedpans, and gowns in the operating room, etc., are now considered absolutely necessary. In a New York medical center, for example, 7000 paper plates, 7000 paper cups, and 3000 bed sheets must be disposed of each day.

Without the use of disposable items, a hospital will generate about 12 lb (5.4 kg) of solid waste per bed per day. With a determined effort to use disposable materials, this figure is easily doubled. Of this amount, 2–4% can be defined as being potentially hazardous. Such wastes include pathological and surgical waste, clinical and biological laboratory waste, needles, syringes, patient-case items (linen, personal, and food service items), drugs, and chemicals.

The special handling necessary for hazardous hospital wastes begins with proper storage of such wastes. Double-strength bags that are impervious to moisture and resistant to ripping or tearing under normal use make excellent containers. In addition, containers for blades, needles, or other sharp waste objects should be puncture-proof. Whenever possible, the wastes should be processed prior to storage. Processing should include sterilization by autoclaving or chemical treatment in order to destroy the disease transmission potential of infectious wastes.

The preferred disposal methods for hazardous hospital wastes are either combustion or landfill. The combustible solids should be completely incinerated followed by disposal of the ash in a sanitary landfill. Noncombustible solids should be buried in sanitary landfills packaged in containers that clearly identify them as hazardous wastes. If operations such as baling or shredding occur at the landfill, they should not include loads of hazardous wastes.

The proper handling of hospital wastes can nullify the hazardous potential of such wastes to patients, staff, or the community. The reader is referred to Refs. 53–55 for further information.

D. Packaging Waste

The ever-increasing generation of solid waste in the United States owes in some measure to the increased consumption of packaging and packaged items. The estimated national per capita packaging consumption was 404 lb (183 kg) in 1958, 475 lb (215 kg) in 1966, and over 650 lb (295 kg) in 1976. Combined with the population increase, the total tonnage of discarded packaging has increased substantially over the past few decades.

Table 23 shows the trend in packaging consumption according to different packaging materials. Paper, glass, metal, and wood are all long established packaging materials with plastics being the most recent addition to the field. The growth in plastics is significant because plastics appear not only as containers, but also as coatings and laminants in combination with metals, glass, and paper, thereby permeating the

Table 23
Packaging Consumption in the United States

	1966		1976	
Material	Million tons	%	Million tons	%
Paper	25.2	55	39.5	57
Glass	8.2	18	13.2	19
Metals	7.1	16	9.0	13
Wood	4.1	9	4.9	7
Plastics	1.0	2	2.8	4
Total	45.6	100	69.4	100

entire packaging industry [56]. Plastics are further discussed in a later section of this chapter.

A discussion of packaging in relation to solid waste management necessitates a review of the entire solid waste disposal problem. Table 24 shows how packaging relates to six prominent aspects of the solid waste problem, namely, collection, processing, littering, soil and groundwater pollution, air pollution, and loss of resources [57].

Collection of solid wastes leads to enormous national expenditures—in excess of 4 billion dollars annually—and the role of packaging materials in this connection is significant. The total annual packaging waste tonnage (assuming 90% of the totals in Table 23 enter the solid waste stream) has increased by over 20 million tons per year from 1966 to 1976. At a modest cost of $8/ton per year, the increase caused by packaging wastes alone is about $160 million per year. Furthermore, packaging materials are becoming less dense owing to the increased use of paper and plastics, and consequently require more space in collection vehicles.

Packaging materials often necessitate development of better disposal technology, such as proper control of incinerator operations, waste reduction by grinding and shredding prior to landfill, etc. These techniques, although readily available, suggest more processing and increased cost.

Packaging also plays a dominant role in littering. On a tonnage basis, packages may not be the greatest contributors to litter; on the other hand, discarded packages account for the majority of items in litter. This subject is dealt with in a later section.

Packages play an indirect role in the contamination of soil and groundwater, because some packaging materials (aluminum, glass, plastics) are not degradable and persist in the environment. Among the

Table 24
Packaging Related to Aspects of Solid Waste Problems[a]

Aspect	Basic problem	Packaging contribution	Possible solution
Waste collection	Collection is labor-intensive, thus costly	Proportion of packaging materials in municipal waste growing	Automation of collection On-site volume reduction and disposal
Waste processing	Disposal technology is relatively backward Insufficient support of research and development Land available for waste disposition is dwindling	Packaging materials are usually nondegradable by natural processes	Retooling of financial support for waste processing Development of new disposal technology Modify packaging materials to make them more degradable
Esthetic blight from littering	Public carelessness and/or indifference	Packaging is major component of litter	Intensive anti-litter publicity Rigorous anti-litter law enforcement Economic incentives for returning containers

Soil and groundwater pollution from decomposition of organics	Inadequate waste processing Poor selection of disposal sites	Packaging plays indirect role; some materials may contain organic residues	Relocation of adequate sites Replacement of dumps by incinerators
Air pollution from waste combustion	Existence of burning dumps Poorly operated or designed incinerators	Role of packaging same as for solid waste in general	Elimination of burning dumps and inadequate incinerators R&D on improved combustion equipment High-quality pollution abatement equipment for incinerators
Loss of potentially valuable raw materials	Solid waste has low value because of contamination by intermixing Low cost of virgin material reduces demand for secondary materials	Many high value raw materials combined in single package Exploitation of concept of throwaway containers	New technology for low cost automated separation of heterogeneous solid waste Incentives for wider industrial use of secondary materials Modifications in package design which would utilize more homogeneous materials

[a] From Ref. 57. Reprinted with permission of the American Chemical Society.

persistent variety of packaging materials there may be some that contain organic residues which could lead to soil or groundwater contamination.

The combustion of packaging materials such as paper, metal, glass, wood, plastics, and textiles does not lead to any unusual gaseous emissions (in comparison with solid waste material in general) except in the case of poly(vinyl chloride) which decomposes into undesirable chlorine compounds. The sulfur content of packaging materials is lower than that of most hydrocarbon fuels; a typical ton of packaging contains about 2 lb of sulfur, a mere 0.1% by weight.

The principal resources squandered in throwaway packaging are paper, metal, glass, and plastics. Any attempts at recovering these resources would first require the separation of the heterogeneous mix of solid wastes; the recovery of the valuable materials would depend on the economic incentives for doing so.

Packaging is a complex activity pursued by a number of interlocking industries that can be classified broadly as material producers, package fabricators, and packagers. Consequently, it must be recognized that a solution to the packaging waste management problem cannot be achieved by any single measure. A spectrum of alternatives must be explored in order to arrive at any reasonable strategy.

The various alternatives (as proposed by McGauhey [58]) available to the packaging industry for reducing packaging wastes are:

1. Less volume/material in the package.
2. Redesign the product to use less packaging.
3. Reuse packages for initial purposes.
4. Design packages with disposability in mind.
5. Use more easily recoverable materials in packaging.
6. Limit variety of materials used in packaging.
7. Design the package to be consumed with the product.

The first two alternatives aim at reducing the package-to-product volume ratio. The ways in which the volume of packaging might be reduced without interfering with the objectives and needs of packaging is a matter which deserves serious consideration by those fabricating packages as well as those implementing the packaging.

The reuse of packages can be implemented two ways: within the industry itself, or in the industry-to-consumer relationship. The former results in less packaging waste to be disposed of by industry, the latter results in less packaging waste to be disposed of by the consumer; the net result is less packaging waste in the environment.

Alternatives 4, 5, and 6 listed above are closely related, each being concerned with improving the disposability of, and/or material recovery from, the discarded packaging. Disposability should take into account both collection and final disposal. To facilitate collection, package design should encourage hand disposal to reduce bulk. Furthermore, design can make packages more collapsible and structurally more compact, taking up less room. The toothpaste tube is perhaps a good example. Why use plastic instead of metal for the tube? The collapsible metal tube is convenient to dispense and becomes smaller as it is used. When disposed, it occupies one-tenth the space of a noncollapsible plastic tube. It may be argued that the plastic tube has a lower initial cost, but this does not take into account the increased cost of collection owing to greater bulk.

From the final disposal standpoint, destructability of packaging material is of some concern. Packaging materials should be amenable to disposal by some of the more common or promising disposal methods such as landfill, incineration, composting, digestion, wet oxidation, or pyrolysis. This means that, for instance, incineration should produce the normal end products of carbon dioxide and water, without the accompanying undesirables such as nitrogen oxides or chlorine compounds; digestion and composting should biodegrade packaging synthetics with the same ease that they do other organic wastes. These requirements obviously cannot apply to all types of packaging; nevertheless, the alternative of deliberate design and materials selection holds some prospects of easing the problem of packaging wastes management although it may not reduce the volume to be hauled.

In order to facilitate resource recovery, of first concern should be the segregation of materials for return to the reprocessing cycle. The technology for separating refuse, prior to materials recovery, typically entails magnetic separation, vacuum pickup, ballistic or cyclone separation of shredded particles, screening, etc. Packaging design should bear in mind the complexities of recovery of resource material from a heterogeneous mixture of solid wastes, and the technology available to achieve such recovery. Packaging materials should be limited to those kinds of chemical compounds, alloys or combinations which can be reprocessed for reuse as resource materials, with a reasonable amount of effort and within technological constraints. A good example is the aluminum beverage can in the seamless version versus the one with a tin seam. The latter variety entails an additional detinning step before the aluminum scrap can be reused.

Alternative 7 is a variation of the ice cream cone concept, and although not widely applicable, is certainly worth consideration.

If the objective of society is to reduce packaging wastes and the attendant waste management problems, the methods for achieving this end must clearly involve not only industry, but also active public concern and governmental action.

Public concern must begin with the consumer's awareness of the price that is paid in packaging waste in return for the conveniences demanded in the marketplace. Furthermore, there must be specific direction as to how the consumer can influence the government and industrial sectors of the economy.

Governmental involvement in the packaging wastes problem may be felt from the impact of such contemplated measures as (a) a national "bottle bill" placing a mandatory deposit on all beverage containers, (b) a tax on all containers according to their weight, (c) a disposal charge relative to the ease of disposal and/or recovery of original materials. All of these are drastic actions and severe restrictions on a supposedly free enterprise system. The overriding question is whether we as a society can afford the luxury of unrestricted packaging waste.

E. Litter

Litter is a specific packaging waste problem. It is usually only a visual affront and seldom a health problem. Because of its noncritical nature, litter collection in the city is one of the least vital services, and a city can often neglect the service without causing a crisis.

Three types of costs are associated with litter, none of which are easily quantifiable. One is esthetic blight. Although there is little disagreement with the fact that more litter is uglier than less litter, little work has been done to translate this fact into dollars and cents. A second type of cost is medical. Approximately 80% of an estimated 300,000 litter-caused injuries in California alone were caused by broken glass and pull-tabs in 1974 [59]. The third type of cost is litter cleanup. Using the existing cost of litter pickup, actual expenditures, which reduce litter only to a limited extent, are currently estimated to range between $500 million and $1 billion annually [60, 61].

In addition to the difficulty in placing a total cost on litter, there is little agreement on how to measure litter. Even if there were agreement on measurement techniques, the question remains as to what litter to count: permanent or accumulated† and what size items should be included.

† A permanent litter measurement would count litter which was removed from an area which had not been cleaned for a long period. An accumulated litter measurement counts the litter which has accumulated over a period of time after an area is cleaned.

The average composition of litter from four surveys is shown in Table 25. These data were obtained by actual survey, although the sampling methods used in such surveys are usually not reported. Consistent techniques of litter measurement must be developed if data from various surveys can be correlated and meaningful inferences drawn therefrom. A reasonable set of guidelines for litter measurement may be the following:

1. Count items larger than a certain size, say a bottle cap.
2. Count as one bottle all glass pieces clearly belonging together.
3. Classify the litter into components such as beverage containers, newspapers, fast food containers, etc.
4. Do not count dirt, rocks, etc.
5. Do not count apple cores and other organics, since these decompose rapidly.
6. Measure litter as (a) visible litter, (b) piece count, and (c) weight.

Interesting correlations between the last three exist. For example, only about 6% (by count) of all litter is visible [63].

Regardless of the sampling techniques used in measuring litter, all studies on the subject show that the beverage container (cans and bottles), and its related items (bottle caps, pull tabs, six-pack carriers) are a significant part of litter in the United States. Beverage containers alone are estimated to average 20–30% of all litter by item count and 40–60% on a volume basis [64]. In addition, their size and visibility make them particularly noticeable. Moreover, the relative permanence of cans and bottles in the environment makes them among the most esthetically damaging form of litter. These esthetic damages are not readily estimated, but the increasing value placed on wilderness, parks, forests, and other recreational areas suggests that they are appreciable.

Two factors have contributed to the beverage container's prominence in today's litter. First, soft drink and beer sales have more than tripled in the last 15 years. Second, the containers used for these beverages have shown a significant trend toward one-way† packaging. As recently as 1960, beer and soft drinks were mostly purchased in refillable glass bottles that carried a refundable deposit. Today over 70% of these beverages are sold in no-deposit glass or metal containers that are used once and discarded.

The trend toward one-way packages in the beverage industry is both praised and criticized. The praise centers around the convenience

† Beverage containers designed for one-time use have been described by many adjectives; throw-away; no-deposit; no-return; disposable; one-way; and nonreturnable.

Table 25
The Composition of Litter: A Summary of Four Surveys,[a] Percent

	Keep America Beautiful[b]			Florida[b]		
	1st Pickup	2nd Pickup	Dallas[c]	1st Pickup	2nd Pickup	Continental Can Company[d]
Cans (total)	28.3	16.3	36.0	35.3	14.3	12.0
Beer	21.7	11.7	28.8	27.4	11.1	6.0
Soft drinks	4.4	3.1	5.0	7.1	3.1	4.3
Food cans	1.0	0.6	1.1	0.3	0.0	—
Other	1.3	0.8	1.7	0.4	0.1	1.7
Bottles (total)	6.9	5.9	5.0	4.9	3.7	2.8[e]
Nonreturnable beer	2.7	2.3	2.6	1.4	0.5	0.5
Returnable soft drink	1.6	1.6	1.6	0.2	1.8	0.4
Liquor	0.8	0.6	0.4	0.5	0.6	0.5
Nonreturnable soft drink	0.8	0.5	0.3	0.4	0.7	0.5
Returnable beer	0.4	0.4	0.1	0.2	0.1	0.2
Food	0.3	0.2	0.1	—	—	0.7
Other	0.4	0.2	0.1	0.0	0.0	—
Paper (total)	48.9	59.5	57.0	36.9	55.4	55.0
Paper packaging or containers	10.7	11.5	12.5	12.5	7.6	24.4
Newspapers, magazines	1.8	1.9	2.3	4.1	2.3	4.3
Other	36.4	46.1	41.6	20.4	45.5	31.3[f]
Auto parts	0.8	0.8	0.3	—	—	0.7
Miscellaneous	15.1	17.5	1.7	22.9	26.6	29.5

[a] From Ref. 62.
[b] Indicates pickup on interstate highway.
[c] Indicates pickup on 50 miles of streets and highways.
[d] Indicates pickup on residential and commercial blocks.
[e] Total does not include broken glass.
[f] Includes 5% paper cups.

that one-way containers provide the consumer; the initial criticism was a reaction to the obvious increase in beverage container litter.

In Vermont, the sale of one-way glass bottles was banned as early as 1953. Since that time innumerable bills on this subject have been introduced into legislative bodies. For the most part, these bills are intended to encourage consumers not to litter their beverage containers, by imposing a mandatory deposit on all beverage containers, including one-way bottles and cans.

The rationale behind a mandatory deposit is economic in nature. The consumer would be required to pay a minimum amount per container. When the container is returned to the retail store, the deposit would be refunded. The consumer by returning the container, would be reacting to the financial incentive, thus making the can or bottle available for multiple use (refilling) or recycling (remelting and manufacture into containers again).

The debate on mandatory deposit legislation has continued for some time and the issues have broadened. Proponents of mandatory deposit legislation view the present system as a symbol of a society that litters and wastes too much, uses energy excessively and depletes resources needlessly. In their view, a mandatory deposit system would help to eliminate these negative aspects of the present predominately one-way system.

Opponents counter that the present beverage system is a direct response to consumer demands. Any deposit legislation would interfere with the orderly working of the market system. They contend that our lifestyle would be deleteriously affected by such a law through higher beverage costs, disruptions in service, and a loss of both convenience and local brands. Furthermore, they maintain that a mandatory deposit system would not reduce litter or solid waste very much nor conserve energy.

The proponents believe that the producers and sellers of goods should bear some responsibility for the waste generated by their products and that the consumer should be more directly aware of the costs of consuming and disposing of one-way packaging. The opponents view a national mandatory deposit system as only a partial solution to the problems of solid waste. In fact, such legislation is seen as a hindrance to implementation of municipal recycling programs because it removes valuable cans and glass from the waste system.

Currently, four States (Oregon, Vermont, Michigan, and Maine) have enacted some type of mandatory deposit legislation and a national mandatory deposit law has been suggested.

The Oregon Minimum Deposit Law went into effect on October 1,

1972. This law required that a deposit be paid on all beverage containers. The results of this law are perhaps the best documented and analyzed of all subsequent State and local mandatory deposit laws. The experience in Oregon has been that the beverage container litter has been reduced by approximately 60% at no net cost in jobs or economic hardships. The only adverse effect is the loss of convenience to the consumer. The question remains whether Oregon's positive experience with a mandatory deposit law can be projected to the national level without a comprehensive assessment of the law's potential impact on the nation's environment, economy and energy [65, 66].

F. Plastic Waste

Probably the fastest growing constituent of municipal solid waste is plastic waste. This is evident from the vigorous growth and diversity exhibited by the plastics industry in recent years, especially in packaging. For more than 20 yr the market for plastics has increased at a faster pace than any other basic material. Each year plastic products contribute to more of our needs in the areas of packaging, building, furniture, automobiles, housewares, appliances, etc.

Plastic materials are divided into two classes, thermoplastics and thermosets, according to the way each behaves under repeated conditions of heating and cooling. A *thermoplastic* requires heat to make it formable and after cooling can be reheated and reformed into new shapes a number of times without significant change in properties. Polyolefins (polyethylene and polypropylene), styrenes (polystyrene, acrylonitrile–butadiene–styrene), vinyls [poly(vinyl chloride), poly(vinylidene chloride)] and nylon are some examples of thermoplastics.

A *thermoset* uses heat to make its shape permanent. After a thermoset is formed into permanent shape, usually with heat and pressure, it cannot be remelted and reformed. Additional heating will only degrade or destroy it. Phenolic, polyester, epoxy and amino polymers are examples of thermosets.

Plastic wastes generally fall in the category of collected solid wastes. Although no precise figures are available on total collected solid wastes, current estimates range between 200 and 450 million tons (180–410 million metric tons) of industrial and municipal wastes. Of this amount, plastics account for approximately 2%, or about 4–9 million tons (3.6–8.2 million metric tons). Of this, packaging accounts for more than half, industrial wastes account for about one-sixth, with the remainder distributed over items such as housewares, construction, toys, appliances, furniture, transportation, footwear, and phonograph records. Polyolefins are estimated to account for approximately

55–70% of total collected plastic wastes, with styrene polymers contributing about 15–20%, and poly(vinyl chloride) about 11–13%.

Waste plastic comes from three main groups of waste generators: (a) the basic resin and polymer producers who recycle part of their waste internally, and send the remainder for reprocessing to scrap processors or to disposal; (b) fabricators of plastic products who recycle most of their own wastes internally; and (c) postconsumer wastes from household, industrial, and commercial sources that are hauled away as part of the municipal solid waste [67].

The present combination of possible feedstock shortages, definitely higher feedstock prices, and even higher resin prices has led to a vigorous search for ways of reusing and recycling plastic waste.

Collected municipal solid waste contributes the largest portion of plastic waste; however, salvaging plastic from these wastes poses a difficult problem technically because of the heterogeneity of such wastes. The recovery and reuse of plastics from municipal wastes has been difficult because there are no established procedures for separating plastics from other wastes and no established markets for contaminated mixed plastics, and techniques for cleaning and separating mixed plastics have not been developed for large-scale economic application.

Techniques for separating plastics from mixed wastes are still in the experimental stage and identification of plastics by type is difficult. Segregation of waste plastics by resin type is further complicated since there are several hundred different grades of plastics and some of these are proprietary formulations that hamper any efforts at categorization. The economics of collecting and shipping plastic wastes to points of reuse are also generally unfavorable.

Currently almost all recycled plastics are recovered from the industrial sector. This recycling activity has concentrated chiefly on thermoplastics, which account for approximately 80% of all plastic waste. In 1974, about 29 billion lb (13 billion kg) of plastics were produced and consumed in the United States of which 14.1 billion lb (6.4 billion kg) ended as plastic waste. Of this waste quantity 8.3 billion lb (3.8 billion kg) were from nonindustrial sources and none of this was recycled. The remaining 5.8 billion lb (2.6 billion kg) were from industrial sources, of which fabricators recycled more than 2.6 billion lb (1.2 billion kg) of their process scrap, and reprocessors handled another 1.5 billion lb (0.7 billion kg), and only 1.7 billion lb (0.7 billion kg) of industrial plastic wastes ended up in the solid waste stream. Low-density polyethylene (LDPE) and poly(vinyl chloride) (PVC) accounted for the bulk of this unrecovered material (37 and 30%, respectively) [68].

It is not difficult to understand why virtually all plastics recycling is done by industry itself. Industrial plastic wastes are more easily separated than municipal wastes, and also less contaminated. Among the plastics in the industrial waste stream that are not being recycled, mixtures of different plastics are a particular problem. The presence of different types of polymers makes reprocessing very difficult because of the lack of compatibility of most polymers. Furthermore, scrap containing plastic mixtures has poor physical properties. Techniques are being developed to separate these waste streams. One approach that avoids the difficult separation problem, is to introduce an additive to improve adhesion between the polymers and thereby produce a blend with improved physical properties [69, 70].

In most fabricating operations, in-process waste material is generally recycled by blending with virgin material, as long as adequate control can be maintained over the quality of the final product. When contamination and degrading of in-process waste are such that it cannot be recycled within the process, the material is often sold to industry reprocessing operations where it is sorted, blended, or modified in such a fashion that it becomes a useful product for some other end use. In this way high specification compounds, such as those used for wire and cable, bottles, high quality film, and some coatings, can be recycled into lower quality compounds used in molded products (shoe soles, bicycle handles), pipe, and structural materials.

Some of the activities that relate to the reclamation of plastics from solid waste, have been covered in a study commissioned by the Society of Plastics Industry [71]. Over 50 promising recycling projects were reviewed and categorized in one of three categories: recycling for reuse as a conventional plastic product, recycling for reuse as a substitute for nonplastic products, and recycling as an energy source. The conclusion drawn from this study was that the use of waste plastic in an unsorted form for the recovery of energy, by pyrolysis or incineration, will be the most successful plastics recycling course of action in the context of present-day technology and economics. However, such technological and economic evaluation are still in a developmental phase.

Since recovery of plastics from municipal waste still appears to be some years away, the two current concerns with plastic waste are (a) Can they be disposed of in an incinerator without causing operational or air pollution problems? and (b) Will they decompose in a landfill?

Incineration of plastics is, on the surface, an ideal solution. Various types of plastics have high heating values; for example, polystyrene, 19,000 Btu/lb; polyester, 12,000 Btu/lb; poly(vinyl chloride), 8500 Btu/lb; polyurethane, 11,500 Btu/lb. When compared with municipal solid waste, which has a heating value of about 5,000 Btu/lb, the worth of plastic in refuse as a source of heat is readily apparent.

The increase in plastics in refuse over the past two decades however, produced a number of difficulties for incinerators. Sudden flare-ups, melting of the plastics on the grates, and corrosion of the flues, all were widely reported. Controlled research, however, showed that many of these claims were exaggerated, and proper incinerator design and operation resulted in the absence of such problems.

The one problem that still exists in some cases is the production of corrosive gases. The major culprit is poly(vinyl chloride), which, when combusted, emits hydrogen chloride. The amount of PVC in refuse is small, less than 2% by weight, and one pound of PVC can theoretically produce only 0.56 pounds of HCl. Thus, the maximum theoretical yield of HCl from the combustion of PVC can be only 0.11% of the total refuse weight. Emission analyses show that the HCl in stack gases is about three times that value, and hence most of the HCl produced must come from the combustion of other chlorine-containing materials such as paper, textiles, rubber, etc.

Plastics are for the most part inert, and will not readily decompose in a landfill. Nevertheless, chopped plastics make a compact inert landfill material without the usually attendant problems of leachate contamination or explosive gas release.

The plastics industry, having learned how to stabilize its products so that they do not degrade during processing and use, has in recent years witnessed a concerted effort to develop additives that will accelerate plastic degradation. Prompted by the apparently growing problems of plastics litter and the difficulties experienced during landfill and composting operations, there has been a rapid development of degradable plastics [72, 73].

There are three approaches to the introduction of degradability to a plastic: (a) the introduction of light-sensitive additives; (b) the addition of biologically degradable components; and (c) the use of a water-soluble inner layer which can dissolve when the protective outer layer is broken.

Because of their latent susceptibility to photolytic degradation, the long-chain structures of plastics can be activated by additives to bring about their degradation when exposed to certain wavelengths of light [74]. Biological degradability on the other hand is not easily imparted to plastics. They must be mixed with organic substances that decompose readily (such as starch) in order to render them biologically decomposable [75]. The multilayer technique has so far been limited to one application for the manufacture of bottles: hydroxylpropyl cellulose is used as the water-soluble layer sandwiched between polystyrene layers.

The use of plastics and the quantities of plastics in the solid waste stream will undoubtedly be increasing. For example, wide use of a plastic soft drink bottle seems to be imminent, and the trend toward smaller and lighter cars will dictate greater use of plastics. Considering the source of plastics (fossil fuels) and energy costs in the production of plastics, it seems reasonable that an economical and environmentally acceptable recovery method will find wide (and profitable) use in the future.

G. Junked Automobiles

Between 9 and 10 million vehicles are junked annually in the United States, and this number is expected to rise to over 13 million units by the late 1980s. Trucks and buses account for about 15%, with automobiles accounting for the remaining 85%.

Although discarded auto hulks constitute only a small fraction of the waste disposal problems in terms of tonnage, they are higher in metal values than most solid waste materials. Consequently, the auto disposal industry is active and vigorous.

A junked automobile may go through as many as four different stages after it has outlived its usefulness and begun its recycling journey:

1. Acquisition (collection) of damaged, junked or abandoned automobiles.
2. Dismantling or stripping of useful parts.
3. Scrap processing.
4. Scrap residue processing.

The collector or tow truck operator, who acquires and sells autos either to a motor vehicle salvage dealer or scrap processor, is often the first link in the recycling chain. Collectors obtain automobiles from owners, and municipal pounds and disposal facilities.

A number of factors influence the rate at which junked autos are acquired. These include junk rate, auto scrap prices, used part prices, state automobile laws, and location of an auto wrecker.

The dismantling process begins after the autos have been collected at some central point by a salvage dealer, frequently referred to as an *auto wrecker* or *dismantler*. The autos are stripped of usable parts such as radiator, wheels, starter, generator, gas tank, and other replacement parts and components. The battery may be sold for its lead content. The motor block and other cast iron parts (differential and transmission housings) may be sold as foundry scrap, whereas heavier steel components (wheels, springs, axles, gear, including the frame) may be removed and sold as No. 2 heavy melting scrap.

The auto wrecker operates by one of several methods: (a) the auto hulks may be parked in a yard, with parts being stripped as required for sale to walk-in trade; (b) the vehicles may be stripped completely, the parts placed in storage, and the hulks sold immediately to scrap processors; (c) the hulks may be parked in a yard where customers can strip parts as necessary; (d) a scrap processor who will take the hulk as it stands may be found, and (e) the wrecker may deal directly with small foundries and steel companies for foundry scrap and No. 2 heavy melting scrap.

Besides depending on the independent collector, auto wreckers also collect many of the junked autos themselves from used car agencies that accumulate trade-in vehicles with little resale value, salvage pools operated by insurance companies, and individual owners. The auto wrecker is vitally concerned with the dates and models of the individual cars he purchases. Many wreckers are reluctant to take old vehicles that have no parts value.

It is estimated that approximately 16,000 companies with a total employment of about 100,000 workers are actively engaged in the auto wrecking business.

At some point, the stripped auto hulks are sold to a scrap processor. The hulks are delivered from auto wrecker to processor by flatbed trucks and may be either flattened or unflattened. Unflattened, a truck load may contain up to eight hulks, whereas up to 25 flattened hulks may be transported in a single truck. The auto flattener (or crusher) facilitates transportation by reducing the standard automobile body to a slab 7 ft (2 m) wide and 7–10 in. (18–25 cm) high. Flatteners can process between 60 and 160 cars per day. At this rate the obsolete inventory of an auto wrecker may be processed in a matter of days.

The *scrap processor* uses machinery such as shredders balers and shears for processing the auto hulks and manufacturing iron, steel, or nonferrous metallic scrap into prepared grades, which in turn are purchased by steel mills and foundries [76].

Shredders, the most significant development in scrap processing in recent years, can process an automobile body and frame into small fragments of pure metal at rates of up to 120 cars per hour. Typically, a crane equipped with an orange-peel bucket picks up the hulk and lowers it into a shredder where swing hammers batter against cutting edges and shred the metal. In less than a minute the car is pounded and sheared into a mixture of iron, steel, aluminum, copper, zinc, plastics, and fabric. The iron and steel are separated by magnetic separators whereas aluminum, copper and zinc are handpicked.

Shredders are economically justified when the annual acquisition rate of junked autos is at least 40,000 because they can cost anywhere from $500,000 for small units to over $3 million for units handling more than 100 tons scrap per hour. Many shredders come equipped with afterburner devices to incinerate such combustible trash as upholstery, insulation, and rubber remaining in the junk body. More than 100 shredders have been installed in scrap yards all over the United States.

For areas that do not discard enough cars to support a shredder, balers (or presses) can be used to process automobile hulks. This is by far the most widespread method of auto scrap processing, and it is estimated that about 1000 balers are in operation in the United States. Balers are long bins equipped with a removable lid and hydraulic rams. Depending on size, a baler can cost from about $50,000 to several hundred thousand dollars. In this operation the hulk, after being stripped to varying degrees either by hand or by incineration, is charged to the bin and then compressed into a cube with the approximate dimensions of a TV console. The cube weighs approximately 1200 pounds (545 kg) and is referred to as a No. 2 bundle, a basic item in auto scrap. A certain degree of quality control has to be exercised in producing No. 2 bundles or else quality hazards may discourage consumers to use them. This necessitates the removal of glass and other nonmetallics and nonferrous metals as completely as possible.

Shears are a relatively new development in the scrap industry. They first process the hulk into a rectangular (20 × 2 × 2 ft) log, which is then fed into guillotine shears which cut and compress the log into predetermined lengths. The resulting sheared slabs are pillow-shaped uniform pieces of very high-density scrap. Total time required for the complete shear operation is about 40–60 man minutes for each ton of product. Depending on its size, the guillotine shear can represent an investment of between $225,000 and $1.5 million.

Incineration is used mainly as an adjunct to hand stripping and is useful in removing combustibles that might contaminate the final scrap. However, in view of the cost, technical problems, and air pollution regulations, this practice is very limited. Most incineration is confined to the smokeless incinerator developed by the U.S. Bureau of Mines. This is a relatively inexpensive operation that can efficiently process as many as 10 cars per hour; the combustion gases are smokeless and meet or exceed most clean air standards [77].

The principal value in automobile scrap lies in its iron and steel content, i.e., ferrous scrap. However, the nonferrous scrap metals such as zinc, aluminum, copper, and lead, are also economically important factors in the total value of the scrap metal in junked automobiles, although they constitute a very insignificant part of the metallic scrap weight. A Bureau of Mines study showed that a representative junk automobile weighing 3600 lb (1635 kg) could yield approximately 2500 lb (1135 kg) steel, 500 lb (225 kg) cast iron, 54 lb (25 kg) zinc, 51 lb (23 kg) aluminum, 32 lb (15 kg) copper, and 20 lb (9 kg) lead. The remaining 443 lb (200 kg) consisted of nonmetallics.

One of the problems that scrap processors face is that copper is a serious contaminant in finished steel ingot. Consequently scrap processors have to eliminate copper from the hulk prior to processing it and this can be a troublesome and time-consuming chore since the copper is usually found in inaccessible places in the form of copper wiring in starters, generators, armatures, etc. A process has been developed whereby copper-containing scrap is dipped in molten salt (calcium chloride) and agitated briefly. The copper quickly melts and collects at the bottom, whereas the iron and steel are unaffected. The molten copper can easily be tapped off and recovered.

At present no practical methods are available for recovering the nonferrous metals from automobile scrap and consequently these metals are wasted as rejects of the scrap processing operations. Air separation methods have been devised that can separate over 95% of the nonferrous metals from shredded auto hulks. This metallic concentrate can be refined into some of its constituent parts by physical concentration methods and pyrometallurgical processing. These techniques await widespread commercial acceptance.

The nonmetallic residue resulting from hulk and scrap processing results in approximately 1.5 million tons of solid waste annually. This waste consists of scrap tires (about 25%) and scrap glass (about 20%). Scrap tires are the subject of the following section.

At present, about one-fourth of the recycled steel scrap is from discarded automobiles. However, the availability of scrapped autos exceeds the demand for them. It is estimated that a backlog of some 12–20 million auto hulks exists, with an additional 8 million being abandoned each year [78].

H. Scrap Tires [18]

More than 200 million old tires containing over 2 million tons of rubber are discarded annually. Used tires pose significant disposal problems for municipalities, industries, and private citizens. Although many junked tires are burned haphazardly in open air dumps, this practice should cease with enforcement of more stringent air pollution codes. Most cities refuse to incinerate scrap tires because they burn so fiercely as to burn out the grates of ordinary incinerators. Even if special incinerators are used, large quantities of particulates, sulfur and nitrogen oxides, and hydrocarbons are released thus creating an air pollution problem.

Because of low bulk density and resistance to biodegradation, tires do not lend themselves well to disposal by landfilling; whole tires resist compaction and burying, and rise to the surface, thus creating other problems such as breeding places for disease vectors and as fire hazards.

The alternatives currently available in dealing with the tire disposal problem are: (a) waste reduction through longer-life tires and retreading, (b) immediate options, (c) short-term options, and (d) long-term options [79].

Although *waste reduction* measures will not solve the tire problem, they can decrease the number of tires being discarded. The tire lifetime option can be implemented by consumer insistence on more durable tires (currently the most durable tires last about 40,000 miles); furthermore, tire lifetime can be increased with proper tire maintenance, i.e., proper tire pressure, wheel alignment, and tire balancing.

Retreading of tires usually faces two possible problems. The shortage of tires suitable for retreading and reservations about the performance of retreaded tires.

There are several *immediate alternatives* available to tire disposal/reuse. Rubber tires properly ballasted and secured in groups make very satisfactory artificial reefs. Approximately 1000 such reefs have been built in U.S. coastal waters. Proponents of tire reefs cite the abundant supply of tires, low total cost ($0.50 to $4 per tire, depending on type of reef construction), nonmagnetic properties, crevices for fish homes, marine organism growth potential, and long-lasting materials. However, because of transportation costs, reef building is feasible only along coastal areas, and even so, only a certain number of reefs can be built. Furthermore, the long-term effects of slowly disintegrating tires are not known.

Intact tires may also be used as crash barriers around obstructions near high-speed freeways, bumpers for docks and towing vessels, and retaining walls for soil erosion control.

Another immediate alternative is the reclaimed rubber industry, which uses scrap rubber in making compounds for manufacturers of new tires and various rubber products. Reclaimed rubber is produced by treating vulcanized waste rubber with chemical agents, heat, and intensive mechanical working to produce a uniform consistency. Tires are first passed through crackers for size reduction. Crackers are machines with two rolls rotating at different speeds to create a friction ratio for shearing. As the tires pass into the cracker, the slower roll holds the tire while the faster roll corrugations shear, slice, crush, and abrade the tire. Bead wire which holds the tire firmly to the rim of the wheel, is removed after the tire passes through the cracker.

Rubber particles from the cracker then pass through a fiber separation process where tire-reinforcing materials such as cotton, nylon, rayon, polyester, fiberglass, and metal are removed from the rubber. This involves use of hammer mills, sifters, and beaters, which complete fiber removal. The fiber waste is baled and the clean rubber particles are finely ground to 30-mesh size. This finely ground rubber can be softened by the digester or wet process, devulcanizer or dry process, and mechanical process.

In digestion, the rubber particles are placed in a vessel with water and softening chemicals, heated, and then discharged as a slurry. The now softened rubber is mechanically dewatered and dried.

In devulcanizing, the rubber particles are premixed with softening chemicals and subjected to steam pressurization in an autoclave after which the rubber is cooled and ready for further processing.

The mechanical process, unlike the above two processes, is continuous rather than batchwise. Finely ground rubber is fed along with softening chemicals, into high-temperature shear machines. The discharged rubber needs no drying and is ready for further processing.

The rubber from all of the softening processes is intimately mixed with compounding ingredients which give the finished reclaimed rubber special physical properties. The softened product is afterward strained and refined into a thin film by huge rollers. It can then be formed into slabs or bales for shipping.

One of the drawbacks of reclaimed rubber is the fact that it is black and consequently cannot be used for brightly colored products. Nevertheless, reclaimed rubber is used in a number of goods (inner tubes; hard rubber battery boxes, steering wheels, etc.; heels, soles, footwear; hose belting; insulated wire; surfacing materials; cements and dispersions).

Tires can also be crumbed, i.e., ground into fine particles, for use as aggregate or in products where devulcanized rubber is not required, such as in asphalt patching mixes, and under house foundations as a cushioning agent.

Short-range alternatives include landfilling and road building. Landfilling is feasible only when tire splitting or shredding equipment is used. Shredding equipment can process from 60 to 1000 tires per hour and vary in cost from about $10,000–100,000. Tire slicers handle about 300 tires per hour and cost about $3000–5000. Both shredders and slicers are available in the portable variety, thus making it possible to bring such equipment to retreaders or tire collection points.

There is increasing interest in using shredded tires as an additive to asphalt in road building and repairing. However, most proved asphalt-additive applications require rubber particles smaller than present shredder output. It is also necessary to remove any steel belt or bead material from the rubber used for road building. Consequently, for most road building applications rubber reclaimers will have to be relied upon.

Asphalt pavement consists of two components, aggregate and binder. Reclaimed rubber enhances the properties of the binder in several ways. The resilience of the road is increased in the winter, and flowing characteristics of asphalt are reduced in the summer.

Rubber reduces the tendency of asphalt to bleed to the surface where it creates a skidding hazard. The road has greater resistance to compaction owing to "rebound" of the rubber in the mix; in addition, increased resiliency reduces surface cracking. Rubber has been used as an asphalt additive in Phoenix, Arizona, and on the New York State Thruway System.

There are several *long-term alternatives* for scrap tire disposal/reuse. All of these are capital intensive and are currently either in the developmental or demonstration phases.

Pyrolysis is a controlled, oxygen-deficient heating process that decomposes organic materials. Tires are typically made up of 83% carbon, 7% hydrogen, and 6% ash, with trace quantities of nitrogen, oxygen, and sulfur. In pyrolysis, chemically complex oils and gases and a solid residue are obtained. The proportions of the products depend on the temperatures used in the process.

A series of pyrolysis tests carried out by the Bureau of Mines (with cooperation and funding from The Firestone Tire and Rubber Co.) showed that at around 500 °C (930 °F) nearly 50% of the charge is converted to oils with the rest being converted to solid residue (about 45%) and gas (about 5%); when the temperature was raised to 900 °C (1650 °F), a larger portion of the charge is converted to solid residue (about 55%) and gas (about 20%) with less oils (about 25%) being produced. In the above tests the charge consisted of approximately 100 lb (45 kg) of shredded scrap tire material. A typical test required

7–14 h of heating to completely devolatilize the charge in a hermetically sealed retort. As many as 50 different chemicals have been identified in the oils that were recovered, and were mainly classified as olefins (about 15%), aromatics (about 55%) and paraffins and naphthenes (about 30%). The gases were a mixture of hydrogen and hydrocarbons and had a heating value of approximately 900 Btu/ft^3 (3.3×10^4 kJ/m^3); the heating value of gases produced at higher temperatures were lower because of the higher percentages of hydrogen and low-molecular weight hydrocarbons. The solid residue has a heating value equivalent to coal, but is relatively high in sulfur content (1.5–3.0%) [80].

Another unusual process pyrolyzes scrap tires in molten salts. Pieces of scrap tires immersed in a lithium and potassium chlorides eutectic mixture at 500 °C (930 °F) yield (by weight) 47% oil, 12% gases, and the remainder as solid residue. The oil composition is about 21% aromatics, 34% olefins, and 45% paraffins; the gases are C_1–C_4 paraffins and olefins. The residue is char, unchanged glass, and steel tire components. Molten salts are excellent heat transfer media, there is good contact of liquid and rubber, and the reactions are rapid. This technique works well with whole or split tires, thus eliminating the costs of shredding. Furthermore, the salt eutectic mixture is not consumed or degraded during the reaction and thus can be reused. The drawbacks are that the technology of separating solids from molten salt is not well developed. Commercial success would depend on being able to use the solid residue as an extender; using pyrolysis products merely as fuel may not be enough economic incentive [81].

Energy can be recovered from tires by shredding and incinerating them as supplemental fuel in conventional coal-fired installations or as the sole fuel in specially designed furnaces for industrial applications. The Goodyear Tire and Rubber Co., for example, has constructed and is operating a cyclone furnace for whole-tire incineration and steam generation. Whole tires are fed continuously by conveyors onto a rotating hearth. As the tires move in a spiral toward the center of the hearth, temperatures up to 2400 °F (1315 °C) oxidize even the wire bead. The furnace can burn up to 3100 lb (1400 kg) of scrap tires per hour and, with the aid of a specially designed boiler, can generate 25,000 lb (11,300 kg) of process steam per hour.

The viability of large-scale pyrolysis or incineration plants hinges on the ability to procure the scrap tires in sufficient quantity and at a reasonable cost. Even though the economics of such processes appeared marginal in the past, it is possible that recent price trends in the energy and petroleum products fields will make the economics appear more favorable [82].

VII. LEGISLATIVE ASPECTS OF SOLID WASTE

There are laws at almost every level of government, federal, state, and local, that pertain in some manner to solid waste. This section will be devoted to the legislative aspects at the federal level only, since the greatest strides toward the regulation and control of solid waste have been taken at that level.

The first significant federal legislation that recognized solid waste as a national problem and strived to set a standard and precedent for the future was the Solid Waste Disposal Act (PL 89-272) of 1965. It incorporated the basic concepts of waste reduction at the source, recovery and reuse, and regional solutions, and established the necessary organizational structure for the future.

The Federal Solid Waste Management Program activities as authorized by this act, included policy formulation, demonstration, research and development, technical assistance, planning solid waste management systems, public information, state-of-the-art studies, and training.

During the initial years of federal activity, major emphasis was placed on learning as much as possible about the practice and technology of solid waste management and assisting states to develop viable solid waste agencies. As a result of the studies in these earlier years much insight was gained into the quantity, quality, and distribution of solid waste in the United States and a rather ominous picture of the solid waste problem began to emerge. These early studies also established an understanding of the state-of-the-art of solid waste management and provided insights into future efforts needed.

New federal solid waste legislation was enacted in October 1970, the Resource Recovery Act of 1970, PL 91-512. It amended the earlier act and came at the same time that the Environmental Protection Agency (EPA) was established in the Federal Government. Thus the Federal Solid Waste Management Program became part of the EPA as the Office of Solid Waste Management Programs.

This act, authorizing a continuation of past work, also authorized several new significant activities and studies: (a) the establishment of guidelines; (b) hazardous wastes; (c) resource recovery; and (d) solid waste management manpower.

The guidelines effort mandated by the act resulted in regulations governing the operation of sanitary landfills and incineration facilities on federal lands. These also served as a guide for state agency standards [83].

The 1970 Act, PL 91-512, required that the EPA "submit to the Congress no later than two years after the date of enactment of the Resource Recovery Act of 1970, a comprehensive report and plan for the creation of a system of national disposal sites for the storage and disposal of hazardous wastes, including radioactive, toxic chemical, biological, and other wastes which may endanger the public health." The report was formally submitted to Congress in 1973 [84]. It pointed out the magnitude of the hazardous waste problem, emphasized the inadequacies of the federal, state and local regulations and management practices to combat hazardous wastes, and urgently called for federal and state action to regulate these wastes.

The resource recovery aspects of the act emphasized the need for, and directed the EPA to study and determine ways to recover materials and energy from solid waste. EPA was required to report on the findings of these studies annually. The first report was submitted in 1973 [85]. This report mainly emphasized the recovery of materials and energy from municipal and "post-consumer" solid wastes.

The act also required the EPA to initiate a study to determine current and future manpower needs for solid waste management in the United States. This study disclosed, among other things, that although over 300,000 people worked in solid waste management, only 350 full-time employees were working at the state level.

In 1976, Congress made broad changes in the legislative mandate for the EPA in solid waste management. The 1976 amendments to the Solid Waste Disposal Act, titled the Resource Conservation and Recovery Act of 1976 (RCRA), PL 94-580, mandate national action for the first time against solid waste management practices that lead to environmental and public health hazards. They also seek to promote resource recovery and conservation as waste management options. The main provisions of RCRA for achieving these goals are as follows:

1. Federal financial and technical assistance to State and local governments is authorized for planning and development of comprehensive solid waste management programs that include environmental controls on all land disposal of solid wastes, regulation of hazardous wastes from point of generation through disposal, and resource recovery and conservation activities.

2. Such state programs would include schedules for upgrading or closing all environmentally unacceptable land disposal sites ("open dumps") identified according to EPA criteria and a nationwide inventory. Open dumping is prohibited except as covered by an acceptable schedule for compliance under the state plan.

3. Where states do not establish hazardous waste regulatory

programs that meet federal standards, EPA will administer regulatory control.

4. A cabinet-level interagency study of resource conservation policies is mandated; findings and recommendations are to be submitted periodically to the President and the Congress.

5. Public participation is required in the development of all regulations, guidelines, and programs under the Act.

6. Research, demonstrations, studies, and information activities related to a wide range of solid waste problems are authorized.

The following are the highlights of the main provisions under RCRA as excerpted from Ref. 86:

Planning and Development of State and Local Programs

The Resource Conservation and Recovery Act authorizes federal technical and financial assistance to state, regional, and local authorities for planning and implementation of comprehensive solid waste management programs pursuant to federal guidelines. Main elements of comprehensive state programs would include regulation of hazardous wastes, environmental controls of land disposal of all solid wastes covered by the act, resource recovery and conservation measures, and public participation and information activities.

To encourage planning on a regional basis, RCRA required issuance of EPA guidelines within 6 months of enactment, i.e., by April 21, 1977, on identification of areas that are appropriate units for such planning (Section 4002a). Within 6 months of the issuance of the guidelines, the governors, in consultation with local elected officials, were to formally identify regions; within 6 months of that, state and local officials would jointly identify the agencies that will develop and implement the state solid waste management plan, specifying which agencies are responsible for which functions (Section 4006).

EPA was required to issue guidelines by April 1978 on the development and implementation of the state solid waste management plans (Section 4002b). To qualify for federal financial assistance, the state plans must meet minimum requirements, including identification of the responsibilities of state, regional, and local authorities in implementing the plan, the prohibition of new open dumps, provision for closing or upgrading all existing open dumps, and provisions for other disposal, recovery, and conservation measures necessary to meet the objectives of the act.

Federal financial assistance to state and local governments for development and implementation of solid waste management plans is authorized. Under Section 3011, financial assistance is authorized for state hazardous waste program development and implementation pursuant to establishing a federally authorized regulatory program.

Authorities for technical assistance in the act include Section 2003, which requires EPA to provide, upon request, assistance to state and local governments through teams of personnel ("Resource Conservation and Recovery Panels") including federal, state, or local government employees or contractors.

A complete study of the manpower and training needs of state and local solid waste programs is required under Section 7007, and grants for training projects are authorized.

Land Disposal

The elimination of environmentally unacceptable land disposal is a prime objective of RCRA. The law directs EPA to issue, within one year of enactment, criteria for the classification of all land disposal sites as either environmentally acceptable or unacceptable (Section 4004). Within one year after promulgation of the criteria an inventory is to be published of all unacceptable sites ("open dumps") identified according to the criteria (Section 4005). Open dumping is prohibited except as covered by an acceptable schedule for compliance under the state plan (Section 4005). Such a schedule must include an enforceable sequence for actions leading to full compliance within a reasonable time (not to exceed 5 years from date of publication of the inventory).

Section 1008 requires EPA to develop and publish suggested guidelines for solid waste management that provide for the protection of public health and the environment. Following publication of the land disposal criteria, such guidelines will be issued on land disposal practices. They will assist the states in their assessment of compliance with the open dump prohibition and specification of remedial measures.

Thus, the criteria define acceptable land disposal, the inventory is a national listing of sites that do not meet the criteria and therefore should be upgraded or closed, and the suggested guidelines describe acceptable operating practices, that is, the means of achieving the performance goals of the criteria. The state plans provide the framework for the regulatory elements to become functional and effective.

Hazardous Waste Management

Subtitle C of RCRA mandates establishment of a regulatory control program that will prevent serious threat to human health and the environment from current practices in managing hazardous wastes. Key provisions are for development of criteria for determining which wastes are hazardous, institution of a manifest system to track wastes from point of generation to point of disposal, and organization of a permit system, based on standards, for hazardous waste treatment, storage, and disposal facilities. These criteria are further discussed under Nonradioactive Hazardous Solid Waste, Section VI, in this chapter.

Resource Recovery and Waste Reduction

Principal RCRA provisions supporting resource conservation are the following:

1. The greatest benefit to resource recovery is likely to come from the Subtitle D provisions relating to land disposal. The removal of environmentally unacceptable land disposal will eliminate an unrealistically low-cost alternative for waste disposal that has limited the attractiveness of resource recovery. Similarly the regulation of hazardous waste management should encourage reduction and recovery of hazardous wastes.

2. Section 4003 on the minimum requirements for approved state plans, besides closing off open dumping, explicitly states that "all solid waste . . . shall be (A) utilized for resource recovery or (B) disposed of in sanitary landfills . . . or otherwise disposed of in an environmentally sound manner." The requirements also include the provision that no local government within the state shall be prohibited from entering into long-term contracts for the supply of solid waste to resource recovery facilities. The provisions encouraging regional planning and requiring identification of responsible agencies should also facilitate planning for resource recovery.

3. Advice to state and local governments on implementation of resource recovery projects and programs will be focus of the technical assistance panels (Section 2003). Recognition of communities' needs for expert and intensive consultation in the complexities of these enterprises was a prime basis for congressional adoption of Section 2003.

4. Section 1008 requires EPA to issue guidelines for solid waste management; this, in effect, continues the guideline-writing authority under which the beverage container, source separation, and other guidelines in the area of resource conservation have been promulgated [87, 88].

5. Under Subtitle H, EPA has wide authority to conduct and support research, demonstrations, and studies relating to resource re-

covery and conservation systems. EPA is directed to enter into contracts to evaluate full-scale solid waste facilities, whether or not they are partially funded by EPA.

6. Section 8002 (j) establishes the cabinet-level Resource Conservation Committee, chaired by the EPA Administrator and comprised of the Secretaries of Commerce, Labor, Treasury, and Interior, as well as the Chairman of the Council on Environmental Quality and a representative of the Office of Management and Budget. The Committee is charged with analyzing and reporting recommendations to the president and the congress on a wide range of incentives and disincentives to foster resource recovery and conservation. This includes analysis of the effect of removing existing subsidies and allowances for virgin materials. It also includes a specific mandate to evaluate the feasibility of solid waste product charges, that is, an excise tax on products that reflects the cost of collection and disposal of the products. The committee is required to report every 6 months over a 2-year period to the president and the congress.

7. Section 8002 also requires other studies on specific aspects of resource recovery; these relate mainly to types of recovery systems and their compatibility, to recoverability of specific materials (glass, plastics, tires), and to research priorities.

8. Section 6002 requires federal leadership in procurement of products manufactured from recycled materials. Agencies are instructed to eliminate any bias in specifications against products containing recycled materials and to procure products containing the highest percentage of recycled materials practicable. States, localities, and contractors must also comply with Section 6002 in purchasing with federal funds. EPA is to write guidelines to aid agencies in complying with this mandate.

9. In Subtitle E the Department of Commerce is assigned various duties to promote resource recovery: publish guidelines for the development of specifications for secondary materials, stimulate development of markets for such materials, promote proven resource recovery technology, and provide for the exchange of technical and economic data on resource recovery facilities.

Public Participation and Information Activities

In view of the nature and complexity of the issues that RCRA addresses, the voluntary changes in institutional and individual habits and attitudes it is intended to stimulate, and the difficult direct and indirect regulatory actions it prescribes, its successful implementation depends

Table 26

RCRA Regulations and Guidelines Issued or in Preparation as of January 31, 1979

Section of the Act	Description	Statutory deadline	Status[a]
1008	Solid waste management guidelines	October 1977 and time to time thereafter	Guidelines on landfill disposal are scheduled for proposal in March 1979, with final issuance in Jan. 1980.
3001	Identification and listing of hazardous waste	April 1978	Proposed Dec. 18, 1978. Final scheduled for Dec. 1979.
3002	Standards for generators of hazardous waste	April 1978	Proposed Dec. 18, 1978. Final scheduled for Dec. 1979.
3003	Standards for transporters of hazardous waste	April 1978	Proposed April 28, 1978. Final scheduled for Dec. 1979.
3004	Standards for hazardous waste treatment, storage, and disposal facilities.	April 1978	Proposed Dec. 18, 1978. Final scheduled for Dec. 1979.
3005	Permits for treatment, storage, of disposal of hazardous waste	April 1978	Proposal scheduled for Mar. 1979. Final scheduled for Dec. 1979.
3006	Guidelines for development of State hazardous waste programs	April 1978	Proposed Feb. 1, 1978. Reproposal scheduled for Mar. 1979. Final scheduled for Oct. 1979.
3010	Notification system regulations	—	Proposed July 11, 1978. Final scheduled for Aug. 1979.
4002 (a)	Guidelines for identification of regions and agencies for solid waste management	April 1978	Interim guidelines published May 16, 1977.
4002 (b)	Guidelines for State plans	April 1978	Proposed Aug. 28, 1978. Final scheduled for June 1979.
4004	Criteria for classification of disposal facilities	October 1977	Proposed Feb. 6, 1978. Final scheduled for July 1979.

(continued)

Table 26 *continued*

Section of the Act	Description	Statutory deadline	Status[a]
6002	Guidelines for procurement practices	—	Proposal of the first guidelines, on cement and concrete, scheduled for April 1979.
7002	Prior notice of citizen suits	—	Final regulations published October 21, 1977.
7004	Public participation guidelines	—	Interim guidelines published Jan. 12, 1978.
	Public participation guidelines for programs under RCRA, Clean Water Act, and Safe Drinking Water Act (will supersede previous guidelines)	—	Proposed August 7, 1978. Final scheduled for Feb. 1979.
	Regulations to implement the Resource Conservation and Recovery Act of 1976; Grants and other financial assistance	—	Interim regulations published October 20, 1977. Amendments published Sept. 25, 1978.

[a] Schedules for issuance of guidelines and regulations in preparation are subject to change.

[b] Financial assistance provisions.

on a high level of public understanding and participation. Fortunately the act contains an array of public information and participation provisions.

Section 7004 requires that public participation in implementation of all parts of the act be provided for, encouraged, and assisted by EPA and the states. EPA, in cooperation with the states, is to develop and publish minimum guidelines for such public participation.

Section 8003 requires EPA to develop, collect, evaluate, and coordinate information in key subject areas, rapidly disseminate this information, implement programs to promote citizen understanding, and establish a central reference library on solid waste management.

In summation, RCRA is committed to the following basic objectives: improved practices in solid waste disposal to protect public health and environmental quality, regulatory control of hazardous waste from generation through disposal, and establishment of resource conservation as the preferred solid waste management approach. The act does not change environmental stewardship to the federal level only, but also clearly spells out state and regional presence in the management of solid waste; it makes especially clear the preeminent role to be assumed by the states in the administration and enforcement of solid and hazardous waste management programs.

At present, few states have a resource conservation and recovery program; it is expected that by 1981 all states will have some form of program to assure that all solid waste is disposed of in an environmentally sound manner.

Table 26 is a list of recent regulations and guidelines issued under RCRA.

VIII. CONCLUDING REMARKS

This chapter has attempted to present an overall picture of the qualitative and quantitative magnitude of the solid waste problem, and its management in terms of three major functions: collection, disposal, and recovery. These are the tangibles that one must have an absolute grasp of in order to engineer an effective solid waste management system.

There are, in addition, certain institutional factors that tend considerably to complicate solid waste management if they are not recognized at the outset. They are the intangibles of a solid waste management system and relate to the inevitable uncertainties of who will administer and operate the system, how it will be financed, and what will be the local public's attitude toward it. These factors are not easily quantified and less easily generalized; consequently each manager faces the difficult task of identifying them in any particular situation. The remainder of this chapter will present a very brief overview of these intangible factors and how they might best be handled if one's object is to improve the overall outcome of a solid waste management project.

Solid waste management differs in one important aspect from air and water pollution in that the bulk of solid wastes is deposited on land; therefore, collection and disposal are regarded as local (as opposed to regional) problems left to be solved locally. As a result, local municipalities generally favor short-range solutions (within the timetable of a 4-year elected term) and have overlooked long-range remedies. This political procrastination effectively prevents the implementation of any cohesive plan for the proper use of many of the alternative methods of disposal and recovery. In addition to the urgent need for careful long-range planning of solid waste management, there lies the need for regionalizing collection and disposal systems.

Many communities are looking toward regional approaches to solid waste management in order to accomplish together what they cannot attain alone. The so-called multijurisdictional approach can be used to hire better management and spread unit costs over a larger population base, thereby taking advantage of economies of scale, and avoiding costly duplication of services [18, 25, 89].

As explained in ref. [18] p. 8, an intergovernmental body created for multijurisdictional management of solid waste can solicit and accept funding from state, federal, and other sources; it can allocate costs fairly among local jurisdictions; it can plan comprehensively for transportation and land use as well as waste processing and disposal; it can make a systems approach to resource recovery more feasible; and it can eliminate the need for direct federal and state controls by being better able to meet standards and laws. The task facing local governments is to determine the conditions under which a multijurisdictional approach is best for them, and which type will best meet their needs. The pros and cons of the different types of multijurisdictional approaches discussed in ref. [18] are summarized in Table 27.

A second intangible to be contended with is the simplistic philosophy that solid waste is something for which one no longer has any use, and of which one wants to get rid as fast and as cheaply as possible. This basic motivation behind waste management limits it in scope, particularly since expenditures for waste collection, disposal, and recovery, must compete with other possible expenditures. The bulk of expenditures (80%) toward solid waste management is in the collection and transportation phases with very little being spent on the disposal and recovery phases. The explanation given for this disproportionate priority by public works officials is that the public will provide money to purchase trucks and to employ men rather than tolerate wastes in their neighborhoods, but the critical phase of disposal and recovery is usually handled by the cheapest means available. In other words, *money* governs the methods of disposal and not public health, conservation of resources, or any of the other factors that should be considered in planning a solid waste disposal facility. Consequently, municipalities have time and again turned to the city dump, and recently to the sanitary landfill.

This again goes back to the fact that solid waste management decisions are made by local communities who have to pay for these

Table 27
Potential Advantages and Disadvantages of Types of Multijurisdictional Approaches[a]

Alternative	Potential advantages	Potential disadvantages	Conditions that favor alternative
Authority	Can finance without voter approval or regard to local debt limit Political influence minimized because board members are private citizens Autonomous from municipal budgetary and administrative constraints Can generate income to make service self-supporting Capital financing is tax exempt	Financing is complex Can become remote from public control Can compete with private industry in some areas, reducing efficiency of both	Debt ceiling prohibits financing by the municipality Voter approval of financing will delay urgent project Political activity has hindered activity in past Autonomy from municipal budgetary and administrative control would mean more efficient delivery of service
Nonprofit public corporations	Tax-exempt status Can finance without voter approval or regard to local debt limit Assets revert to community after bonds are paid	Political influence may be exerted because board members are government officials Difficult to dismantle even if better service can be provided by other sources Financing is not backed by full faith and credit of community	City wishes to shift financing requirements to an organization outside municipal bureaucracy City wishes to avoid administrative details of providing solid waste management services

Multicommunity cooperative	Tax-exempt status is available Does not require state approval	Member communities lose some autonomy Ability to raise capital depends on lead community's debt capacity and financing strength Lead community can be hurt financially unless contracts with other communities are written properly	One city is willing to take lead in securing financing
Special districts	Constituency is a distinct group of residents, not scattered bond-holders Local autonomy can be protected by having county officials serve on board	Powers limited by state statute Must rely on special tax levies requiring voter approval Creates an additional unit of government not directly elected by citizens	No other governmental unit can provide service
Governmental agreements	Flexible and enforceable method of cooperation Basic governmental structures are not changed Can be implemented quickly and easily	May be difficult to raise capital since each community must borrow No single corporate body, so all communities must agree on any decision If contracts are not carefully written, misunderstandings may arise	Service or function to be provided is not costly or complex

[a] From Ref. [18].

Table 28

Potential Advantages and Disadvantages of Taxes and User Charges as Sources of Operating Revenues, and the Conditions that Favor Each[a]

Alternative	Potential advantages	Potential disadvantages	Conditions that favor alternative
Property tax	Simple to administer—no separate billing and collection system necessary If part of local property tax, it is deductible from federal and state income taxes	Solid waste management is often a low-priority item in the budget and receives inadequate funds Costs are hidden—less incentive for efficient operation Commercial establishments pay taxes for service they may not receive	Tradition of tax financing for most public services
Sales tax	Simple to administer	Variable monthly income Requires voter approval Income may not be adequate Commercial establishments pay taxes for services they may not receive	Recreation areas with high tourist trade

Municipal utility tax	Simple to administer More equitable than ad valorem taxes Can be instituted without voter approval	Variable monthly income Income may be inadequate	Ceiling on property tax rates Tradition of tax financing for most public services
Special tax levies	Voter approval usually not required	Amount limited by statute	Ceiling on property tax rates Tradition of tax financing for most public services
User charges	Enables localities to balance the cost of providing solid waste services with revenues Citizens are aware of costs of service and can provide impetus for more efficient operations	More complex to administer Can cause problems for users on fixed incomes	Ceiling on property tax rates

[a] From Ref. [18].

decisions themselves. Many communities are unwilling or even unable to pay the price necessary to take advantage of the best currently available solid waste disposal systems or devices. This brings up the next intangible facet of solid waste management, namely, modes of financing.

A major problem confronting many local governments is that the increasing volumes of waste collected and the need to meet more stringent regulatory standards in disposal operations are demanding greater reserves of capital and operating revenues.

Traditionally, municipally operated solid waste systems have been funded from property tax revenues. Pressed for more operating revenues for solid waste services, many communities have sought other sources of revenue such as a sales tax, municipal utility tax, or some type of special assessment (see Table 28). With increasing pressure on municipal budgets, there has been a trend toward raising new or increased revenues through "user" or "service" charges. Funding solid waste services through user or service charges is an equitable method if properly administered. It allows a community the opportunity to establish fees on the basis of actual costs of collection and disposal. Although many communities charge a uniform rate, others base their charges on the amount and kind of service rendered.

Regardless of the funding arrangement, three things are essential in deriving maximum benefits from it: First, accurate cost accounting is needed to establish cost-effective operations. Second, funds collected for financing the system should be set aside in a dedicated fund so that they are available as needed. Third, the revenues received should be reflective of the cost of the service provided. Otherwise, not only will the operating revenue be insufficient for the service expected, but awareness of the cost of service among the users is not fostered [18, 90].

Municipalities commonly draw from two basic sources to obtain capital for facilities and equipment: borrowed funds and current revenues. A third alternative is to contract with private firms for the service and shift the capital-raising burden onto them. The decision facing the local government is which of these alternatives is best for the local conditions and needs (see Table 29). The decision will be controlled by such factors as the financial status of the city, voter attitude, legal constraints on debt limits or long-term contracts, and the magnitude of the project to be undertaken [18, 91–93].

As demands from environmental regulations lead to more capital-intensive waste management programs, the burden on many local

Table 29

Potential Advantages and Disadvantages of Different Capital Financing Methods and the Conditions that Favor Each[a]

Alternative	Potential advantages	Potential disadvantages	Conditions that favor alternative
Borrowing:			
General obligation bonds	One of the most flexible and least costly public borrowing methods Requires no technical or economic analysis of particular projects to be funded Small projects may be grouped to obtain capital Least difficult to market	Requires voter approval, and elections may be expensive Must not exceed municipality's debt limit Issuing jurisdiction must have power to levy ad valorem property tax Transaction costs impose a bench minimum of $500,000 Capital raised becomes part of general city treasury, thus other city expenditures could draw on amount, unless specifically earmarked for solid waste Since careful project evaluation is not required, decision-makers may be unaware of technological and economic risks Ease of raising capital is a deterrent to change in existing public/private management mix, little incentive for officials to consider use of private system operators	Size of community is small or medium Voter approval likely

(continued)

Table 29 *continued*

Alternative	Potential advantages	Potential disadvantages	Conditions that favor alternative
Municipal revenue bonds	Projected revenues guarantee payment Can be used by institutions lacking taxing power, such as regional authorities and nonprofit corporations Does not require voter approval Is not constrained by municipality's debt limitations	Effective minimum issue of $1 million, thus only useful for capital-intensive projects Information requirements of the bond circular are extensive Technical and economic analysis of project must be performed by experts outside the municipal government Cost is higher than general obligation bonds Can be used only for specific projects	Capital-intensive projects Regional facilities desired Municipality's debt limit has been reached Initiating institutions lack taxing power
Bank loans	Small-scale capital requirements for short-term funding (5 years or less) Some medium-term funding applicability since notes may be refinanced as they expire Relatively low interest cost because interest paid by municipality is tax-free to bank Source of funds on short notice No external technical or economic analysis required Essentially no minimum Relatively inexpensive	Low maximum Short term Not useful for capital-intensive projects	Capital requirement is small Funds needed on short notice

	Voter approval generally not required No debt ceilings Can be used by institutions lacking taxing power		
Leasing	Useful as interim financing for equipment needed before appropriations or long-term capital arrangements can be made Negotiating agreement is simple and fast Only certification required is assurance of municipality's credit standing Reduces demand on municipal capital outlays since original capital raised by private corporation	Relatively high annual interest rate (9–18 percent) Amount of capital is usually limited Lease terms are generally 5 years or less Some states prohibit municipalities from entering multi-year, noncancellable contracts City will not own asset unless it purchases facility upon completion of lease period	Equipment needed before appropriations available Municipality has good credit rating
Current revenue capital financing	Least complex mechanism available No consultant or legal advice required No need for formal financial documents	No cost in the conventional sense (but higher taxes result) Communities' ability to generate surplus capital is frequently lacking Current taxpayers should not have to pay for a system that will be used far into the future Solid waste projects must compete with others municipal demands	Amount of capital necessary is small

(*continued*)

Table 29 *continued*

Alternative	Potential advantages	Potential disadvantages	Conditions that favor alternative
Private financing	Municipality need not borrow capital Provides long-term flexibility for municipality	Municipality must locate acceptable firm and negotiate contract Higher cost of capital reflected in system charges There may be legal constraints which prevent signing of long-term contract Displacement of city employees	Municipality's debt limit has been reached Municipality wishes to avoid administrative details of operating solid waste facility
Leveraged leasing	Reduces demand on municipal capital funds Interest rate on entire financial package may be lower than general obligation bonds	Legally complex City will not own asset unless it purchases facility upon completion of leasing period	

[a] From Ref. [18].

municipalities could get to the point where federal assistance will be a necessity. Thus far such assistance has been slow in coming, caused no doubt by the fact that a clean environment is not yet high enough on our national value system.

Perhaps the most significant intangible factor and at the same time the most difficult to assess is public opinion and attitude toward solid waste management. To be effective, solid waste management must take into consideration people and their attitudes.

Every individual's motives are arranged in a certain pattern, which defines one's attitudes and hence one's actions. One can think of this set of motives as a triangle, with the strongest motive at the apex of the triangle, and others below it. Under certain circumstances these motives shift, i.e., they rise or fall, with the ascendant motive being strongest and influencing general attitude and decisions. In order to predict how people will act in a given situation, it is absolutely necessary to identify possible rises or drops in motives. Failing to do so may result in disastrous mistakes [94].

To a great many people, the solid waste problem is most apparent in its esthetic dimensions—the esthetics of uncollected garbage and trash, dump sites, garbage washed up on beaches, and littered streets and landscapes. Consequently, billions of dollars are spent annually by state and local governments on garbage, and litter pickups. Such expenditures signify that society places a considerable value (motive) on the esthetic quality of the environment. What is lacking however, is a similar or greater value placed on the ecological damage and public health aspects of the environment. Consequently, once out of sight, the disposal and recovery aspects of solid waste are handled in the cheapest possible manner.

Solid waste managers must try to utilize their knowledge of our value system in order to explain to the public that the cheapest possible waste management may be much more expensive in reality, considering the ecological and public health consequences. It is only through such realization that active public support for effective solid waste management will become a reality.

Too often solid waste problems are considered as simple people–machine–money systems, with waste collection and disposal therefore considered chiefly as a problem of optimizing the use of employees, machines, and money. It is imperative that public attitudes be taken integrally into consideration in formulating optimal technological approaches, lest only suboptimal solutions ultimately emerge. To this end, public education and the applied psychology of solid waste management clearly will need much work in the years ahead [95].

REFERENCES†

1. J. A. Michener, *The Source*, Random House, Inc., New York, 1965 (also in paperback by Fawcett Publications).
2. C. G. Gunnerson, "Debris Accumulation in Ancient and Modern Cities," *J. Envir. Eng. Div., ASCE*, **99**, EE3 (1973).
3. A. J. Mukich, A. J. Klee, P. W. Britton, *Preliminary Data Analysis—1968 National Survey of Community Solid Waste Practices*, USEPA, Washington, D.C., 1968.
4. *Resource Recovery and Waste Reduction*, Fourth Report to Congress, USEPA-OSWMP, 1976.
5. W. R. Niessen and S. H. Chansky, *The Nature of Refuse*, Proc. Natl. Incin. Conf., Cincinnati, ASME, 1970.
6. W. R. Niessen and A. F. Alsobrook, *Municipal and Industrial Refuse: Composition and Rates*, Proc. Natl. Incin. Conv., New York, ASCE, 1972.
7. F. Smith, *A Solid Waste Estimation Procedure: Materials Flow Approach*, USEPA-OSWMP SW-147, 1975.
8. A. J. Klee and D. Carruth, "Sample Weights in Solid Waste Composition Studies," *J. San. Engr. Div., ASCE, SA 4*, **96** (1970).
9. F. Smith, *Comparative Estimates of Post-Consumer Solid Waste*, USEPA-OSWMP, SW-148, 1975.
10. P. A. Vesilind, J. F. McAlister, and J. A. Nissen, "Data Generation for Solid Waste Collection and Disposal Models," *Solid Waste Manag.*, November (1977).
11. *Third Report to Congress: Resource Recovery and Waste Reduction*, USEPA-OSWMP, SW-161, 1975.
12. National Center for Resource Recovery, *Municipal Solid Waste—Its Volume, Composition and Value*, NCRR Bulletin, Vol. III, no. 2 (Spring 1973).
13. P. H. McGauhey and C. G. Golueke, "Control of Industrial Solid Waste," in *The Industrial Environment—Its Evaluation and Control*, U.S. Dept. of Health Education and Welfare, Washington, D.C., 1973.
14. *Wealth Out of Waste*, U.S. Bureau of Mines, Washington, D.C., 1968.
15. *A Statewide Comprehensive Solid Waste Management Study*, Roy F. Weston, Inc., New York Dept. of Health, Albany, 1970.
16. J. C. Saxton and M. Narkus-Kramer, "EPA findings on Solid Wastes from Industrial Chemicals," *Chem. Eng.*, Apr. 28 (1975).
17. R. Stone, *The Effects of Air and Water Pollution Controls on Solid Waste Generation, 1971–1985—Executive Summary*, U.S. Dept. of Commerce, NTIS PB #240739, Washington, D.C. (Dec. 1974).
18. *Decision-Makers Guide in Solid Waste Management*, USEPA-OSWMP, SW-500, 1976.
19. E. S. Savas, Unpublished Report, NSF Grant No. SSH-74-02061-AC1, Washington, D.C., 1975.
20. ACT Systems Inc., *Residential Collection Systems*, Vol. 1, Report Summary, USEPA-OSWMP, SW-97c.1, 1974.

† USEPA-OSWMP = U.S. Environmental Protection Agency, Office of Solid Waste Management Programs, Washington, D.C.

21. Booz, Allen and Hamilton, *Cost Estimating Handbook for Transfer Shredding and Sanitary Landfilling of Solid Waste*, USEPA-OSWMP, SW-124c, Aug. 1976.
22. R. A. Boettcher, "Pipeline Transportation of Solid Wastes," in *Chemical Engineering Applications in Solid Waste Treatment*, G. E. Weismantel, ed. AIChE Symposium Series, No. 122, Vol. 68, 1972.
23. I. Zandi and J. A. Hayden, "Are Pipelines the Answer to Waste Collection Dilemma," *Envir. Sci. Technol.* **3** (9), Sept. (1969).
24. Ralph Stone & Co., *A Study of Solid Waste Collection Systems Comparing One-Man with Multi-Man Crews*, USPHS Publication 1892, Washington, D.C., 1969.
25. K. C. Clayton and J. Hvie, *Solid Waste Management: A Regional Approach*, Ballinger, Cambridge, Massachusetts, 1973.
26. L. H. Hickman, Jr., "Planning Comprehensive Solid Waste Management Systems," *J. San. Eng. Div., ASCE* **94**, SA 6 (1968).
27. R. M. Clark, "Measures of Efficiency in Solid Waste Collection," *J. Envir. Eng. Div., ASCE* **99**, EE4 (1973).
28. D. D. Smith, *Marine Disposal of Selected Solid Wastes—A Major Beneficial Use of Ocean Space*, 69th Annual AIChE Meeting, Cincinnati, May 1971.
29. Committee of Sanitary Landfill Practice of the Sanitary Engineering Division, *Sanitary Landfill*, ASCE Manual of Engineering Practice No. 39, 1959.
30. D. R. Brunner and D. J. Keller, *Sanitary Landfill Design and Operation*, USEPA-OSWMP, SW-65ts, 1972.
31. National Commission on Materials Policy, *Towards a National Materials Policy—Basic Data and Issues*, Washington, D.C., April, 1963.
32. *Hazardous Waste; Proposed Guidelines and Regulations and Proposal on Identification and Listing, USEPA*, *Federal Register* **43** (243), 58954–58967, Dec. (1978).
33. "Hazardous Waste Fact Sheet," *EPA J.* **5** (2), 12, Feb. (1979).
34. *Alternatives for Managing Wastes from Reactors and Post-Fission Operations in the LWR Fuel Cycle*, ERDA-76-43, Washington, D.C., May 1976.
35. Wolfe, R. A., "An Identification of the Wastes Generated Within the Nuclear Fuel Cycle," *AIChE Symposium Series*, **72** (154), 1–3 (1976).
36. "Nuclear Fuel Reprocessing: U.S. Outlook Dim," *Chem. Eng. News*, p. 9, March 20 (1978).
37. *Energy and the Environment—Electric Power*, Council on Environmental Quality, Washington, D.C., Aug. 1973.
38. *The Safety of Nuclear Power Reactors and Related Facilities*, U.S. Atomic Energy Commission, WASH-1250, Washington, D.C., July 1973.
39. E. D. North, "Solid Waste Generation in Reprocessing Nuclear Fuel," *AIChE Symp. Ser.*, **72** (154), 5–9 (1976).
40. G. L. O'Neill, "Purex Process," in *Reactor Handbook*, Vol. 2, Chap. 4. Interscience Publishers, New York, 1961.
41. A. W. DeAgazio, "Fuel Reprocessing and the Environment," University of Arizona Press, Tucson, 1973.
42. B. D. Devine, "Department of Transportation Regulations for Nuclear Waste Shipments," *AIChE Symp. Ser.*, **72** (154), 42–44 (1976).
43. M. J. Jump, "Current Practices for Disposal of Solid Low Level Radioactive Waste," *AIChE Symp. Ser.*, **72** (154), 65–68 (1976).

44. R. A. Wolfe, "The Research and Development Program for Transuranic-Contaminated Waste Within the U.S. Energy Research and Development Administration," *AIChE Symp. Ser.*, **72** (154), 28–32 (1976).
45. G. H. Daly and O. P. Gormley, "Handling, Storage, and Disposition of Solid Low Level Wastes," *AIChE Symp. Ser.*, **72** (154), 10–16 (1976).
46. K. J. Schneider and A. M. Platt, *High Level Radioactive Waste Management*, Vol. 1, Battelle Northwest Laboratories, BNWL-1900, Richland, Washington, May 1974, p. 1.8.
47. W. A. Freeby, "Fluidized Bed Calcination of Simulated Commercial High Level Radioactive Wastes," *AIChE Symp. Ser.*, **72** (154), 140–144 (1976).
48. C. C. Chapman, H. T. Blair, and W. F. Bonner, "Experience with Waste Vitrification Systems at Batelle Northwest," *AIChE Symp. Ser.* **72** (154), 151–160 (1976).
49. W. F. Bonner, J. F. McElroy, and J. E. Mendel, *Waste Solidification USA*, presented at the International Symposium of Management of Wastes from the LWR Fuel Cycle, Denver, July 1976.
50. R. Bonniaud et al., "French Industrial Plant for Continuous Vitrification of High Level Radioactive Wastes," *AIChE Symp. Ser.* **72** (154), 145 (1976).
51. J. P. Cagnetta, "Spent Fuel Storage," *Combustion*, June 27–28 (1978).
52. W. Lepkowski, "New Mexicans Debate Nuclear Waste Disposal," *Chem. Eng. News*, Jan. 1, 20–24 (1979).
53. I. Kiefer, *Hospital Wastes*, USEPA SW-129, 1974.
54. R. D. Singer et al., *Hospital Waste: An Annotated Bibliography*, USEPA, Distributed by National Technical Information Service, Springfield, Va., as PB-227-708.
55. A. F. Iglar and R. S. Bond, *Hospital Solid Waste in Community Facilities*, USEPA, 1973. Distributed by National Technical Information Service, Arlington, Va., as PB-222-018.
56. A. J. Darnay and W. E. Franklin, *The Changing Dimensions of Packaging Wastes*, First National Conference on Packaging Wastes, San Francisco, September 1969.
57. A. J. Darnay, "Throwaway Packages—A Mixed Blessing," *Envir. Sci. Technol.* **3** (4), 328, April (1969).
58. P. H. McGauhey, *Developing Strategies for Packaging Waste Management*, First National Conference on Packaging Wastes, San Francisco, September 1969.
59. D. B. Syrek, *California Litter: A Comprehensive Analysis and Plan for Abatement*, Institute for Applied Research (for the California State Legislature), Carmichael, California, May 1975.
60. R. W. Powers, President, Keep America Beautiful, Inc., Testimony before the California Senate Committee on Natural Resources, Washington, D.C., January 1974.
61. *Community Litter Prevention Guide*. U.S. Brewers Association Inc., Washington, D.C., 1972.
62. W. C. Finnie, "Field Experiments in Litter Control," *Envir. Behav.* **5** (2), 262 (1973).
63. P. A. Vesilind, *Measuring Roadside Litter*, Duke Environmental Center Publication, Duke University, 1976.
64. *Fourth Report to Congress: Resource Recovery and Waste Reduction*, USEPA-OSWMP, SW-600, 1977, p. 69.

65. *Committee Findings and Staff Papers on National Beverage Container Deposits*, Second Report to the President and Congress of the United States Mandated by the RCRA of 1976, Resource Conservation Committee, Washington, D.C., Jan. 1978.
66. *Potential Effects of a National Mandatory Deposit on Beverage Containers*, Comptroller General's Report to the Congress, Washington, D.C., December 1977.
67. L. E. Sheftel, *Recycled and Restored Scrap Combats Pollution and Inflation*, 30th Annual Technical Conference, Society of Plastics Engineers, Chicago, May 1972.
68. E. V. Anderson, *Chem. Eng. News*, September 22, p. 16 (1975).
69. D. R. Paul et al., *Pol. Eng. Sci.* **12** (3), 157, May (1972).
70. D. R. Paul et al., *Pol. Eng. Sci.* **13** (3), 202, May (1973).
71. International Research and Technology Corp., *Recycling Plastics—A Survey and Assessment of Research and Technology*, Report to The Society of the Plastics Industry, Inc., New York, June 1973.
72. L. J. White, *Chem. Eng.* Oct. 15, p. 68 (1973).
73. J. E. Guillet, *Plastics Eng.*, August, p. 48 (1974).
74. G. E. Sheldrick and O. Vogl, *Pol. Eng. Sci.*, **16** (2), 65, February (1976).
75. F. Rodrigues, *Chem. Tech.* p. 409, July (1971).
76. K. C. Dean, "Recycling Automotive Scrap," in *Effective Technology for Recycling Metal*, National Association of Secondary Material Industries, Inc., New York, 1970.
77. C. J. Chindgren et al., *Construction and Testing of a Junk Auto Incinerator*, Bureau of Mines, Salt Lake City, Utah, Feb. 1970.
78. *The Automobile Cycle: An Environmental and Resource Reclamation Problem*, USEPA-OSWMP, SW-80ts.1, 1972.
79. C. C. Humpstone et al., *Tire Recycling and Reuse Incentives*, USEPA, SW-32c, 1974.
80. J. Beckman, "Destructive Distillation of Used Tires," *Poll. Eng.* **2** (5), 22, Nov./Dec. (1970).
81. J. W. Larsen and B. Chang, *Scrap Tires Pyrolyzed in Molten Salts*, American Chemical Society, Division of Rubber Chemistry Meeting, Minneapolis, April 1976.
82. "Finding Offbeat Uses for Scrap Tires," *Chem. Eng.* August 14, p. 88 (1978).
83. "U.S. Environmental Protection Agency, Proposed Guidelines for Thermal Processing and Land Disposal of Solid Waste," *Fed. Reg.* **39** (158), 29327–29338, August 14 (1974).
84. *Report to Congress on Hazardous Waste Disposal*, USEPA, June 1973.
85. *First Report to Congress on Resource Recovery*, USEPA, February 1973.
86. *EPA Activities Under the Resource Conservation and Recovery Act of 1976: Annual Report to the President and Congress, Fiscal Year 1977*, USEPA, SW-663, February 1978.
87. "U.S. Environmental Protection Agency, Solid Waste Management Guidelines for Beverage Containers," *Fed. Reg.* **41** (184), 41202–412-5, September 21 (1976).
88. "U.S. Environmental Protection Agency, Source Separation for Materials Recovery Guidelines," *Fed. Reg.* **41** (80), 16950–16956, April 23 (1976).
89. R. O. Toftner and R. M. Clark, *Intergovernmental Approaches to Solid Waste Management*, USEPA, SW-47ts, 1971.

90. E. R. Zausner, *Financing Solid Waste Management in Small Communities*, USEPA, SW-57ts, 1971.
91. R. M. Clark and R. O. Toftner, *Financing Municipal Solid Waste Management Systems, J. Sanit. Eng. Div., Proc. ASCE* **96** (SA4), 945–954, August (1970).
92. Resource Planning Associates, *Financial Methods for Solid Waste Facilities*, USEPA, SW-76c, 1974.
93. R. Randol, Resource Recovery Plant Implementation Guide: Financing, USEPA, SW-157.4, 1975.
94. B. Bocciarelli, *Waste Management and the Socio-Economic System*, 30th Annual Technical Conference, Society of Plastic Engineers, Chicago, May 1972, p. 199–201.
95. A. J. Klee, "The Psychology of Solid Waste Management," *APWA Reptr.* **36** (5), 14–15, 18, 20, May (1969).

2

Mechanical Volume Reduction

Eugene A. Glysson

Department of Civil Engineering, College of Engineering, University of Michigan, Ann Arbor, Michigan

I. INTRODUCTION

Solid waste management consists of three phases. The first is concerned with collection, the second with processing, and the third with disposal. This chapter will deal with the second of these phases, processing. The method employed for processing has a great influence on the final or disposal phase. Methods employed may range from none at all, whereby the refuse is disposed of much in unchanged form to thermal processing where only the noncombustible residues remain for disposal. Other methods include biological and mechanical processing. This chapter is devoted to an in-depth discussion of mechanical volume reduction.

The importance of volume reduction relates to the disposal of solid wastes and the difficulty in obtaining suitable disposal sites. This difficulty, which is many-faceted, requires that the maximum amount of refuse be put into the space available. Volume reduction may also be essential when refuse must be transported over long distances or for other handling operations.

Mechanical as opposed to thermal or biological reduction methods make use of machines rather than chemical (thermal) or living organisms (biological) to accomplish the reduction. The mechanical processes employ compression, tension, and shearing (combining

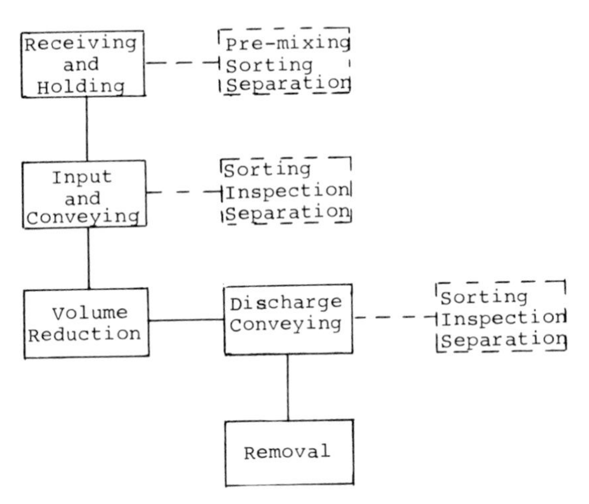

FIG. 1 Volume reduction facility flow chart.

tension and compression). Actually these forces are usually combined in such a way as to make it very difficult to theoretically analyze the various components. Therefore most sizing of equipment is based on experience.

A typical flow chart of a volume reduction facility is shown in Fig. 1. This chart applies to both shredding and baling operations. The operations shown by the dashed lines as subsystems are optional. In some instances premixing, presorting, and separation improve the subsequent operations. Activities such as recycling of certain materials (cardboard for instance) may be desirable in some cases.

Solid wastes are classified in several broad categories (see Table 1) for purposes of selecting the minimum horsepower requirements for shredders.

When applied to solid waste processing the words grinding, shredding, and milling all generally apply to the same process. The equipment to accomplish the size reduction may vary somewhat from hammer mills to chippers, depending on the character of the waste but the results are much the same, that is, the waste is reduced to a more or less uniform size, depending on the need and the constituents are blended.

The Waste Equipment Manufacturers Institute (WEMI) defines shredding equipment as "a mechanical device used to break up solid waste and recoverable materials into smaller pieces." WEMI has designated the term "shredder" to cover *all* such equipment that fulfills this definition. "Shredding" is the term recommended by WEMI for the mechanical process of solid waste size reduction and supersedes such terms as milling and grinding. [2]

Table 1
Solid Waste Categories [1]

Category	Composition	Minimum horsepower required
Light	Sorted packer truck wastes, such as paper, cardboard, bottles, cans, garbage, and lawn trimmings	250
Medium	Normal packer truck wastes, such as the above plus small crating, small appliances, small furniture, bicycles, tree trimmings, and occasional auto tires	400
Combined light and medium with some bulky		600
Bulky	Oversize and bulky wastes of the above plus items such as stoves, refrigerators, washers, dryers, doors, large furniture, springs, mattresses, tree limbs, and truck tires	800
Heavy	Large and dense materials, all of the above plus items such as demolition rubble, automobiles,[a] logs, and stumps	2000

[a] Automobiles are usually shredded separately by a scrap processor for recovery of the metal.

The forces at work are indicated in Fig. 2. These forces, i.e., tension, compression, and shear, are combined in the actual shredding process resulting in the size reduction. It is very difficult, if not impossible, to derive a mathematical expression that can accurately describe the combined effect of these forces as it actually occurs in the shredding process, especially in the case of solid wastes that contain such a diverse mixture

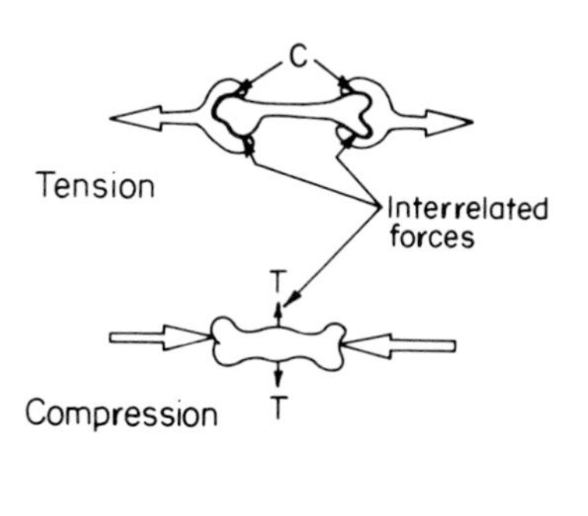

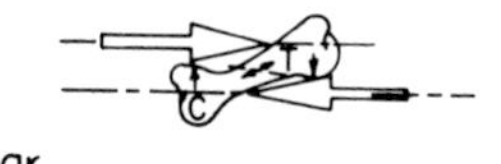

FIG. 2 The application of forces in size-reduction operations. From Ref. 3.

of materials. The design of the machine components therefore is based on a combination of theory and experimentation. No attempt is made here to discuss these theories; further information may be obtained from the following sources:

G. C. Lowrison, *Crushing and Grinding*, C.R.C. Press, Cleveland, Ohio, 1974.
W. L. McCabe and J. C. Smith, *Unit Operations of Chemical Engineering*, 2nd Ed., McGraw-Hill, New York, 1967.

The other process to be described here is baling, which may be defined as the process of applying pressure to loose, compressible material within an enclosure in such a way as to reduce the voids in the material and to increase its density. In some cases the deformation of the material may be such as not to require binding, whereas in other cases binding with wire, etc., may be necessary to hold the material in its confined condition.

II. SHREDDING

The shredding of refuse can be generally classified in two categories, depending on whether the refuse is shredded dry or wet. Several different machines can be used for the more common dry shredding [16, 18, 19].

A. Dry Processes

1. *Hammer Mills*

To date hammer mills have been the most common type of machines used for shredding mixed municipal refuse. The principal reason is that the machine is capable of doing a reasonably good job on the very heterogeneous material found in mixed refuse. It can tolerate almost any type of material and is chosen for this reason rather than for its high efficiency at handling any one type.

Basically the hammer mill consists of a heavy rotor with hammers attached to its outer circumference, enclosed in a housing. As the rotor turns the hammers strike the material to be shredded. The hammers are sometimes aided by stationary breaker plates or cutter bars. The material may be required to exit through a grating at the bottom of the machine that determines the final size of the shredded material. Material fed into the machine is reduced in size by impact, shear, and abrasion forces caused by the hammers.

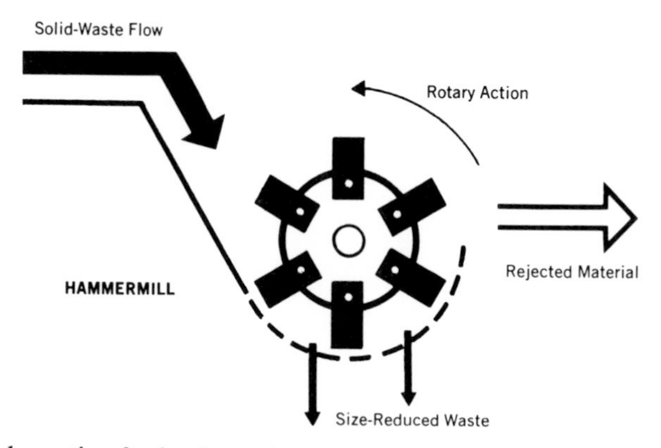

FIG. 3 Schematic of a horizontal swing-hammer hammer mill. From Ref. 4.

The most common configuration of hammer mills are the swing hammer type in which the hammers are pivoted on the rotor so that they can swing back should they encounter a very hard object that lessens the chance of breaking the hammer or stalling the machine (see Fig. 3).

The shapes of the hammers themselves range from rectangular to hammer-headed, depending on the manufacturer. The hammers range widely in weight, from a few kilos in small machines to 226.5 kg (500 lb) in automobiles' shredding units (see Fig. 4).

In addition to rigid or fixed hammer machines, the swing-hammer type is the most common because the hammer is free to pivot on pins in a way that reduces damage to the machine from impact with a very hard or strong piece of input material. A balance problem may arise if the hammers become entangled with the input material (e.g., sheet plastics or wire) and do not swing freely.

Hammer mills are also classified by the position of the shaft as horizontal or vertical, the former being the more common. In this case the rotor shaft is supported at each end. Incoming material enters at the top and flows through the machine by gravity. Control of the particle size of the discharge material is accomplished by a grating through which the material must pass to leave the machine. The size of the openings—usually rectangular—in this grating primarily determines the finished particle size.

The vertical-shaft hammer mills use a rotor shaft in an upright (vertical) position. Material also enters at the top and flows through the machine by gravity. Control of the particle size is commonly accomplished by decreasing the cross-sectional area between the external diameter of the hammer and the stationary housing. As a result the particle

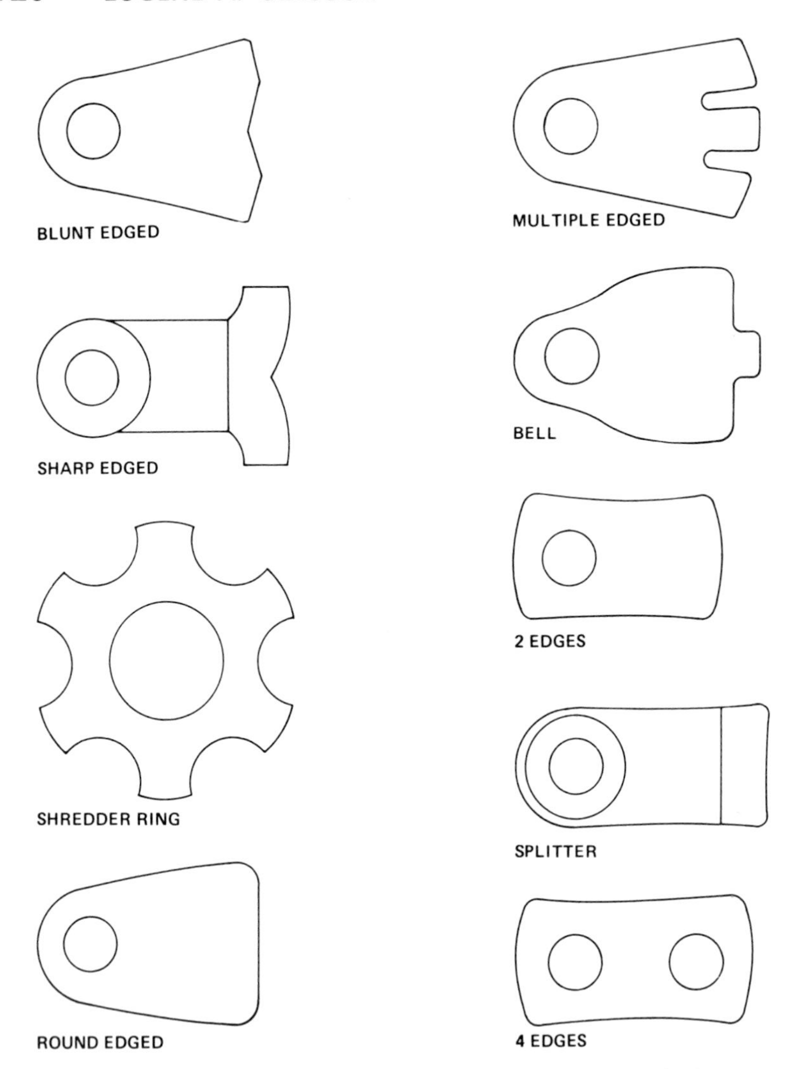

FIG. 4 Types of hammers used on hammer mills—depicting their shapes. From Ref. 2.

size progressively diminishes as the material passes through the machine, and no exit grating is required. The lower shaft bearing of this type of shredder must be capable of supporting the weight of the rotor (see Fig. 5).

A modification of the vertical type of mill is a machine that uses rolling star wheels or gears positioned around the periphery of the rotor

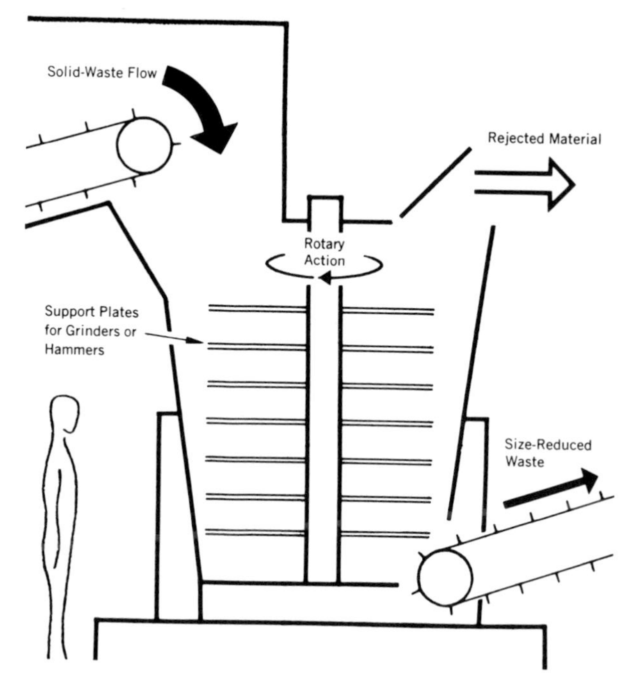

FIG. 5 Schematic of a vertical type size-reduction unit. From Ref. 4.

instead of pivoted hammers. These wheels or cogs grind the material against the side walls of the shredder to reduce its size. The diameter of the housing progressively decreases so that the material is ground smaller and smaller until it reaches the discharge outlet of the machine.

In all of the types of shredders indicated the particle size varies according to the wear on the internal components, moisture content of the refuse, number of hammers being used, feed rate, and solid waste composition.

2. *Chippers*

Special, single-component, solid wastes are more efficiently served by devices especially designed for that purpose. Chippers designed for handling tree limbs and trunks are of this type. These machines contain sharp knife edges positioned on a large rotating disk in such a way that the tree can be fed butt first into it. The knives cut the tree into approximately 1.9-cm ($\frac{3}{4}$-in.) size chips that can be disposed of in a number of ways. Smaller versions are commonly employed by various utility companies for handling tree limbs and other brush. The life of the knives is directly related to the amount of extraneous material (i.e., nails, wire, and concrete) that happens to be included in the wood being processed. The largest of this type of equipment can accommodate a tree which has 61 cm (24-in.) diameter at the stump. They commonly are trailer mounted, which allows them to be moved to accommodate an accumulation of material to be chipped.

3. *Shears*

Bulky wastes (i.e., furniture and timber from building demolition) can be prepared effectively for further processing, such as incineration, by a multiblade hopper-type shear (see Fig. 6). These machines operate like the jaws of an alligator crushing the items between the components

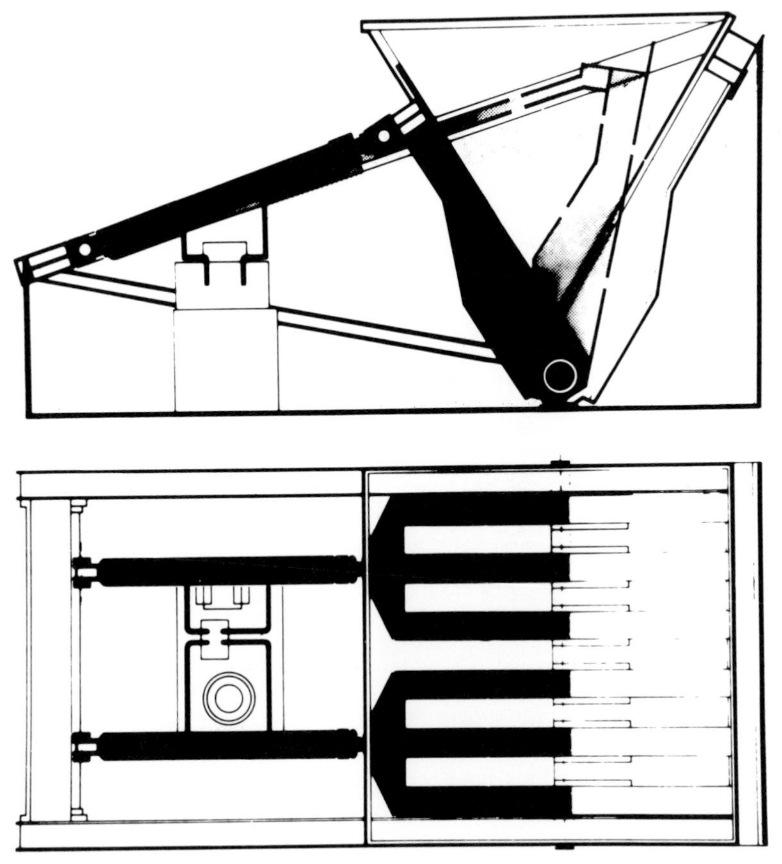

FIG. 6 Von Roll bulky waste shear. From Ref. 5.

of the jaws as they are closed. Material to be crushed at the incinerator is usually handled and delivered to these machines by a traveling bridge crane normally needed to charge refuse into the incinerator. Other than for this rather specialized use this type of crusher is very rarely used in processing mixed municipal refuse.

Traditionally the single-blade shear has been used in scrap metal yards to reduce the size of metal objects to a more uniform and manageable dimension. Crushed and baled automobile bodies, scrap-steel items such as bars, angles, and pipe are cut into shorter lengths by using the shear. This operation is very slow so that shears are rarely used in any other application of solid waste processing.

B. Wet Processes

1. *Rasp Mills*

Rasp mills are not commonly used at present in the United States. They are large cylindrical machines in which an internal rotor moves heavy arms that sweep the refuse around against rasping points on the bottom plate. Holes in this plate serve to control the size (about 5 cm, 2 in.) of the material discharged from the unit. Any material which cannot be reduced in size is rejected tangentially. The rotor turns slowly (5–6 rpm), and has a diameter of about 6.1 m (20 ft). Refuse is introduced through the shroud over the top (see Fig. 7).

Water is usually added to the refuse being processed by the rasp mill since moist refuse tears apart more effectively; thus process efficiency is increased. In Europe, these machines have been used mostly in conjunction with composting operations.

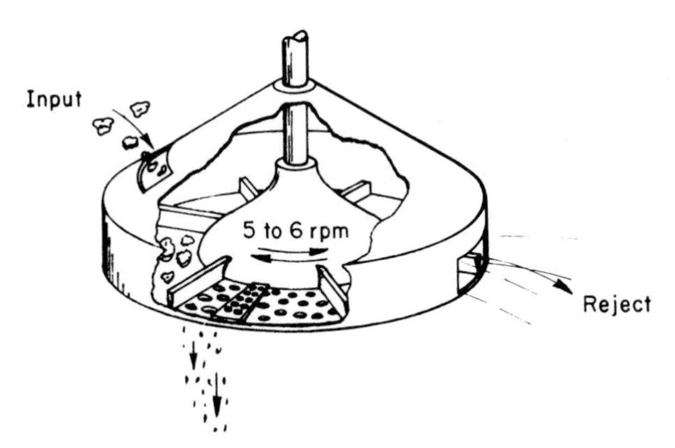

FIG. 7 Rasp mill. From Ref. 1.

2. *Hydropulper*

A central piece of equipment in the recycled paper industry is the hydropulper, which transforms the paper fiber into a slurry for re-forming into new paper. This machine has been developed and used successfully for many years. It consists of a large circular tank filled with water and equipped with a high-speed impeller at the bottom that causes the liquid to swirl and form a vortex. The paper to be slurried is added to the swirling water and is swept downward by the vortex. The combination of the wetting and of the shear of the paper against itself causes the paper to be disintegrated into a fibrous slurry. This slurry is drawn off through holes in the bottom plate of the hydropulper, discharge being possible only when the mixture is of such nature as to pass through these openings (see Fig. 8).

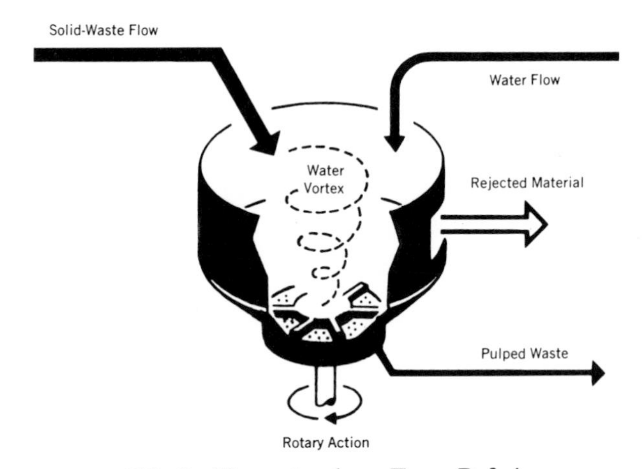

FIG. 8 The wet pulper. From Ref. 4.

This method of size reduction has been applied to mixed municipal refuse as a means of separation for reclaiming the fiber content and some of the other constituents of refuse such as steel, aluminum, and glass. Refuse that is free of large objects (furniture, white goods, tires, etc.) is introduced to the hydropulper. The rotating impeller at the bottom of the machine produces the vortex that draws the refuse down into the water in the tank. In addition to the impeller sharp cutting edges and tips often aid in reducing the size of the components of the mixture. Bottles are broken, and cans are crushed and expelled tangentially through a "junk extractor." The remaining slurry leaves the hydro-pulper through a perforated plate beneath the rotor plate. The diameter of these openings is approximately 2.2 cm ($\frac{7}{8}$ in.). Subsequent screening and dewatering operations recover the fiber and other materials for reuse or disposal.

III. BALING

Another method of volume reduction of solid wastes is the use of heavy duty balers. The technique is simply one of reducing the voids and increasing the weight per unit volume (density) of the material by compression [17].

The baling of loose material for shipment has long been a common practice. Balers for baling waste paper, cotton, etc., are available from several manufacturers. Many iron and steel scrap processors have equipment capable of crushing and baling whole automobile bodies into blocks known as "Number 2 bundles."

Applying the baling concept to mixed municipal refuse using these various types of balers has been considered for several years. Operations to study such techniques have been carried out in San Diego, Chicago, and St. Paul, Minnesota [6–8], as well as in Japan.

Baling is basically accomplished by compressing a given amount of refuse into a small volume. The baler employs a series of rams that force the refuse into a confined space within the baler in a specific sequence. A typical three-stroke baling machine is shown in Fig. 9. After the refuse is charged into the baler the top of the baler is closed. Each of the three strokes is then applied sequentially from the smallest remaining surface of the compartment. In the first or gathering stroke the compression ram (No. 1 in Fig. 9) is extended horizontally or from the side, the second or tramping stroke (ram No. 2), is extended vertically or from the top, and the third or final compression stroke (ram No. 3) is extended from the back. After completing its final stroke this ram is held in its final position for a few seconds to allow the bale to "set," and then the chamber is opened and the finished bale is expelled by the last ram. The cycle can then be repeated. The newer baling machines complete a cycle in 90 sec. Compressing the refuse too quickly is not efficient since time must be provided for air to escape from within the refuse mass and holding the refuse under pressure for a short interval produces a more stable bale. It has been found that holding the pressure during compaction always leads to a decrease in bale volume. This indicates that the expansion (spring back) forces created in the refuse during compression can be partially overcome by applying pressure for a longer time since the refuse may be classified as a semielastic material [7]. By maintaining

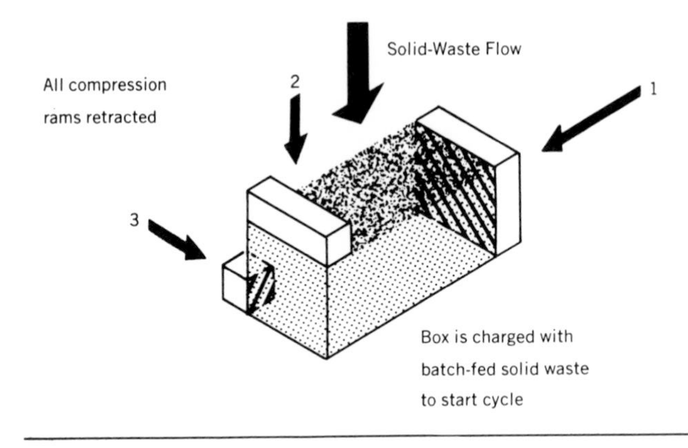

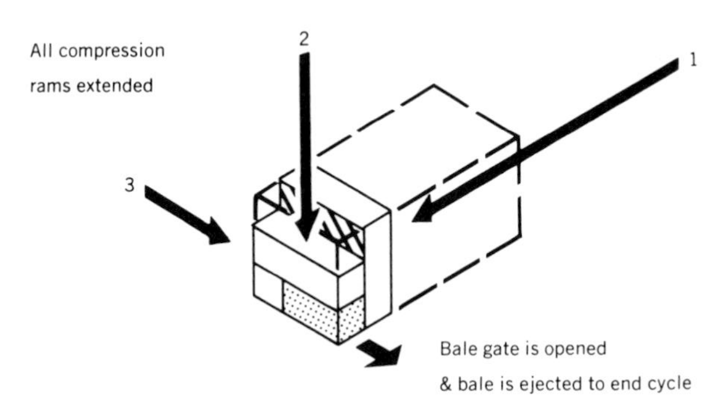

FIG. 9 A three-stroke baling process. From Ref. 4.

the pressure on the refuse after compression the fibers within the material become more permanently deformed and spring back is somewhat reduced. A period of 5–10 seconds is considered adequate, and an 8-s period is used by one baler [8].

Expansion of the finished bales (spring back) occurs immediately after the bale is released from the baler. This has been measured and on the average a volumetric increase of about 36% can be expected (St. Paul, Minn. study) with the length of the bale expanding up to 25% [9].

It has been established by various investigators that if a baler is loaded with charges of equal weight the resultant bales are of similar size whereas if the baler is fed on a volume basis the resultant bales vary greatly in size. In the St. Paul operation the waste is conveyed onto a weighing platform that accumulates refuse until the desired weight is reached and then the conveyor is stopped long enough for the platform to be raised and emptied into the baler. Bales produced by this type of equipment reach densities on the average of between 970.7 and 1132.5 kg/m^3 (60–70 lb/ft^3), being subjected to about 19,320 kN/m^2 (2800 psi) at the face of the final ram. These bales normally do not require binding or strapping in any way. If the bale does not remain intact upon removal from the baler, the material is simply put back onto the conveyor and baled again. Problems with bale stability have occurred when large quantities of a similar material such as grass or leaves constitute the entire charge. Furthermore, a moisture content in excess of 30% by wt results in bales that tend to be unstable [7].

A type of baler similar to that used for baling hay is available for baling shredded refuse. This baler is fed continuously rather than in batches as the previously described type. Friction within the enclosure surrounding the compressed material holds it in the machine until the desired size is produced. This type of bale usually requires binding with wire or straps since the greater compressive forces needed to produce a stable unbound bale are not developed. Bales obtained from this type of baler are also less dense, 403.6–485.7 kg/m^3 (25–30 lb/ft^3).

Balers of smaller sizes such as are normally employed to bale waste paper or corrugated cardboard have been considered for use with mixed refuse. This type of baler has not been very successful in baling unprocessed (as received) mixed municipal refuse. "The major factor[s] found to affect the stability of compacted bales of loose residential wastes are compaction pressure, time of pressure application, and the moisture content of the waste." [7] Small balers which cannot produce compaction pressures of more than 3450–6900 kN/m^2 (500–1000 psi) are more likely to produce fragile and unstable bales. Maximum stability is obtained at pressures between 13,800 and 20,700 kN/m^2 (2000 and 3000 psi). No significant additional improvement in stability occurs above 24,150 kN/m^2 (3500 psi) [7].

A special adaptation of the baling concept is embodied in the small household appliance type of refuse compactor. These devices provide some compression of the household trash and convenient containerization, but are in no way comparable to the baling operation being described here.

IV. SIZE REDUCTION APPLICATIONS

The technique of size reduction can be applied at any of the many stages of solid waste handling. These include transfer stations, resource recovery processing, and the final disposal of the material by incineration or landfilling.

A. Transfer Stations

In many instances, for economic or other reasons, it is desirable to transfer the collected refuse from the collection vehicle, which is a very uneconomical transporting device, to another vehicle that is designed for economical transport over longer distances [15, 20]. Such a transfer frees the collection vehicle and its crew to return to collecting, for which it was designed, and precludes the need for additional collection vehicles since there is no time lost in long hauls. Additional savings are realized by using more efficient transporting methods such as rail haul or large trailer trucks. Refuse is transferred from one vehicle to another at a transfer station.

Volume reduction as a part of the transfer operation may play an important part in realizing the benefits of a more economical transportation medium. To be economical, shipment by rail requires that maximum densities be obtained and that the refuse be transported in a form that can be quickly and efficiently loaded and unloaded. It is also important to assure that the refuse not be allowed to escape along the route resulting in litter, traffic hazards, etc.

Baling of refuse lends itself very well to transfer operations. The bales can be readily loaded and unloaded, they are stable during transportation, and only a minimum of enclosure is required for esthetic reasons. Several studies involving the shipment of baled refuse by rail indicate that such volume reduction lends itself well to this type of transportation [7, 10]. A railroad car capable of carrying 72,576 kg (80 t) can be loaded to its full weight allowance with baled refuse. If unprocessed refuse were to be shipped in this manner its volume to weight relationship would allow only about 36,288 kg (40 t) to be placed in the car. In addition, the baled refuse can be shipped on a bulkheaded flat car or in a gondola and does not require a closed car.

Tests on rail shipment have shown that the bales do not break up and refuse does not flake off to any significant degree during shipment. The test included switching where impact between cars produced severe jolts to the cargo [7, 10]. The same advantage applies to hauling the bales on a flat-bed truck.

Shredding as a part of the transfer operation improves the handling characteristics of the refuse as well as reduces its volume. Shredded refuse must be shipped in a closed vehicle. This is generally accomplished by using a compacting loading device that compacts the shredded material into a large (e.g., 57.4 m^3; 75 yd^3) transfer trailer that is equipped to discharge the shredded material at the disposal site.

B. Resource Recovery

A process that is almost universally applied in resource recovery is shredding. The reason is that the separation necessary for resource recovery by almost any method, is most effectively and efficiently carried out on as nearly a homogeneous material as possible. Shredding results in a more nearly uniform particle size, provides for better access to the various constituents, and improves the handling characteristics of the refuse.

Certain types of reclamation processes depend on shredding to allow the required technology to be effective; these processes include glass segregation, wood fiber reclamation, and metal separation. Separation by air classification and most composting and pyrolysis processes also include shredding as a requirement.

Resource recovery in the form of heat energy resulting from the combustion of solid wastes requires shredding if suspension burning and pneumatic handling and feeding of refuse is the method to be employed. This method of energy recovery seems to have great potential; however, shredding to improve the quality of refuse as a fuel is essential.

Special types of combustion process, such as the vortex furnace and the fluidized-bed reactor, require that the refuse being burned be shredded prior to its being fed into the combustion chamber. Certain types of wastes, such as old tires and junk automobiles, may be greatly benefited by shredding. Tires have potential as fuel and can be handled more easily if they have been shredded. Even if no recovery is contemplated, land disposal is much easier if the tires are broken down. Shredded scrap from old automobiles has become a readily saleable commodity in the secondary metals market. This owes to the high quality resulting from the separation that can be carried out as a result of the shredding operation.

C. Disposal Operations

Although incineration should more properly be classified as processing rather than as a disposal method, the volume reduction carried out by incineration indicates that much of the solid waste material is actually being disposed of. Size reduction in the form of shredding is a very important step in the operation of mass-burning incinerators. The necessity for shredding in relationship to the less conventional types of incinerators such as the fluidized bed or vortex type has been mentioned. The combustion equipment for mass burning is generally not designed to accept large bulky items such as furniture, large crates, pallets, tree limbs, and demolition wastes, for example. These items, although combustible, must be reduced in size before being charged into the furnace because of the limited size of the charging hopper. The shredder used for this purpose is usually the hammer mill or the jaw-type crusher. Units of this sort are being installed in most of the modern incinerator plants near the refuse-receiving area to accommodate the oversize wastes.

The final disposal is the sanitary landfill. The size reduction process makes a very profound impact on this operation.

Shredding refuse prior to land filling has the following effects:

1. Improved esthetics owing to the homogenizing effect of blending the paper and other constituents of refuse so that none of them are any longer distinguishable.

2. Paper is not blown around because of the small particle size and entangling effect of shredding. Wind cannot pick up and carry the paper away as is the case with unshredded refuse. (Plastic films pose a continuing litter problem because of less efficient shredding of such films in the hammer mill.)

3. There are no flies or rodents on a shredded landfill. Research has shown that fly larvae do not survive the shredding operation and the garbage content is so well distributed throughout the refuse mass that rodents cannot find an adequate food supply [11].

4. Odors are eliminated as a consequence of blending the refuse components by shredding.

5. Shredded refuse offers very little attraction to birds.

6. Shredded refuse, owing to its more nearly uniform particle size and freedom from voids spreads and compacts evenly and easily on the landfill.

7. Because it lacks voids and is easily compacted, the in-place densities of shredded refuse can be in excess of 710.6 kg/m^3 wet wt (1200 lb/yd^3).

8. Shredded refuse has a reduced fire hazard compared to unprocessed refuse.

9. Experience has shown that loaded trucks can easily travel over shredded refuse, even in inclement weather, with little additional tire wear.

10. The need for daily earth cover over the compacted refuse is eliminated by shredding so that omitting daily cover has been allowed in several cases. Consequently more refuse can be placed in the available landfill volume and the cost of the landfill operation reduced since

cover material does not have to be obtained and transported. This is not to say that final cover may be omitted but that the day-to-day operation can be modified and made more flexible.

A problem that is associated with the omission of daily cover, is the unhampered percolation of rain water into the refuse, which adds to the moisture available to form leachate. The cover material, if properly selected and applied, aids in preventing such percolation.

An additional advantage of shredding associated with the landfill disposal is its ready conversion to resource recovery operations if the market and economic conditions should become favorable.

Baling refuse prior to landfilling also has some definite advantages that accompany the reduction in volume or result from it. So-called balefills have been introduced in a number of cities with the following effects:

1. Control of litter and blowing papers is excellent since the waste material is confined in the bale throughout the delivery and placing operation.
2. A minimum of equipment and manpower is needed at the landfill site since no spreading or compacting of the refuse is required and no daily cover material is usually needed. An unloading device (front-end loader, for example) is all that is required for day-to-day operation.
3. When loose refuse is discharged from collection or transfer vehicles, considerable dust may be produced. Baling prevents any such problem from occurring at the landfill.
4. The baled refuse does not attract or harbor flies or rodents since the density is such that they cannot gain access to the material.
5. The baling operation encapsulates the odor-emitting substances, in most cases in such a way that odors at the landfill are eliminated.
6. Fire hazard associated with loose refuse is overcome by baling.
7. One of the greatest advantages of baling is the high density resulting from the high compressive forces employed during the baling process. Densities that allow 724.8 kg (1600 lb) of refuse to be put into 0.765 m^3 (1 yd^3) of available fill volume increases landfill capacity (perhaps as much as 50%).
8. As a result of the maximum density mentioned above, heavy truck traffic can easily move over the balefill immediately.
9. Many of the reasons for daily cover are fulfilled by the baling operation, thus eliminating the need for daily cover. This reduces the operation costs and increases the capacity of the available volume. Intermediate and final cover, of course, are necessary to provide for ultimate use of the area.

10. As a result of the high density and elimination of voids settlement of the balefill is minimized, allowing the area to be put to use immediately after completion.

11. The problem of leachate and gas generation relative to balefills remains undetermined at this time but biological degradation within the bales is slowed as a result of the high density and reduced surface area. This should lead to a slower rate of gas production and a minimum of leachate generation if any occurs at all.

Disadvantages common to both shredding and baling of solid wastes prior to disposal by landfilling are principally economic. The increased costs are due to the equipment and its operation and maintenance. Furthermore, additional manpower is needed to operate this equipment. However, if the environmental effects are considered along with savings resulting from many of the advantages described earlier, the total costs may be less than anticipated.

V. ECONOMICS

Cost information is difficult to provide because it changes so rapidly and is affected by so many factors. Much of the information is in the

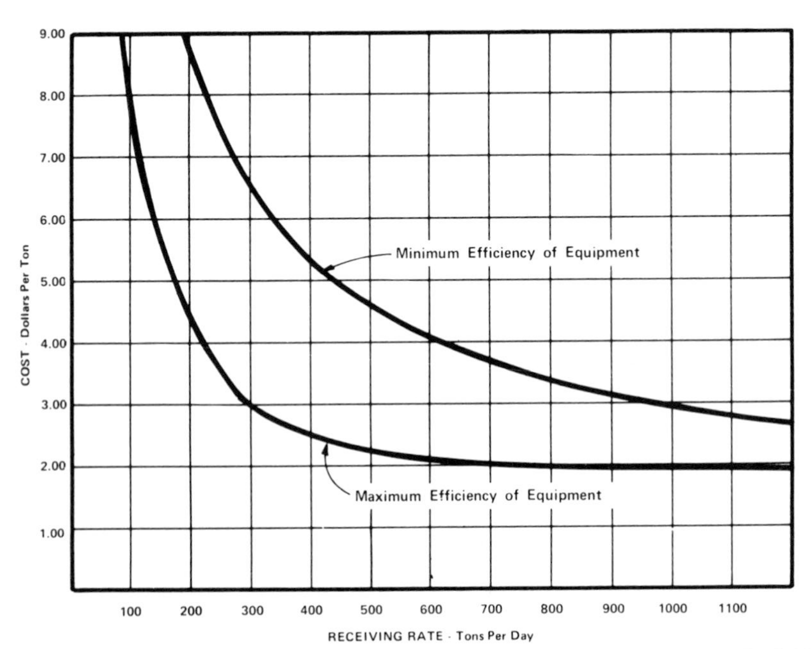

FIG. 10 Operating and maintenance costs of a shredding facility. From Ref. 12.

Table 2
Refuse Shredding Facility Costs [12] Costs projected for 1976

Item	Unit Cost dollar	Unit
Site		
Land	3,800	Acre
Site preparation	2,000	Acre
Scale (remote-computerized)	90,000	Each
Utility construction	35,000	Site
Roadways and miscellaneous construction	180,000	
Building (excluding machinery and equipment)	45	SF
Design and legal fees, and contingencies	35% [a]	
Equipment		
Receiving pit conveyor	63,000	Each
Shredder, including feed, discharge conveyors, and auxiliary equipment		
20 t/h capacity	260,000	Each
40 t/h capacity	600,000	Each
Air classifier	280,000	Each
Magnetic separator	30,100	Each
Front loader, 22,000 lb class	59,000	Each
Compactor	41,400	Each
Truck, trailer enclosed (75 cubic yards)	37,400	Each
Tractor	38,400	Each
Shop tools, miscellaneous equipment	25,000	
Labor		
Superintendent	30,700	Year
Clerical	11,900	Year
Scale attendant	17,300	Year
Loader operator	18,600	Year
Shredder operator	18,600	Year
Mechanic, welder	18,600	Year
Laborer	16,100	Year
Truck driver	18,600	Year

[a] Of total cost.

possession of equipment manufactures and is held to be proprietary or confidential. Some representative costs are provided to indicate the general level of costs for the installation and operation of both shredding and baling facilities. The pertinent dates for these data are indicated so that they can be updated by means of a cost index such as that developed by *Engineering News Record.*

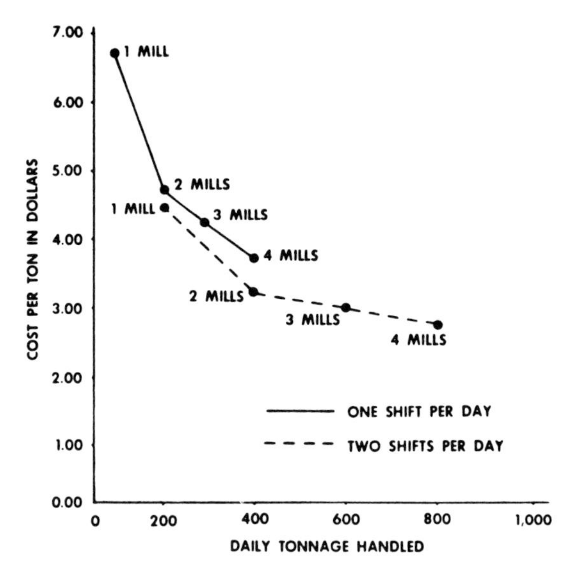

FIG. 11 Projected costs per ton from Madison, Wisconsin, demonstration milling project including the cost of landfilling and a half-mile haul to the landfill. From Ref. 13.

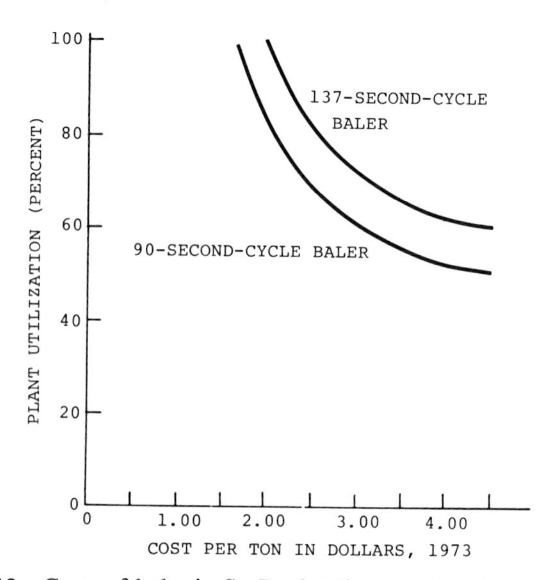

FIG. 12 Costs of baler in St. Paul, Minn., study. From Ref. 13.

Table 3
Baling Plant and Landfill Capital Costs[a] [14]

Item	Cost, $
Baling plant	
Baler SWC-2528A (50 t/h)	676,200
Conveyors	122,500
Hydraulic bale loader	25,000
Overhead crane lift	35,000
Building, including special equipment	580,000
Paving	50,000
Equipment	
2 Rubber tired front end loaders	90,000
2 Fork lift	20,000
Miscellaneous	25,000
Tire shredder	100,000
Incinerator (small animal)	30,000
Automatic scales	40,000
Subtotal	1,793,700
Landfill	
Hauling equipment	
4 Tractor–trailer units	100,000
Fork lift	10,000
Front end loader	45,000
Maintenance building	21,000
Site improvements	30,300
Subtotal	206,300
Legal, administrative, engineering and contingencies (20%)	400,000
Total capital investment	2,400,000
Annual operation	
Salaries[b]	168,500
Heat, light, power	60,000
Insurance	10,000
Accounting, administrative	20,000
Building, equipment maintenance	20,000
Trucks operating expense[b]	125,000
Miscellaneous and contingencies	35,000
Total	435,000

[a] Average cost/t $195.
[b] Director, and 20 men at 44 h/week.

Table 2 gives the costs for a refuse shredding facility including air classification and ferrous metal separation.

The chart in Fig. 10 provides a means of estimating the operating and maintenance costs of a shredding facility (projected to 1976) based on the amount of refuse received per day.

Table 4
Dimension and Operating Properties of Automatic Baling Press,[a] 25 t/h

	Capacity and A Rating	Property
A1	Press box, in.	60 wide × 28½ deep × 176 long
A2	Charging box opening, in.	60 wide × 111 long
A3	Compression chamber, in.	48 wide × 30 deep × 60 long
A4	Approx. expanded bale, in.	52 wide × 32 deep × 64 long
A5	Bale wt (av.), lb	2000–2200[b]
A6	Bale vol, ft^3	50
A7	Bale vol, expanded, ft^3	62
A8	Baling cycle,[c] s	120
A9	Hourly capacity, t	25

[a] For shredded or unshredded solid waste material.

[b] Solid waste material.

[c] A8 is based on solid waste material with loose density of 15 lb/ft^3, requiring three strokes to make a bale. Material can vary from 2 to 30 lb/ft^3; 30 cycles/h.

Table 5
Projected Costs of Baler[a] Rated at 25 t/h

Item	Cost, $	
Capital investment		
Building (80 ft × 120 ft)	182,500	
HRB-SWC Baler installed	143,500	
Conveyor installed (30 ft Z)	86,000	
Wheel loader	40,000	
Lift truck	15,000	
Overhead, legal, and engineering (20%)	93,400	
Total	560,400	
Capital cost[b] per ton		0.98
Interest (7%)	207,978	
Interest cost[b] per ton with ten-year amortization		0.36
Operation		
Labor (3 men @ $5/h)	0.61	
Fringe benefits (25%)	0.16	
Maintenance	0.23	
Heat, power, and light	0.16	
Wire cost	1.05	
Supplies	0.07	
Contingencies	0.10	
Total per ton		2.38
Grand total per ton		3.72

[a] Based on a tonnage capability of 80% utilization or 57,000 t of solid waste annually operating 44 h/week.

[b] With 10-year amortization.

Note: Transportation and balefill costs should be included for total disposal costs, depending on each situation.

Table 6

Dimensions and Operating Properties of Automatic Baling Press,[a] 50 t/h

Capacity and A Rating	Property
A1 Ram push weigh hopper (693 ft^3), Loading dimensions, in.	154 wide × 36 deep × 216 long
A2 Charging box opening, in.	68 × 228
A3 Press box, in.	68 wide × 60 deep × 280½ long
A4 Press box, ft^3	680
A4.1 Bale chamber, in.	70 wide × 123 deep × 36 long
A4.2 Chamber capacity, ft^3	179
A5 Bale size, in.	36 × 36 × variable
A6 Baling forces, compression ram face, psi	
A6.1 1st	172
A6.2 2nd	1450
A6.3 3rd	2740
A7 Bale weight, av, lb.	2500–3500
A7.1 compressed, in.	2900, 36 × 36 × 48
A7.2 expanded, in.	2900, 38 × 38 × 60
A8 Baling cycle,[b] s	90

[a] For solid waste material suitable for packer truck handling.
[b] 40 cycle/h.

Table 7

Approximate Cost of Baler and Conveyor Installation[a]

Item	Price, $
1. Rigid frame metal building, 8 bays at 20 ft with 24 ft eave height, 4 bays at 20 ft with 32 ft eave height, including all concrete work, foundations, concrete floor, foundation for baler with piling. Power room to house power unit and office, 10 overhead doors, 4 main doors, conveyor pit, fire protection system (1 sprinkler head/100 ft^2), general lighting, plumbing, and operator's tower.	835,000
2. Electrical work for baler and conveyor installation including main switchgear.	35,000
3. Baler and power unit, including bale diverter and power unit, load cell assembly with round dial indicator 0–50,000 capacity; baler capacity 50 t/h.	922,460
4. Piping and heat exchanger installation including crane time.	15,000
5. Unloading of baler at site.	6,000
6. Setting and assembly of baler, including crane rental.	16,500
7. Conveyor system horizontal and inclined conveyors.	200,000
Total	2,029,960

[a] In 120 ft × 240 ft Rigid Frame Building.

Table 8
Baling Costs at St. Paul, Minn., Rated Capacity, 50 t/h, July 1972–July 1974

Expenses	Total cost $	Actual cost/t at 137 s, $	Projected cost/t at 90 s, $
Wages[a]	255,403	1.34	0.88
Depreciation[b]	136,135	0.72	0.47
Supplies	7,785	0.04	0.03
Fuel and Oil	12,505	0.07	0.05
Repairs and Maintenance			
Baler	46,893	0.25	0.16
Conveyor	33,494	0.18	0.12
Mobile Equipment	37,515	0.20	0.13
Plant	16,077	0.08	0.05
Equipment Rental	62,378	0.32	0.21
Electricity	45,634	0.24	0.17
Property Insurance	12,365	0.06	0.04
Fringes (Insurance; FICA)	37,775	0.20	0.13
Property Tax	38,812	0.20	0.13
Other	14,604	0.08	0.05
Total	757,375	3.98	2.62

[a] Average rate of $5.23/h. 1st shift, 6 men; 2nd shift, 3 men.

[b] Baler and conveyor, 10 yr amortization at $56,056; building, 40 yr at $11,437. Land costs not included. Total tonnage handled July 1972–1974: 190,323. Recycling expenditures or profits not included.

A general curve relating costs of shredding vs. tons of refuse handled based on the shredding project in Madison, Wisconsin, is shown in Fig. 11. Note the sharp decline in costs as shifts and number of mills are increased.

Costs reported for a baling facility in Cobb County, Georgia (as of 1975) handling 408.2 kg/day (450 t/day) 6 days per week are indicated in Table 3 (1974 prices).

It should be noted that the baling plant in Cobb County includes facilities for handling and baling junk automobiles and shredding old automobile tires that may not be included in other situations.

Additional cost data on baling facilities have been provided by the American Hoist and Derrick Co. It describes two sizes of balers, one rated at 22,680 kg/h (25 t/h) (see Tables 4 and 5), and the other at 45,300 kg/h (50 t/h) (see Tables 6 and 7).

The costs per ton for the operation of the St. Paul baling facility, (45.360 kg/h, 50 t/h) are listed in Table 8 and shown in Fig. 12.

VI. OPERATION AND MAINTENANCE

Shredders, particularly hammer mills, require extensive and regular maintenance because of the rugged type of operation they are subjected to. In a three-shift operation, for example, one shift would be devoted entirely to maintenance of the mill. Necessary maintenance involves component wear, particularly wear on the hammers, mill liners, and grates. Improved performance has been obtained by reversible rotors, improved hammer tipping materials, and easier access to internal components. Wear on bearings is also a problem and calls for careful attention to lubrication.

Explosions within the housing of the shredder arise from various causes (aerosol cans and "empty" gasoline cans) and may result in damage to the shredder and injury to workmen. Antiexplosive devices are available and should be considered.

The maintenance requirements of heavy duty balers are mainly for the hydraulic systems, equipment lubrication, and motor upkeep. Wear does occur but not to the same extent as in shredding.

Another important maintenance problem common to all systems is the feed conveyor system. Refuse must be fed to the size reduction process and removed from it in a reliable manner. This is usually done by a system of conveyors of one type or another and these devices must be adequately maintained. If a weighing device is included in the system, it also requires maintenance.

VII. ILLUSTRATION OF SHREDDER SELECTION

As a means of illustration the following discussion deals with the selection of the general shredder requirements for a shredding facility.

The criteria for the designing of size reduction facilities are based on experience and data gathered by equipment manufacturers.

The following information is taken from the study conducted by the Midwest Research Institute [1].

The shredder size and power required for shredding solid wastes are determined by at least three items:

1. The size and nature of input material.
2. The processing rate (tons per hour) desired.
3. The output particle size required.

The output size required establishes the degree of size reduction needed and the minimum energy requirements. The size and nature of the input material establishes the minimum horsepower required to provide the minimum performance without frequent jams and damage to the machine. The processing rate determines the physical size of the machine and the total horsepower required.

Based on the categories of solid waste described in Table 1, the following nomographs have been developed by discussion with manufacturers and others [1]. (These values and other data and nomographs provided here must be considered tentative and should be used only as guides.)

Figure 13 relates output particle size to horsepower requirements and production rate for the various classes of solid wastes. As expected, it shows that the smaller the particle size, the larger the motor required for a given capacity. The output particle size is the size of the square opening through which the particle passes. The importance of particle size varies with the subsequent destination of the shredded refuse. The least constraint on variations and allowable maximum size would probably be established by landfilling whereas much more stringent size requirements would be dictated by boiler firing. A 20.3 cm (8-in.) particle size might be acceptable for landfilling whereas use as a fuel might require that 95% of the output be of 3.8 cm (1.5-in.) size and 100% of 12.7 cm (5.0-in.).

Since it appears that the power requirements increase exponentially as particle size decreases, there may be a particle size below which it becomes more economical to use two-stage size reduction. This would imply a primary and secondary shredder in series. This critical particle size may be in the range of 5.1 cm (2 in.) to 10.2 cm (4 in.) when shredding combines bulky and medium wastes [1]. (This is not well established however.)

Many of the larger shredder plants are designed to operate in two shifts which provides ample time for receiving refuse and to handle the peaks that occur during the day. Daily maintenance is required, as discussed earlier.

> General experience indicates that there is a practical limit to the capacity of a single size reduction operation. A capacity of 36,388 kg/h (40 t/h) to 54,432 kg/h (60 t/h) begins to approach the limit for reasonable conveying and feeding systems. The *volume* of material represented by 36,388 kg (40 t) to 54,432 kg (60 t)/h begins to present such severe logistical problems that efficient operation may require subdividing into two or more units if larger capacities are desired. [1]

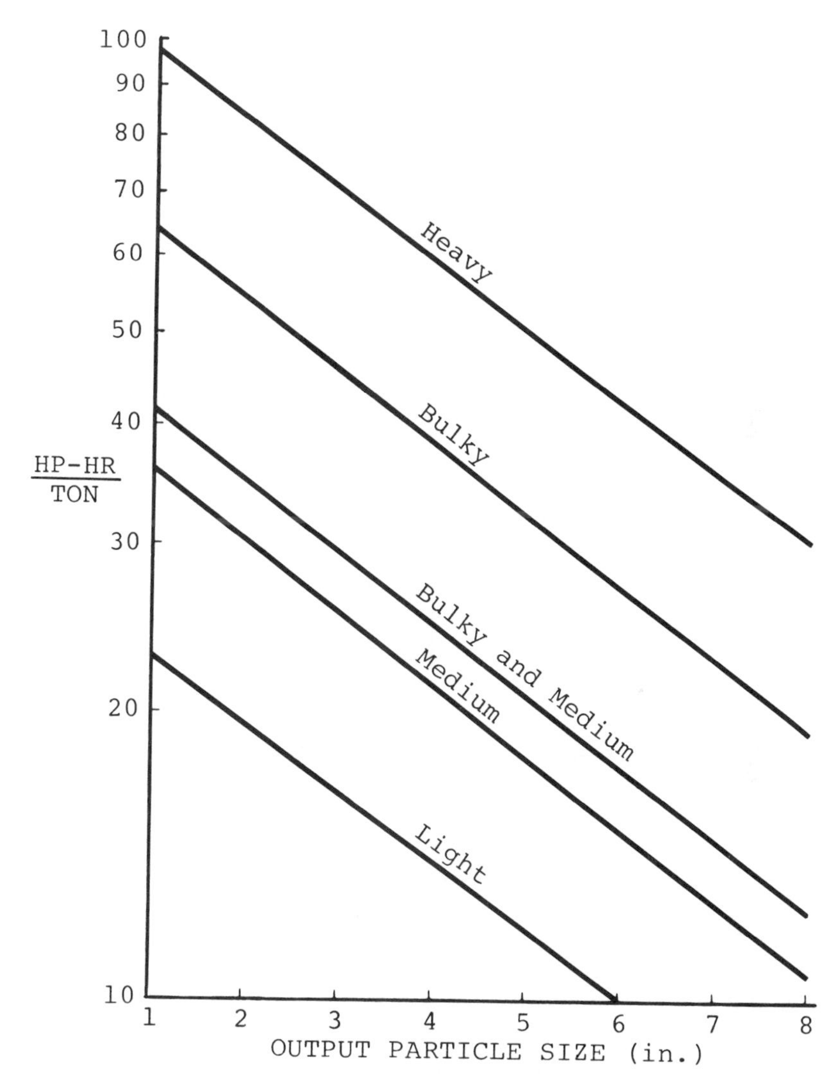

FIG. 13 Solid waste size reduction power requirement. From Ref. 1.

Considerations of reliability and availability also dictate additional capacity and standby equipment.

As much of the necessary information and procedures as could be accommodated have been used to establish the nomograph, in Fig. 14, that can be used as follows:

1. Starting with the population served extend a straight line

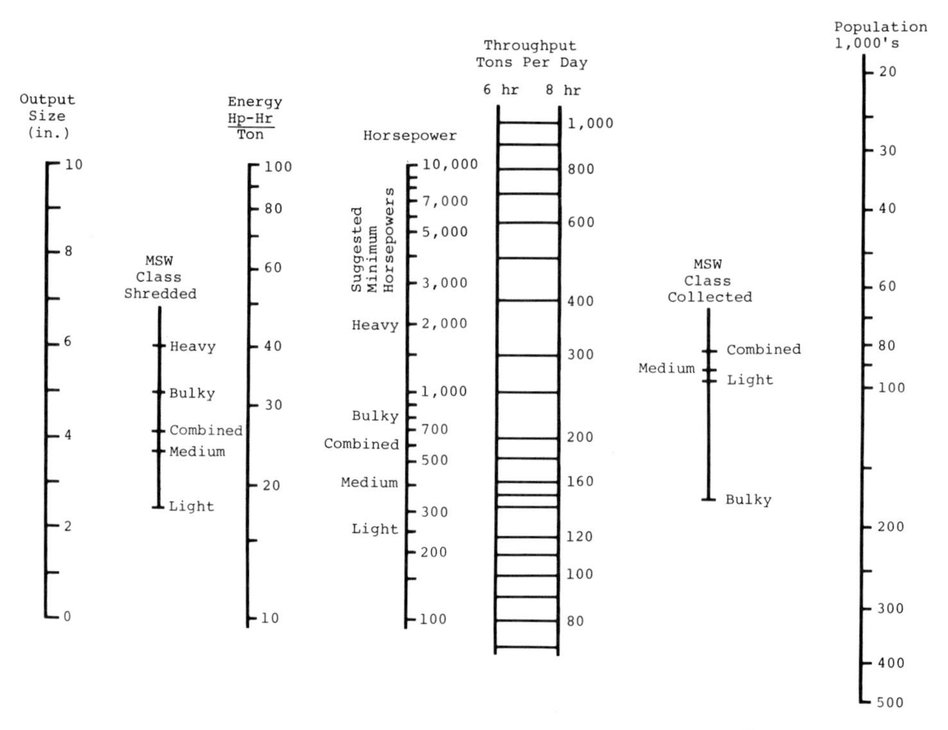

FIG. 14 Shredder selection nomograph. From Ref. 1.

through the type of waste collected to determine the throughput tons per day on an 8-h basis.

2. If an 8-h per day shift is to be used, use the value found in Step 1. If a 6-h per day shift is to be used transfer to the 6-h axis using a horizontal line extending from the 8-h axis.

3. Starting from the desired output particle size extend a straight line through the type of municipal solid waste being shredded to the energy (hp-h/ton) axis.

4. Determine the horsepower required by connecting the two points just established (one on either the 8-h or 6-h throughout axis, and the other on the energy axis).

An example of the use of this nomograph for a community of 150,000 population planning to shred combined solid waste to a 4-in. particle size with an 8-h day and a 5-day week follows:

1. A straight line extended from 150,000 through the combined

refuse point on the MSW class collected axis intersects the 8-h per day throughput axis at 570 t per day.

2. A straight line extended from the 4-in. output size through the combined refuse point on the MSW class shredded axis indicates an energy requirement of 27 hp-h/t.

3. Connecting this last point with the one determined first (on the 8-h per day throughput axis) suggests a total horsepower requirement of 2000 hp.

This would suggest that two shredder systems in parallel might be more suitable for a plant of this size. The particle size specified is also reaching the point at which two stages of shredding may be indicated.

REFERENCES†

1. *Size-Reduction Equipment for Municipal Solid Waste*, Midwest Research Institute, USEPA-OSWMP, SW-53C, 1973.
2. H. W. Rogers and S. H. Hitte, *Solid Waste Shredding and Shredder Selections*, USEPA-OSWMP, SW-140, 1975.
3. N. L. Drobny, H. E. Hull, and R. F. Testin, *Recovery and Utilization of Municipal Solid Waste*, USEPA-OSWMP, SW-10C, 1971.
4. E. A. Glysson, J. R. Packard, and C. H. Barnes, *The Problem of Solid Waste Disposal*, College of Engineering, University of Michigan, Ann Arbor, 1972.
5. J. Skitt, *Disposal of Refuse and Other Waste*, Halstead, New York, 1972.
6. *Baling Solid Waste to Conserve Sanitary Landfill Spaces*, USEPA-OSWMP, SW-44d, 1973.
7. K. W. Wolf and C. H. Sosnovsky, *High Pressure Compaction and Baling of Solid Waste*, USEPA-OSWMP, SW-32d, 1972.
8. K. W. Wolf and C. H. Sosnovsky, "Investigate Baling," *American City* **87** (2), 60, February (1972).
9. F. R. Zettel, "High Density Baling as a Solution to Solid Waste Disposal," presented 1974 Waste Management Control Conference, University of South Wales, Sydney, Australia.
10. F. R. Zettel, "Standard Rail Cars Haul Baled Refuse Without Problems," *American City* **88** (4), 91, April (1973).
11. J. J. Reinhardt and R. K. Ham, *Milling of Solid Waste*, Final Report on a Demonstration Project at Madison, Wisconsin, Vol. 1, 1966–1972, USEPA-OSWMP, 1973.
12. *Washtenaw County Plan for Management of Solid Wastes*, prepared by Jones and Henry Engineers, Ltd., Toledo, Ohio, 1975.
13. R. A. Colonna, C. McLaren, *Decision-Makers Guide in Solid Waste Management*, USEPA-OSWMP SW-127, 1974.

† U.S. Environmental Protection Agency, Office of Solid Waste Management Programs, Washington: USEPA-OSWMP.

14. M. A. Oberman, "High Density Solid Waste Baler is Installed in Cobb County, Georgia," *Waste Age* **6** (7), 46, June (1975).
15. M. Oberman, "Transfer Station Cuts Costs for Myrtle Beach, So. Carolina," *Waste Age* **8** (5), 103 May (1977).
16. P. J. Franconeri, "Size Reduction of Industrial Solid Wastes," *Pollution Eng.* **8** (3), 54, March (1976).
17. M. Oberman et al., "The Evolving Refuse Truck: A Baler on Wheels," *Waste Age* **8** (8), 12, August (1977).
18. J. E. Butler, "Brevard County, Florida Launches a Shredder and Resource Recovery Facility," *Waste Age* **8** (2), 46, Feb. (1977).
19. M. Oberman et al., "Continued Interest in Shredding Wastes," *Waste Age* **8** (7), 10, July (1977).
20. F. Fregerio, "Estimating the Cost of A Transfer Station," *Waste Age* **5** (4), 6, July (1974).

3

Combustion and Incineration

Walter R. Niessen
Camp Dresser and McKee, Boston, Massachusetts

I. INTRODUCTION TO INCINERATION

Solving the problem of volume is often the key to successful solid waste management. The low bulk density of solid waste requires costly storage containers at the point of generation, greatly influences the cost and difficulty of collection and handling, and is the primary factor in setting the cost and scale of landfill and other ultimate disposal operations.

Incineration is often an attractive processing step to reduce the volume of solid waste. Properly, incineration should not be considered as an ultimate disposal method since it cannot function without a landfill or other means to receive its solid residues.

In addition to volume reduction, incineration also offers the potential advantages of detoxification of combustible toxins or pathologically contaminated material, reduction of environmental impact related to leaching of organic material from raw refuse landfills, and provision of an energy resource through the use of a boiler enclosure.

The disadvantages of incineration include high cost relative to many other waste management options, operating problems owing to the sensitivity of the process to changes in refuse character, staffing problems made more critical in light of the complexity of the process, secondary environmental impacts such as air and noise pollution and discharge of highly polluted wastewater, adverse public reactions in many cases, and technical risk reflecting the complexity of incineration systems.

Nonetheless, incineration persists as an important concept in solid waste management. Indeed some 20% of municipal waste and a somewhat smaller fraction of commercial and industrial waste was incinerated in 1970. The future of municipal incineration will depend on the changing balance between several conflicting influences. These are, principally, the availability of landfills near areas of high population density, the value of energy recovered using boiler-type incinerators, and the burgeoning capital and operating costs of incineration systems.

II. PROCESS ANALYSIS OF INCINERATION SYSTEMS

The designer or analyst of an incineration system faces a formidable challenge. Incineration processes are complex, involving the interplay of chemical reactions, fluid flow, and heat transfer in a nonisothermal, nonhomogeneous, reacting system. This already complex physicochemical process is made less tractable to rigorous analysis by the ever-changing nature of the waste and the irregular pattern of major and minor adjustments of key parameters owing to the action of the plant operators or automatic control instrumentation.

This pessimistic picture, however, should not lead the engineer to abandon theoretical analysis as a powerful tool in design and problem solving tasks related to incineration. In the following sections of this chapter and in other texts [1, 2], the interested student or practicing engineer is introduced to analytical tools which give the understanding and insight to cope with this important sector of environmental engineering.

The analytical methods presented here draw heavily on the disciplines of chemical and mechanical engineering, beginning with a review of the fundamentals of process analysis (heat and material balances, chemical kinetics, and equilibrium). This is followed by several topics relating to the fundamental processes occurring in burning and pyrolyzing systems.

A. Stoichiometry

Stoichiometry is the discipline of tracking matter (particularly the elements), partitioned in accord with the laws of chemical combining weights and proportions, and energy (in all its forms) in either batch or continuous processes. Applying the laws of conservation of mass and energy, supplemented by the consideration of chemical kinetic and equilibrium relationships, provides great insight into the behavior of incineration systems.

In the paragraphs below, it will be advantageous to use the kilogram mole (or kilogram atom)—the molecular (atomic) weight expressed in kilograms—as the unit quantity. The advantage arises because one molecular (atomic) weight (mole) of any compound (element) contains the same number of molecules (atoms) and, for gases, occupies approximately the same volume (at similar pressures and temperatures). The approximation holds exactly if the gases are "ideal," an assumption acceptably accurate for gases at atmospheric pressure and elevated temperatures.

1. *Gas Laws*

The relationship between absolute pressure P, absolute temperature T, and volume V for ideal gases is given by

$$PV = nRT \tag{1}$$

where n is the number of moles of the gas and R is the universal gas constant (Table 1).

Example 1. To generate carbon dioxide for process use, 3000 kg/d of a waste containing 80% carbon, 7% ash, and 13% moisture is to be burned. The combustion gases leave the furnace at 1000 °C and pass through a gas cooler, exiting at 80 °C. How many kilogram-moles and how many kilograms of CO_2 will be formed per day? How many cubic meters of CO_2 are produced per day at the furnace outlet and at the gas cooler outlet at 1.04 atm?

The number of moles of carbon (atomic weight = 12) in the waste is (3000)(0.80)(1/12) = 200. Noting that with complete combustion each

Table 1
Values of the Gas Constant R for Ideal Gases

Pressure	Volume	Moles	Temperature	R
atm	m^3	kg-mol	K	0.08206 m^3-atm kg-mol^{-1} K^{-1}
psia	ft^3	lb-mol	°R	1543 ft-lb lb-mol^{-1} °R^{-1}
atm	ft^3	lb-mol	°R	0.729 ft^3-atm lb-mol^{-1} °R^{-1}
—	—	kg-mol	K	1.986 kcal kg-mol^{-1} K^{-1}

mole of carbon yields 1 mole of CO_2, 200 mol/d of CO_2 are produced. The weight flow of CO_2 (molecular weight = 44) is 200(44) = 8800 kg/d. From $PV = nRT$

$$V = \frac{nRT}{P} = \frac{200(0.08206)T}{1.04} = 15.78T$$

At 1000 °C (1273 K), V = 20,090 m³. At 80 °C (353 K), V = 5570 m³.

2. *Material Balances*

A material balance is a quantitative expression of the law of conservation of matter:

$$\text{Input} = \text{output} + \text{accumulation} \tag{2}$$

This expression is always true for *elements* flowing through combustion systems, but is often not true for *compounds* participating in combustion reactions.

The basic data used in calculating material balances can include analyses of fuel or waste, gases in the system, etc. (e.g., see Appendix) and some rate data (usually feed rate). Coupled with these data are fundamental relationships that prescribe combining proportions in molecules (e.g., two atoms of oxygen to one of carbon in carbon dioxide) and those which indicate the course and heat effects of chemical reactions.

Balances on elements in the fuel or waste allow one to calculate the amount of air theoretically required to completely oxidize the carbon, hydrogen, sulfur, etc. (recognizing that a portion of the oxygen required may be supplied by the oxygen contained in the material being burned). This quantity of air (known as the *theoretical* or *stoichiometric air* requirement) is often insufficient in a practical combustor, and *excess air*, expressed as a percentage of the stoichiometric air quantity, is usually supplied. For example, an incinerator operating at 50% excess air denotes a combustion process to which 1.5 times the stoichiometric air requirement has been supplied.

Example 2. Calculate the air requirement and products of combustion when burning, at 50% excess air, 100 kg/h of a waste having the ultimate analysis: 12.2% moisture, 75% carbon, 5.2% hydrogen, 2.4% sulfur, 2.1% oxygen, 0.5% nitrogen, and 1.6% ash. The combustion air is at 15.5 °C and 70% relative humidity. The sequence of computations is shown in Table 2.

Table 2
Calculations for Example 2

Line	Component	kg	Atoms or moles[a]	Combustion product	Theoretical moles of O_2 required	Moles formed in stoichiometric combustion					
						CO_2	H_2O	SO_2	N_2	O_2	Total
1	Carbon, C	75.0	6.250	CO_2	6.250	6.250	0.0	0.0	0.0	0.0	6.250
2	Hydrogen, H_2	6.2	3.100	H_2O	1.550	0.0	3.1	0.0	0.0	0.0	3.100
3	Sulfur, S	2.4	0.075	SO_2	0.075	0.0	0.0	0.075	0.0	0.0	0.075
4	Oxygen, O_2	2.1	0.066	—	(0.066)	0.0	0.0	0.0	0.0	0.0	0.0
5	Nitrogen, N_2	0.5	0.018	N_2	0.0	0.0	0.0	0.0	0.018	0.0	0.018
6	Moisture, H_2O	12.2	0.678	H_2O	0.0	0.0	0.678	0.0	0.0	0.0	0.678
7	Ash	1.6	N/A	—	0.0	0.0	0.0	0.0	0.0	0.0	0.0
8	Total	100.0			7.809	6.250	3.778	0.075	0.018	0.0	10.121
9	Moles of nitrogen in stoichiometric air[b]			$\frac{79}{21}$ (7.809)					29.377		29.377
10	Moles of nitrogen in excess air			(0.5) $\frac{79}{21}$ (7.809)					14.688		14.688
11	Moles of oxygen in excess air			(0.5) (7.809)						3.905	3.905
12	Moles moisture in combustion air[c]						0.604				0.604
13	Total moles in flue gas					6.250	4.382	0.075	44.083	3.905	58.695
14	Volume (mole) percent in wet flue gas					10.648	7.466	0.128	75.105	6.653	100.0
15	Orsat (dry) flue gas analysis, moles					6.250		0.075	44.083	3.905	54.313
16	A. With selective SO_2 testing, volume percent					11.508	N/A	0.138	81.164	7.180	100.0
17	B. With alkaline CO_2 testing only, volume percent					11.646	N/A	N/A	81.164	7.190	100.0

[a] The symbol in the component column shows whether these are kg-mol or kg-atom.
[b] Throughout this chapter, dry combustion air is assumed to contain 21.0% oxygen by volume and 79.0% nitrogen.
[c] Calculated as follows: (0.008/18)[(29.377 + 14.688)(28) + (3.905)(32)] based on the assumption of 0.008 kg water vapor per kg bone-dry air; found from standard psychrometric charts.

Several elements of the analysis should be noted:

a. *Line 1.* Carbon is assumed here to burn completely to carbon dioxide. In practice, some carbon may be incompletely burned (forming carbon monoxide), and some may end up as unburned carbon char in solid residues or as part of the particulate matter leaving (as soot or char fragments) in the effluent gas.
b. *Line 2.* Hydrogen in the waste (other than the hydrogen in moisture) increases the amount of combustion air, but does not appear in the Orsat analysis (Lines 16 and 17).
c. *Line 3.* Sulfur in the waste as sulfide or organic sulfur increases the amount of combustion air required in burning to SO_2. Inorganic sulfates may leave as ash or be reduced to SO_2. If selective analysis is not used for SO_2 (Line 17), it is usually reported out as carbon dioxide.
d. *Line 4.* Oxygen in the waste reduces the amount of required combustion air.
e. *Line 12.* Moisture entering with the combustion air can be seen to be small and is often neglected. Although this problem considered only waste components of C, H, O, N, and S, the analyst should review waste composition thoroughly and consider the range of possible secondary reactions:

- *Carbon Monoxide.* CO is formed in appreciable quantities in solids burning systems.
- *Chlorine.* Chlorine appearing in the waste as inorganic salts will, most likely, remain as the salt. Organic chlorides, however, result primarily in the formation of hydrogen chloride.
- *Metals.* Metals usually burn to the oxide, although, in burning solid wastes, a large fraction of massive metal feed (e.g., tin cans, sheet steel, etc.) is unoxidized.
- *Thermal Decomposition.* Some compounds may decompose at combustor temperatures. Carbonates, for example, may dissociate to form an oxide and CO_2, and sulfides may "roast" to form the oxide and release SO_2.

In many instances, the analyst is called upon to evaluate an operating waste disposal system. In such studies, accurate data on the flue gas composition are readily obtainable and offer a low cost means to characterize the operation and the feed waste.

One important combustor and combustion characteristic which can be immediately computed from the Orsat (dry) flue-gas analysis is the percentage excess air:

$$\text{Percentage excess air} = \frac{[O_2 - 0.5(CO + H_2)]100}{0.266N_2 - O_2 + 0.5(CO + H_2)}$$

where O_2, N_2, etc., are the *volume percentages* of the gases on a dry basis.

Example 3. The flue gas from a waste incinerator burning a waste believed to have nitrogen or oxygen has an Orsat analysis (using alkaline CO_2 absorbent) of 11.6% CO_2, 7.2% O_2, and the rest nitrogen and inerts. From these data, calculate the weight ratio of hydrogen to carbon in the waste, the percent carbon and hydrogen in the dry waste, the kilograms of dry air used per pound of dry waste, the percent excess air used, and the moles of exhaust gas discharged from the unit per kilogram of dry waste burned. (Note that this example is derived from Example 2.)

Basis: 100 mol dry exhaust gas

Component	Moles	Mol O_2
CO_2 (+SO_2)	11.6	11.6
O_2	7.2	7.2
N_2	81.2	—
Total	100.0	18.8

Considering all N_2 to have come from the combustion air, a total of 81.2 × (21/79) = 21.6 mol O_2 entered with the N_2. The difference, 21.6 − 18.8 = 2.8 mol O_2 may be assumed to have been consumed in burning hydrogen.

H_2 Burned: 2(2.8) = 5.6 mol	11.2 kg
C Burned: 12(11.6) mol	139.2 kg
	150.4 kg

a. Weight ratio of hydrogen to carbon: (11.2/139.2) = 0.08.
b. Percent (by weight) C in dry fuel: (139.2/150.4)(100) = 92.55.
c. Kilogram of dry air per kilogram of dry waste.
First, calculate the weight of air resulting in 1 mol dry exhaust gas from a nitrogen balance:

$$\frac{1}{100}(81.2 \text{ mol } N_2)(1/0.79 \text{ mol } N_2/\text{mol air})(29 \text{ kg air/mol}) = 29.81 \text{ kg air/mol dry exhaust gas}$$

then,

29.81(100/150) mol dry exhaust gas/kg waste
= 19.87 kg dry air/kg dry waste

d. Percent excess air:

The oxygen *necessary*† for combustion is:	11.6 + 2.8 = 14.4 mol
The oxygen *unnecessary*† for combustion	= 7.2 mol
The *total* oxygen	= 21.6 mol

The percent excess air (or oxygen) may be calculated as:

$$\frac{(100)(\text{unnecessary})}{\text{total} - \text{unnecessary}} = \frac{100(7.2)}{21.6 - 7.2} = 50\% \quad (4a)$$

$$\frac{(100)\ \text{unnecessary}}{\text{necessary}} = \frac{100(7.2)}{14.4} = 50\% \quad (4b)$$

$$\frac{(100)(\text{total} - \text{necessary})}{\text{necessary}} = \frac{100(21.6 - 14.4)}{14.4} = 50\% \quad (4c)$$

e. Moles of exhaust gas per kilogram of dry waste:

Noting that 5.6 mol water vapor must be added to the dry gas flow, (100 + 5.6)/150 = 0.702 mol/kg fuel.

Lessons to be learned from comparison of the results of Example 3 with the "true" situation from Example 2 are:

- Waste analysis data are important in calculating combustion air requirements for design.
- Waste moisture data are necessary to determine total flue gas rates.
- Insight into the nature of the waste can be gained from stack gas analysis.
- If data are available, all data—for both fuel and flue gas—should be used to cross-check for consistency.

† Note that the *necessary* oxygen increases and the *unnecessary* oxygen decreases if incompletely burned components (such as CO) are present.

3. *Heat Balances*

A heat balance is a quantitative expression of the law of conservation of energy. In waste incineration, five energy quantities are of prime interest:

- *Chemical Energy.* The heat of chemical reaction; especially the heat of combustion.
- *Latent Heat.* The heat effect of changes in state, especially the heat of vaporization of moisture.
- *Sensible Heat.* The heat content (enthalpy) related to the temperature of materials.
- *Useful Heat.* The heat available for use, especially the sensible heat available to generate steam.
- *Heat Loss.* The heat lost through furnace walls by conduction, convection, and radiation.

In heat of combustion and sensible heat calculations, 15.6 °C (60 °F) is often used as a reference point for "zero energy." Most values of heat of combustion reported in the incineration literature are the "Higher Heating Value" (HHV) which includes the latent heat of vaporization of the water formed in combustion (10,520 kcal/kg-mol at 20 °C). See Appendix for HHV for refuse and refuse components. The "Lower Heating Value" (LHV) does not include the latent heat.†

The sensible heat content (Δh) at a temperature T may be calculated relative to the reference temperature T_0 by:

$$\Delta h = \int_{T_0}^{T} Mc_p{}^0\, dT \quad \text{kcal/kg mol} \tag{5}$$

where $MC_p{}^0$ is the molar heat capacity (kcal mol^{-1} °C^{-1}). The calculation may be carried out using an empirical equation describing the functional relationship of $Mc_p{}^0$ on temperature [1]. Also, one may use a graphical presentation (Fig. 1) of the average molal heat capacity ($Mc^0_{p,\text{avg}}$) between a reference temperature of 60 °F (15.6 °C) and the abscissa temperature. Thus:

$$\Delta h = (T - T_0)(Mc^0_{p,\text{avg}}) \tag{6}$$

Example 4. If the 100 kg of waste described in Example 2 has a heat of combustion of 7500 kcal/kg (HHV) and the combustion air is preheated to 300 °C, what is the temperature of the flue gases? How much steam can be generated if the gases are cooled to 180 °C (about 350 °F) in a boiler? Assume 5% heat loss in the furnace and 5% in the

† By international agreement, the Joule has been selected as the preferred energy unit. One kcal is equivalent to 4190.02 joules.

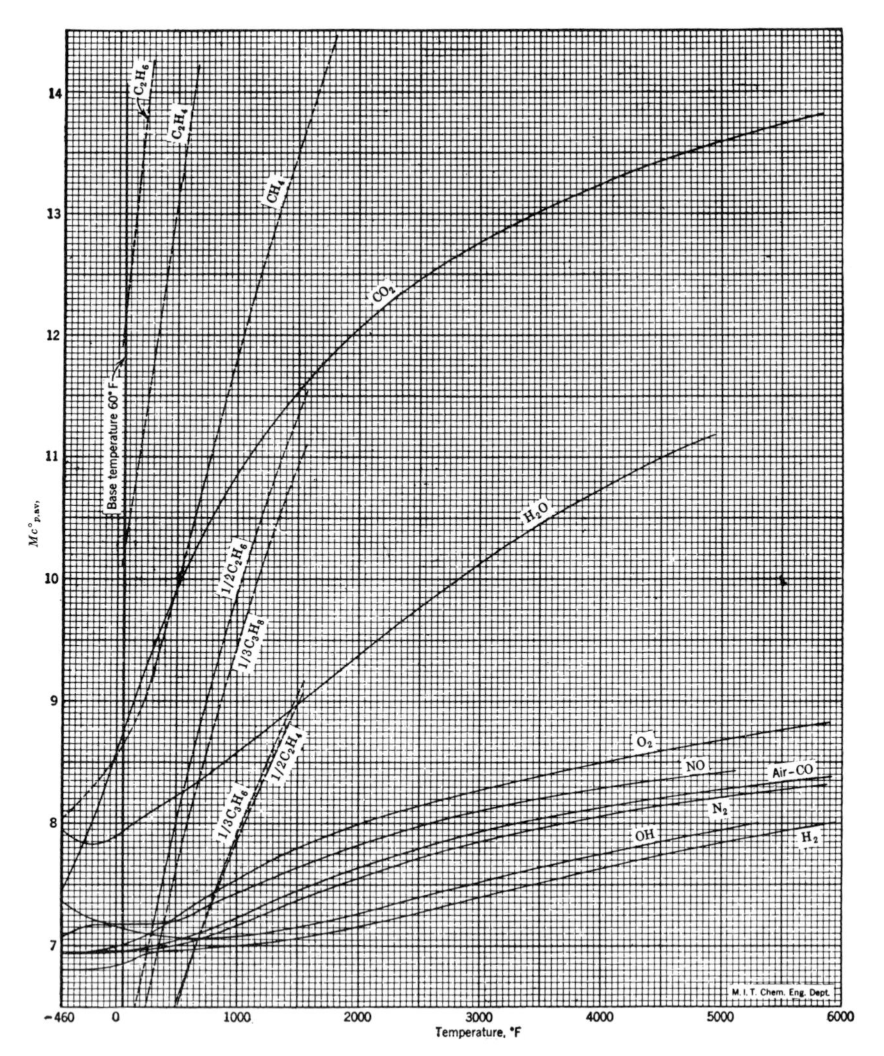

FIG. 1 Average molal heat capacity of fuel and combustion gases, $Mc^0_{p,\text{avg}}$, at zero pressure between 60 °F (15.6 °C) and abscissa temperature.

boiler, and 570 kcal/kg enthalpy change from boiler feedwater to product steam at 204 °C. Basis: 1 h operation.

The total combustion air supplied to the system is 29.377 + 14.688 + 3(3.905) + 0.604 = 56.384 mol (see Table 2). From Fig. 1, the heat content of the preheated air at 300 °C is:

$$(56.384)(7.08)(300 - 15.5) = 113{,}572 \text{ kcal}$$

Table 3

Computation of Heat Content of Flue Gases from Combustion of Benzene Waste at 10% Excess Air

A. Assumed temp., °C	180	500	1,000	1,500	2,000
B. A − 15.5 °C	164.5	484.5	984.5	1,484.5	1.984.5
C. $Mc^0_{p,\mathrm{avg}}N_2$ [a]	7.00	7.13	7.48	7.76	8.00
D. $Mc^0_{p,\mathrm{avg}}O_2$	7.10	7.50	7.92	8.20	8.40
E. $Mc^0_{p,\mathrm{avg}}H_2O$	8.10	8.52	9.23	9.90	10.50
F. $Mc^0_{p,\mathrm{avg}}CO_2$	9.42	10.75	11.90	12.60	13.08
G. Ash [b]	0.2	0.2	0.2	0.2	0.2
H. 44.083(B)(C)	50,750	152,250	324,480	507,925	699,850
I. 3.905 (B)(D)	4,740	14,230	30,565	47,430	65,350
J. 4.382 (B)(E)	5,825	18,125	39,870	64,340	91,265
K. 6.25 (B)(F)	9,680	32,540	73,210	116,910	162,210
L. 1.6 (B)(G) + 85 [c]	190	290	450	610	770
M. 4.382 (10,595) [d]	46,430	46,430	46,430	46,430	46,430
N. (H + I + J + K + L + M) [e]	117,615	263,865	515,005	783,645	1,065,875
O. kcal/mol gas	2,004	4,495	8,774	13,351	18,160

[a] Source: Fig. 1 (kcal/kg mol °C).

[b] Specific heat of the ash (kcal/kg °C) for solid or liquid.

[c] The latent heat of fusion of the ash (85 kcal/kg) is added at temperatures greater than 800 °C, the assumed ash fusion temperature.

[d] Latent heat of vaporization at 15.5 °C of free water in waste and from combustion of hydrogen in waste (kcal/kg mol).

[e] Total heat content of gas stream (kcal).

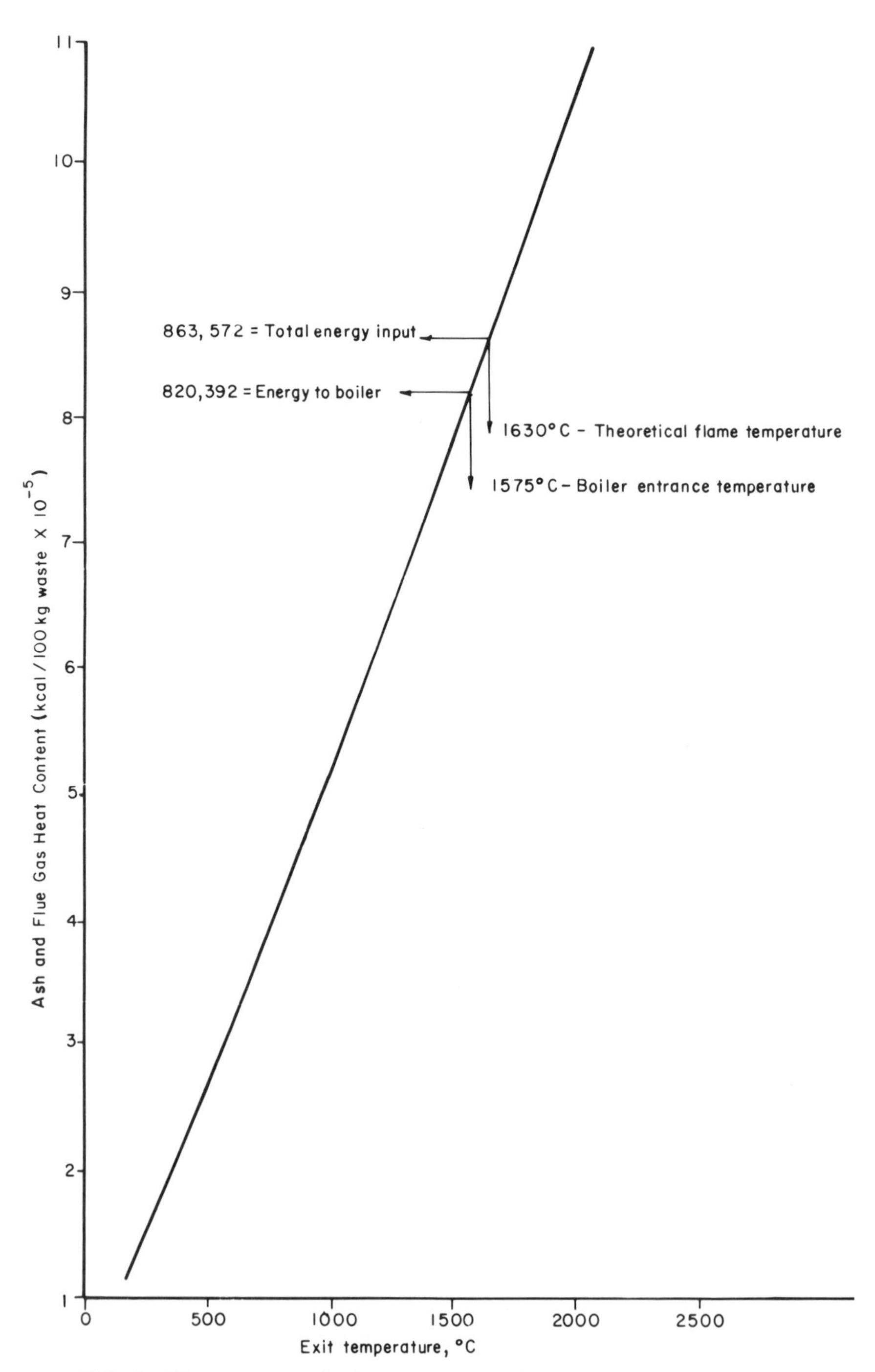

FIG. 2 Heat content of exhaust gases relative to 15.5 °C in Example 4.

therefore, the total energy impact is

$$7500(100) + 113{,}572 = 863{,}572 \text{ kcal energy addition}$$

To find the exit temperature of the combustion chamber and the steaming rate, it is useful to construct a plot of the heat content of the gas stream as a function of temperature, computed as shown in Table 3 and presented in Fig. 2. The flows of thermal energy are then:

Energy flows	kcal	Temperature, °C
Energy into system		
Heat of combustion	750,000	15.5
Air preheat	113,572	300
Total	863,572	1,630[a]
Heat loss (5%) from combustion chamber	(43,180)	
Energy into boiler	820,392	1,575
Heat loss (5%) from boiler	(41,020)	
Heat loss out stack	(117,615)	180
Net energy into steam	661,757	204

[a] The theoretical (adiabatic) flame temperature for this system (the temperature of the products of combustion assuming no heat loss.

For feedwater (at 100 °C and 15.8 atm) changing to saturated steam at 15.8 atm, the enthalpy change is 567.9 kcal/kg, so that the resulting steaming rate for a burning rate of 1100 kg/h is:

$$\frac{661{,}757}{567.9} = 1165 \text{ kg/h}$$

4. *Equilibrium*

No chemical reactions go "to completion." Always, some fraction of the reactants remain in the reaction mass. For the gas phase reaction:

$$aA + bB \rightleftharpoons cC + dD \tag{7}$$

where the reactant and product concentrations are expressed as partial pressures, the equilibrium constant K_p, which is a function (only) of temperature, is given by

$$K_p = \frac{p_C{}^c p_D{}^d}{p_A{}^a p_B{}^b} \tag{8}$$

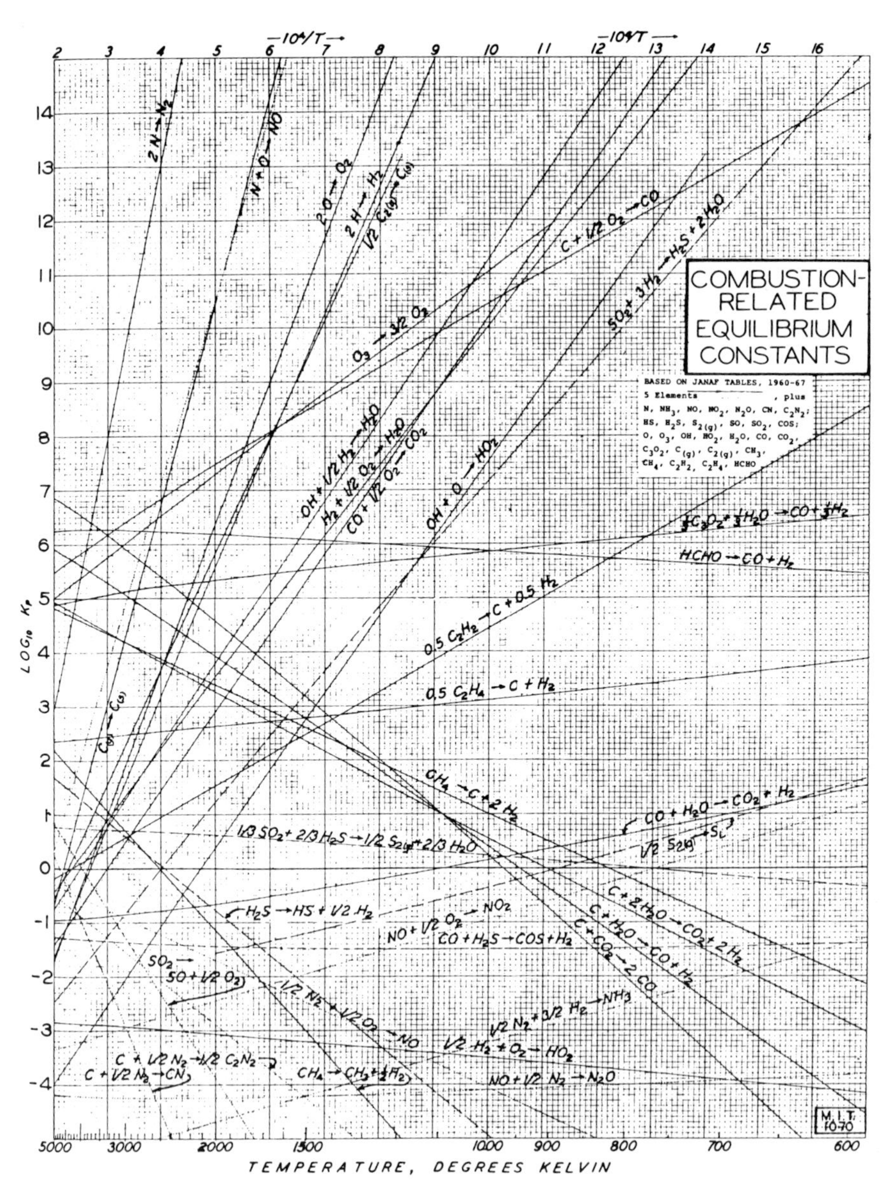

FIG. 3 Equilibrium constants of combustion reactions (partial pressure in atm).

Where the units of K_p depend on the stoichiometric coefficients a, b, c, and d such that if $(c + d - a - b)$ is zero, K_p is dimensionless. If the total is nonzero, K_p will have the units of pressure raised to the appropriate integer or fractional power.

Figure 3 shows the temperature dependence of reactions of interest. Note that when solid carbon is a product or reactant, no partial pressure term for carbon is entered into the mathematical formulation.

Example 5. At the furnace outlet temperature in Example 4 and at a total pressure of 1 atm, what is the emission rate of nitric oxide (NO) formed by the reaction

$$\tfrac{1}{2}O_2 + \tfrac{1}{2}N_2 \rightleftharpoons NO \tag{9}$$

Component	Moles	Partial pressure (atm)
NO	x	$x/58.695$
N_2	$44.083 - 0.5x$	$(44.083 - 0.5x)/58.695$
O_2	$3.905 - 0.5x$	$(3.905 - 0.5x)/58.695$
Total[a]	58.695	1.00

[a] For this reaction, the total number of moles does not change.

From Fig. 3 at 1575 °C, log $K_p = 1.9$ ($K_p = 79.43$) where

$$K_p = p_{O_2}^{1/2} p_{N_2}^{1/2} / p_{NO}$$

At equilibrium, then:

$$(3.905 - 0.5x)^{1/2}(44.083 - 0.5x)^{1/2} = 79.43x$$

Solving this equation gives $x = 0.164$ mol NO at equilibrium, or 0.279 mol% or 2794 ppm. Note, however, that kinetic limitations usually result in NO concentrations substantially below those predicted by equilibrium alone.

5. *Kinetics*

All chemical reactions proceed at a finite rate depending upon the concentration of the reactants, the static pressure (for reactions in the gas phase) and, importantly, the temperature. At combustion temperatures, reactions are usually very fast. Exceptions of importance are the oxidation reactions for carbon monoxide (CO), soot (carbon), and chlorinated hydrocarbons. The reaction rate behavior (chemical kinetics) of CO and soot burning will be discussed here.

a. *Kinetics of Carbon Monoxide Oxidation.* Carbon monoxide (CO) is an important air pollutant, a poisonous gas in high concentrations, and represents unavailable combustion energy if found in stack gases. The rate expression by Hottel *et al.* [3] for the rate of change of the CO mol fraction (f_{CO}) with time is given by:

$$\frac{-df_{CO}}{dt} = 12 \times 10^{10} \exp -\left[\frac{16{,}000}{RT}\right] f_{O_2}^{0.3} f_{CO} f_{H_2O}^{0.5} \left[\frac{P}{R'T}\right]^{1.8} \quad (10)$$

where f_{CO}, f_{O_2}, and f_{H_2O} are the mole fractions of CO, O_2, and water vapor, respectively, T is the absolute temperature K, P is the absolute pressure (atm), t is the time in seconds, and R and R' are the gas constant expressed as 1.986 cal g mol^{-1} K^{-1} and 82.06 atm cm^3 g mol^{-1} K^{-1} respectively.

The term $(-16{,}000/RT)$ is the heart of the kinetic expression, functionally providing a strong sensitivity to temperature through the exponentiation of the ratio of 16,000 (the Arrhenius "activation energy") to the absolute temperature.

It is instructive to note that the reaction rate is dependent upon the water vapor concentration, a reflection of the role of hydrogen (H) and hydroxyl (OH) free-radicals in the combustion reactions. Indeed, bone-dry CO is *very* difficult to burn whereas even a trace of moisture is sufficient to assist in ignition and rapid combustion.

b. *Kinetics of Soot Oxidation.* When carbon-bearing wastes are burned, the existence of regions where the oxygen concentration falls to zero often results in the formation of soot (finely divided carbon). The high optical density of such black smoke can lead to violation of opacity regulations applying to stack discharges and creates system problems by fouling boiler tube surfaces, reducing the collection efficiency of electrostatic precipitators, etc.

Soot burn-out is relatively slow in comparison to many other combustion reactions, owing in part to the slower pace of heterogeneous reactions and the possibility of diffusion limitations (*viz.*, diffusion of oxygen to the surface of the soot particle).

For spherical particles, Field [4] suggests that the rate of carbon consumption q(g cm^{-2} s^{-1}) is related to the oxygen partial pressure in atmospheres (p_{O_2}) by:

$$q = \frac{p_{O_2}}{1/k_s + 1/k_d} \quad (10)$$

where k_s is the kinetic rate constant for the consumption reaction and

k_d is the diffusional rate constant. For particles of diameter d (cm) at a temperature T (K);

$$k_d = \frac{4.335 \times 10^{-6} T^{0.65}}{d} \tag{11}$$

$$k_s = 0.13 \exp\left[(-35{,}700/R)\left(\frac{1}{T} - \frac{1}{1600}\right)\right] \tag{12}$$

where R is the gas constant (1.986 cal g-mol^{-1} K^{-1}). For a particle of initial diameter d_0 and an assumed specific gravity of 2, the time t_b in seconds to completely burn out the soot particle is given by:

$$t_b = \frac{1}{p_{O_2}}\left[\frac{d_0}{0.13 \exp\left[\left(\frac{-35{,}700}{R}\right)\left(\frac{1}{T} - \frac{1}{1600}\right)\right]} + \frac{d_0^2}{8.67 \times 10^{-6} T^{0.75}}\right] \tag{13}$$

B. Thermal Decomposition (Pyrolysis)

The thermal decomposition or pyrolysis of carbonaceous solids in the absence of air or under limited air-supply conditions occurs in most burning systems. Several solid-waste processing systems currently under advanced development exploit this process to effect gasification of refuse. Each produces a low heat-content gas stream containing volatilized water, a mixture of CO, hydrogen, and hydrocarbons, and a solid char which, often, is burned completely in a specialized region of the "pyrolyzer."

Both physical and chemical changes occur in solids undergoing pyrolysis. The most important physical change is a softening effect, resulting in a plastic mass, followed by resolidification. Cellulosic materials increase in porosity and swell as volatiles are evolved.

As cellulose pyrolysis begins (at about 200 °C), complex, partially-oxidized tars are evolved. As the temperature increases, these products further degrade forming simpler, more hydrogen-rich gaseous compounds and solid carbon. The solid residue approaches graphitic carbon in chemical composition and physical structure.

The rate-controlling step in pyrolysis can be either the heat transfer rate into the solid or the chemical reaction rate. Below 500 °C, the pyrolysis reactions appear rate-controlling for waste pieces less than 1 cm in size. Above 500 °C, pyrolysis reactions are fast and both heat transfer and product diffusion are rate-limiting. For pieces greater than

5 cm, heat transfer probably dominates for all temperatures of practical interest.

1. *Pyrolysis Time*

The time required for pyrolysis of most wastes may be estimated by assuming that the rate is controlled by the rate of heating. Neglecting

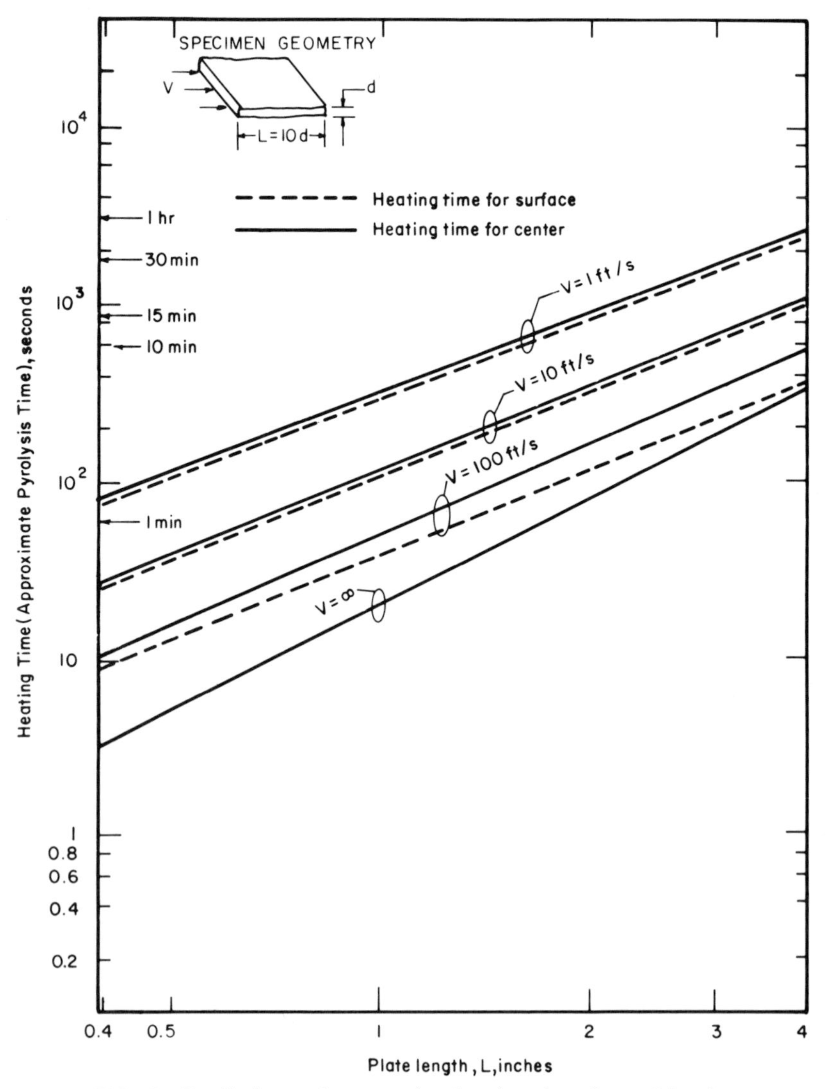

FIG. 4 Radiative and convective heating time for a thin plate.

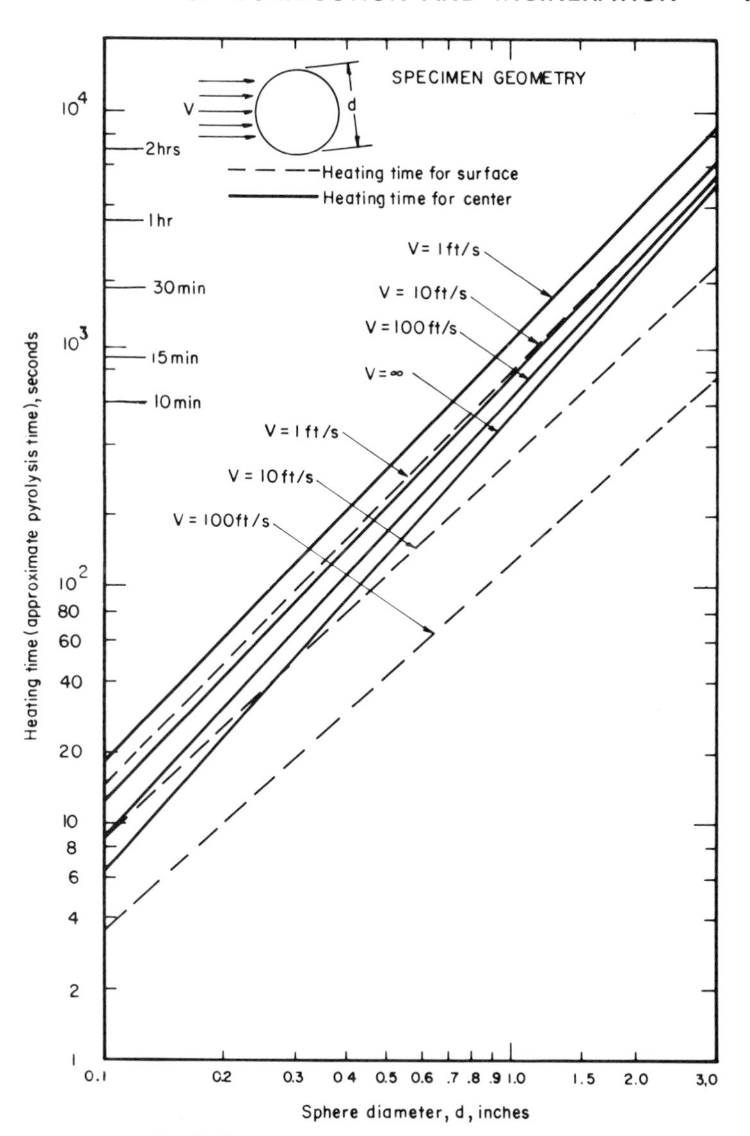

FIG. 5 Radiative and convective heating time for a sphere.

energy absorption or generation by the reactions, Figs. 4 and 5 allow estimation of the time for the center temperature of plates and spheres to rise by 95% of the initial temperature difference between specimen and surroundings. A thermal diffusivity of 3.6×10^{-4} m^2/h has been assumed, roughly equal to that of paper or wood [5]. The heating time at infinite cross-flow velocity (V_∞) corresponds to radiant heating.

2. *Pyrolysis Products*

The products of pyrolysis reactions include ash and carbonaceous char in the solid phase; liquids (at room temperature) which include water, various alcohols, ketones, acetic acid, methanol, 2-methyl-1-propanol, 1-pentanol, 3-pentanol, 1,3-propanediol, and 1-hexanol; and gases which include carbon dioxide, carbon monoxide, hydrogen, and a variety of low molecular weight hydrocarbons.

The distribution between these products is known to be related to the heating rate and the ultimate temperature based on laboratory experiments with muffle furnaces. One suspects that another parameter affecting the degree of char gasification relates to the degree to which moisture, often evaporated and exhausted early in or prior to the test, is permitted to contact the char and thus engage in the *water gas reaction*

$$H_2O + C \rightleftharpoons CO + H_2 \tag{14}$$

Table 4 shows the yield of pyrolysis products from different substrates.

Table 4
Yields of Pyrolysis Products from Different Refuse Components by Weight Percent of Refuse[a]

Component	Gas	Water	Other liquid	Char (ash-free)	Ash
Cord hardwood	17.30	31.93	20.80	29.54	0.43
Rubber	17.29	3.91	42.45	27.50	8.85
White pine sawdust	20.41	32.78	24.50	22.17	0.14
Balsam spruce	29.98	21.03	28.61	17.31	3.07
Hardwood leaf mixture	22.29	31.87	12.27	29.75	3.82
Newspaper I	25.82	33.92	10.15	28.68	1.43
II	29.30	31.36	10.80	27.11	1.43
Corrugated box paper	26.32	35.93	5.79	26.90	5.06
Brown paper	20.89	43.10	2.88	32.12	1.01
Magazine paper I	19.53	25.94	10.84	21.22	22.47
II	21.96	25.91	10.17	19.49	22.47
Lawn grass	26.15	24.73	11.46	31.47	6.19
Citrus fruit waste	31.21	29.99	17.50	18.12	3.18
Vegetable food waste	27.55	27.15	20.24	20.17	4.89
Mean Values	24.25	23.50	22.67	24.72	11.30

[a] Refuse was shredded, air-dried, and pyrolyzed in a retort at 815 °C [6].

Table 5
Percent Yields of Pyrolysis Products from Refuse at Different Temperatures by Weight of Refuse Combustibles[a]

Temperature °C	Gases	Liquid (including water)	Char
480	12.33	61.08	24.71
650	18.64	59.18	21.80
815	23.69	59.67	17.24
925	24.36	58.70	17.67

[a] From reference [7].

Tables 5 through 8 show the effect of ultimate temperature and heating rate on product mix. Table 9 shows the significant differences in gas composition and heat content for differing feed materials, while Table 10 shows the distribution in the products of the elements comprising a mixed municipal refuse.

The yield of liquid (including water) is approximately 50–60% of air-dried, ash-free refuse, decreasing with increasing ultimate pyrolysis temperature. The heat content of the liquid per pound of refuse decreases as the pyrolysis temperature increases.

For typical refuse, gas yield ranges from 15 to 35% of the air-dried

Table 6
Effect of Heating Rate on Yields of Pyrolysis Products and Heating Value of the Pyrolysis Gas from Newspaper

Time taken to heat to 815 °C, min	Yield of air-dried newspaper, wt%				Heating value of gas, kcal/kg of newspaper
	Gas	Water	Other liquid	Char (ash-free)	
1	36.35	24.08	19.14	19.10	1136
6	27.11	27.35	25.55	18.56	792
10	24.80	27.41	25.70	20.66	671
21	23.48	28.23	26.23	20.63	607
30	24.30	27.93	24.48	21.86	662
40	24.15	27.13	24.75	22.54	627
50	25.26	33.23	12.00	28.08	739
60	29.85	30.73	9.93	28.06	961
71	31.10	28.28	10.67	28.52	871

[a] Reference [6].

Table 7
Calorific Value of Pyrolysis Gases Obtained by Pyrolizing Refuse at Different Temperatures[a]

Temperature, °C	Gas yield per kg of refuse combustibles,[b] m^3	Calorific value	
		Gas, kcal/m^3	Refuse combustibles, kcal/kg
480	0.118	2670	316
650	0.173	3346	581
815	0.226	3061	692
925	0.211	3124	661

[a] From reference [7].
[b] At 15 °C, 1 atm.

feed with yield decreasing then increasing as the ultimate pyrolysis temperature is raised from 480 to 925 °C. In general, 1 kg of refuse combustibles yields 0.125–0.185 m^3 of gas with a calorific value of about 3000 kcal/m^3.

The solid char formed from refuse pyrolysis is an impure carbon, similar to coal in proximate analysis. Chars formed at 480° and 925 °C are comparable to bituminous and anthracite, respectively. Char yields range from 17 to 32% of the air-dried, ash-free feed, decreasing with increasing heating rate and ultimate temperature. Char heating value is around 6600 kcal/kg of air-dried char and decreases slowly as the ultimate pyrolysis temperature increases.

Table 8
Composition of Pyrolysis Gases Obtained by Pyrolizing Refuse to Different Temperatures[a]

Temperature, °C	Gas composition, vol%					
	H_2	CH_4	CO	CO_2	C_2H_4	C_2H_6
480	5.56	12.43	33.50	44.77	0.45	3.03
650	16.58	15.91	30.49	31.78	2.18	3.06
815	28.55	13.73	34.12	20.59	2.24	0.77
925	32.48	10.45	35.25	18.31	2.43	1.07

[a] From reference [7].

Table 9
Produced Pyrolysis Gas Analysis[a]

Waste Material	Gas analysis (dry basis), vol%								Heating value[f]	
	H_2	CO_2	CH_4	CO	C_2H_2	C_2H_4	C_2H_6	C_3H_8	BTU/scf	kcal/scm
MSW[b]	44.47	15.78	6.96	24.76	4.97	1.49	0.66	0.91	421	6750
Sawdust[c]	29.32	12.13	11.04	43.79	3.12	0.36	0.36	NM[e]	398	6380
Chicken manure	35.91	29.50	8.31	21.37	2.22	NM	0.61	NM	308	4940
Cow manure[d]	31.07	20.60	7.70	38.06	1.86	NM	0.31	NM	328	5260
Animal fat	11.57	27.63	18.12	14.72	25.05	NM	2.91	NM	683	10,950
Tire rubber	33.81	15.33	29.09	5.67	12.94	NM	3.17	NM	661	10,600
PVC plastic	41.02	19.06	14.51	20.76	4.02	0.21	0.43	NM	412	6600
Nylon	45.38	6.03	15.47	34.64	0.0	NM	0.0	NM	403	6460
Bituminous coal	46.88	11.68	16.63	21.72	2.08	NM	1.01	NM	435	6980
Sewage sludge (digested)	47.01	22.88	11.22	15.57	3.12	NM	0.21	NM	360	5770

[a] From reference [8].
[b] Average of five tests.
[c] Average of three tests.
[d] Average of two tests.
[e] NM = not measured.
[f] scf = standard cubic feet (60° F, 1 atm); scm = standard cubic meter (15 °C, 1 atm).

Table 10
Dry-Basis Yields from Pyrolysis of Refuse in Weight Percent[a]

	C, wt%	H, wt%	O, wt%	N, wt%	S, wt%	Ash, wt%	Total, wt%
Feed composition	30.85	3.84	22.32	(0.4)	(0.1)	42.49	100.00
CO	8.01	—	10.68	—	—	—	18.69
CO_2	4.32	—	11.52	—	—	—	15.84
H	—	2.05	—	—	—	—	2.05
CH_4	2.25	0.76	—	—	—	—	3.01
C_2H_2	3.22	0.27	—	—	—	—	3.49
C_2H_4	0.95	0.16	—	—	—	—	1.11
C_2H_6	0.43	0.11	—	—	—	—	0.54
C_3H_6	(0.52)	(0.09)	—	—	—	—	0.61
C_3H_8	(0.35)	(0.08)	—	—	—	—	0.43
Liquids	3.45	(0.32)	(0.12)	(0.1)	—	—	3.99
Ash	—	—	—	—	—	42.49	42.49
Char	7.35	—	—	(0.3)	(0.1)	—	7.75
Pyrolysis product totals	30.85	3.84	22.32	(0.4)	(0.1)	42.49	100.00

[a] From reference [9]. Parentheses indicate estimated values.

3. *Decomposition Kinetics*

Pyrolysis of cellulose appears to follow a two-step process. The first step involves breaking of the C–O–C bond to yield a mixture of sugar-like molecules, which subsequently degrade by further breaking of C–O–C bonding.

Studies by Kanury [10] using an X-ray technique to monitor density changes during the pyrolysis of wooden cylinders provides useful insight into pyrolysis kinetics. Kanury's data showed the pyrolysis reaction to follow:

$$\frac{d\rho}{dt} = -10^6(\rho - \rho_c)\exp\left(\frac{-19{,}000}{R_T}\right) \tag{15}$$

where ρ is the instantaneous density (g/cm^3) and the subscript c denotes char, t is time (min), R is the gas constant (1.987 cal/mol K), and T is the absolute temperature (K).

Kanury's data showed little reaction up to 350 °C, but rapid reaction above this point. Shivadev and Emmons [11], studying the pyrolysis of filter paper, show an ignition-like, rapid increase in reaction rate above 407 ± 15 °C, in rough agreement with Kanury.

C. Mass Burning

In mass-burning incinerators, solid wastes are burned in a relatively thick bed. In an idealized conceptualization of the bed processes (after ignition down to the grate line)

- Complete combustion is occurring at and near the grate, consuming the oxygen in the undergrate air to form CO_2 and H_2O.
- As the gases pass upward, CO_2 and H_2O react with char to form CO and H_2 in an endothermic reaction to a degree controlled by the water-gas shift equilibrium.
- Above this point, the only reaction which occurs is pyrolysis of refuse in the hot gases from below.

In the idealized mass-burning model described above, it was postulated that in thick beds the upper regions could behave as a true pyrolyzer. Evidence [12, 13] for coal and refuse beds would tend either to discount the existence of the pyrolysis zone or, more probably, to suggest that this zone does not appear in beds of practical thickness. Thus, the off-gas from a bed would be a mixture of gases from both burning and gasification zones.

Recent studies [13] on the off-gas from refuse burning in a municipal incinerator have confirmed the earlier data of Kaiser [14] and the hypothesis [5] that the off-gas composition is controlled by the water-gas shift equilibrium. This equilibrium describes the relative concentration of reactants in the following:

$$H_2O + CO \rightleftharpoons CO_2 + H_2$$

$$K_p = \frac{p_{H_2O}p_{CO}}{p_{CO_2}p_{H_2}} \qquad (16)$$

The importance of this equilibrium in mass burning is the incremental gasification potential given to the underfire air. In tests at Newton, Massachusetts [13], refuse and off-gas stoichiometry were studied. Average refuse was given the mole ratio formula: $CH_{1.585}O_{0.625}$-$(H_2O)_{0.655}$. For this formula, and assuming that the water-gas shift equilibrium holds, over 1.5 times as much refuse can be gasified as would be predicted for stoichiometric combustion to CO_2 and H_2O. Burning rate data showed rates 1.7 to 2.1 times that corresponding to stoichiometric combustion with the combustion air supplied beneath the refuse bed. Entrainment of air from above the refuse bed was assumed to occur to account for relative burning rates in excess of the 1.5 factor due to the water gas shift reaction. A second result coming from the water-gas shift reaction is that a definite and relatively large combustion-air requirement will necessarily be placed on the overfire volume. The air requirement for the CO, H_2, distilled tars, and light hydrocarbons can, indeed, be as much as 30–40% of that expended in gasification, thus creating a need for effective overfire air injection and mixing.

D. Suspension Burning

In suspension burning, a particle of refuse is suddenly thrust into an environment of hot gases and intense radiation flux. The particle undergoes rapid drying and ignition while airborne, and proceeds to burn in an oxygen-rich atmosphere. Depending on the particle shape and weight, the velocity of the gas medium, and the geometry and dimensions of the combustion chamber, the particle may be partially or entirely burned while still suspended in the gas stream.

In general, the chemistry and heat transfer environment of the furnace and the details of the particle characteristics (moisture content, thermal and mass diffusivities, shape factors, etc.) are poorly defined, so that a rigorous analysis is difficult. Even for the somewhat simpler case of pulverized coal combustion, many simplifying assumptions are required to predict flame length, minimum air requirements, etc. [4].

For refuse, the second and third stages of the combustion process (heat-up of the dry solid and pyrolysis) may be analyzed using Figs. 4 and 5.

E. Air Pollution from Incineration

Perhaps no aspect of the incineration of municipal solid waste generates so much public concern as that of air pollution. Although particulate emissions usually receive the greatest attention, concern is also shown about gaseous emissions, particularly hydrogen chloride.

In analyzing the mass-burning incineration process, the full spectrum of pollutants associated with the combustion (and incomplete combustion) of a heterogeneous organic and inorganic mixture merit some, if occasionally passing, attention. The pollutants include:

- Mineral (inorganic) and combustible particulate matter
- Carbon monoxide
- Unburned or partially-burned organic compounds, including the full spectrum of "pyrolysis products" and the several polycyclic organic compounds with which adverse health effects

are associated (e.g., polycyclic organic matter (POM), polyhalogenated hydrocarbons (PHH), and polychlorinated biphenyls (PCB)
- Sulfur dioxide and trioxide
- Hydrogen chloride
- Nitrogen oxides
- Trace elements

1. *Mineral Particulate*

The great majority of the emissions of mineral particulate are associated with the carryover of mineral matter introduced with the waste. The most important mechanism leading to these emissions is the entrainment of ash fragments by the flow of air and combustion products passing as underfire air through the burning refuse mass.

Undergrate air velocities typically range between 0.05 $sm^3/s/m^2$ of grate area to 0.5 $scm/s/m^2$. An analysis of the entrainment process [1, 5] suggests that this range of velocities would entrain particles ranging in maximum diameter from 70 to 400 μm at a mean temperature of the entraining fluid of 1100 °C.

The importance of the underfire air entrainment mechanism on emissions is supported by the observed range of particle sizes of fly ash, the increases in particulate emission with undergrate air velocity and the similarity in the chemical analyses of fly ash and that of the refuse ash [1].

Data by Eberhardt and Mayer [15] and Nowak [16] indicate that from 10 to 15% of the refuse ash can be expected to appear in the gases leaving the furnace. Data by the U.S. Public Health service (PHS) on an experimental incinerator [17] showed emission rates from 8 to 22% of that in the refuse for underfire air rates comparable to those used in municipal incinerators.

The relationship between underfire air rates and emissions were correlated by the PHS as:

$$W = 4.35V^{0.543} \tag{17}$$

where W is the emission factor (kg/ton of refuse burned) and V is the undergrate air flow ($sm^3\ s^{-1}\ m^{-2}$). Data on three municipal incinerators obtained by Walker and Schmitz [18] showed general agreement with Eq. (17) but scattered $\pm 20\%$.

Small effects on emission rate are also shown by incinerator size (emissions increasing as size increases, perhaps owing to the natural convection effects on gas velocity); burning rate relative to design capacity (Rehm [19] noted a 30% reduction in furnace emissions with a 25% reduction in throughput); and grate type (lower emission being associated with grates where sizable air openings pass a large fraction of the fine ash out of the flow as siftings); and several still lesser effects [1].

2. *Combustible Solids, Liquids, and Gases*

By definition, the appearance of combustible pollutants (soot, char, CO, hydrocarbons) in the effluent of a combustion system reflects an inadequacy in the combustion efficiency. This indicates:

- Inadequate residence time to complete combustion and/or
- Inadequate temperature levels to speed combustion reactions to completion and/or
- Inadequate oxygen concentrations in intimate conjunction with the combustibles to allow oxidation to proceed to completion, often indicating incomplete mixing rather than an oxygen deficiency in the overall flow.

The solution of the combustible pollutants problem is, most generally, associated with the attainment of intense mixing in the hot regions of the furnace and provision of sufficient residence time for combustion reactions to be completed. The soot and CO kinetics discussed in Section II.A.1 allow estimation of the order of magnitude of time and temperature required after an adequate degree of mixedness is attained.

3. *Pollutants Related to Waste Chemistry*

Sulfur dioxide and trioxide, hydrogen chloride, and trace elements of concern (e.g., mercury, selenium, lead, cadmium, zinc) are emitted in proportion to their concentrations in the waste fired. Estimation of their importance, therefore, is to some extent dependent upon the results of a comprehensive waste analysis for the area in question.

Sulfur oxides arise primarily from the oxidation of sulfur or sulfides, or from organic or inorganic sulfur-based acids. Owing to some absorption by alkaline fly ash, only about 70% of this sulfur will appear as SO_2 or SO_3 in the flue gases, by analogy with coal burning plants [20]. About 97% of the sulfur appears as SO_2 and 3% as SO_3.

Hydrogen chloride arises most importantly from the incineration of municipal refuse components such as polyvinyl chloride (PVC, 59% chlorine) and poly (vinylidene chloride) (73.2% chlorine). Conversion of the organically-derived chlorine to hydrogen chloride in the furnace gases is almost quantitative.

4. *Nitrogen Oxides*

Nitrogen oxides (NO, NO_2) are formed in all air-oxidized combustion reactions although nitrogeneous fuels produce significantly higher concentrations than fuels barren of nitrogen. Based on work by the PHS on fossil fuels [21], an estimate of the nitrogen oxides emission (expressed as NO_2) for refuse combustion is given by:

$$\text{kg NO}_x/\text{h} = [(\text{kcal/h})/2.26 \times 10^6]^{1.18} \tag{18}$$

F. Fluid Mechanics in Furnace Systems

Furnace fluid mechanics are complex, with flows driven by jets and by buoyancy interacting in swirling, recirculating eddys, and all the while traversing complex geometrical sections. Yet a basic understanding of furnace flow, particularly the behavior of jets and of buoyancy effects, is of great assistance in the design and analysis of incineration systems.

1. *Jet Behavior*

The behavior of jets in furnaces is of particular importance in incinerator design. This importance reflects the function of jets in:

- The controlled addition of *mass* to contribute to the oxidation process, to act as a thermal sink to temper gas temperatures (air jets) or to convey refuse into the incinerator (suspension burning).
- The controlled addition of *momentum* to promote mixing (turbulence) of the furnace gases to assist in attaining complete combustion (air or steam jets).

The behavior and design of jets in combustion situations is covered in more detail elsewhere, e.g., in refs. [1, 13]. The simplest jet system (the round, isothermal, turbulent subsonic jet) will be reviewed here since it embodies many of the basic characteristics important in understanding the behavior of the sidewall air jets commonly used in mass-burning incinerators. The complexities of crossflow, combustion, nonisothermal flow and buoyancy effects should be considered, however, in any final design calculations [1, 22, 23].

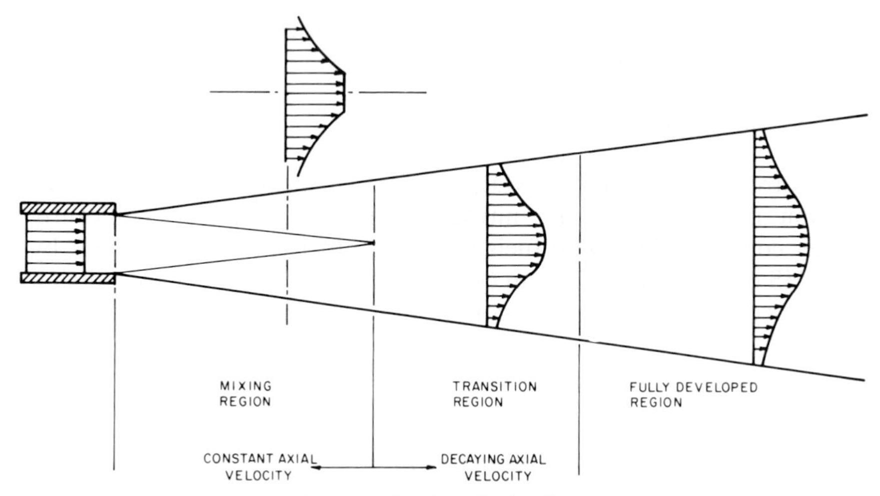

FIG. 6 Regions in jet flow.

The round jet (Fig. 6) shows three characteristic flow regions:

- The *mixing region* adjacent to the nozzle and extending about 4–5 nozzle diameters from the discharge plane which contains, as a distinguishing feature, an undisturbed flow near the axis of the jet (the "potential core") showing a relatively-flat velocity profile surrounded by a region (of high velocity gradient) where the rapidly moving jet gases are mixing with the surrounding fluid.
- The *transition region* extending 4–8 diameters downstream where the velocity profile acquires a stable shape.
- The *fully developed region* where the normalized velocity profile about the jet axis remains of constant or "self-preserving" shape.

Important jet characteristics include (*1*) the change in centerline velocity and concentration (of nozzle fluid) with distance, (*2*) the shape of the radial velocity and concentration profiles in the fully developed region, and (*3*) the rate of entrainment of ambient fluid into the jet.

The following functional relationships [4] describe these characteristics using the nomenclature $\bar{u}$ and $\bar{c}$ for mean velocity and concentration, ρ for density, x for distance from the nozzle, d_0 for the nozzle diameter, r for the radial distance from the jet centerline, and $\dot{m}$ for the mass flow rate. Subscripts 0 denote nozzle conditions, x conditions at a distance x from the nozzle, a conditions in the ambient fluid, and m conditions on the jet centerline.

- Velocity

$$\frac{\bar{u}_m}{\bar{u}_0} = 6.3\left(\frac{\rho_0}{\rho_a}\right)^{1/2} \frac{d_0}{(x + 0.6d_0)} \tag{19}$$

$$\frac{\bar{u}}{\bar{u}_m} = \exp\left[-96\left(\frac{r}{x}\right)^2\right] \tag{20}$$

- Concentration

$$\frac{\bar{c}_m}{\bar{c}_0} = 5\left(\frac{\rho_0}{\rho_a}\right)^{1/2} \frac{d_0}{(x + 0.8d_0)} \tag{21}$$

$$\frac{\bar{c}}{\bar{c}_m} = \exp\left[-57.5\left(\frac{r}{x}\right)^2\right] \tag{22}$$

- Entrainment

$$\frac{\dot{m}_x}{\dot{m}_0} = 0.23\left(\frac{\rho_a}{\rho_0}\right)^{1/2}\left(\frac{x}{d_0}\right) \qquad (\text{valid } x/d_0 > 6) \tag{23}$$

These functional relationships indicate that the jet flow expands inside a cone-shaped envelope. Defining the boundary as that corresponding to a velocity one-half of that on the jet centerline, a cone of half-angle 4.85° is defined. The corresponding half-angle for concentration is 6.2° [4].

These relationships have utility in predicting the flow behavior of side-wall air jets. Indeed, the Bituminous Coal Research (BCR) method of overfire air-jet design [24] is based on an assumed jet penetration depth (clearly, somewhat less than the width of the chamber) corresponding to a centerline velocity of 2.5 m/s as calculated using Eq. (19).

2. *Buoyancy*

Since furnace gases are hot, buoyancy effects can result in substantial flow acceleration. Although often overlooked in furnace analysis, these effects can be of sufficient magnitude as to cause severe erosion damage or to greatly change the velocity field, thus influencing the penetration distance of sidewall jets.

The acceleration of gases from an initial velocity u_0, elevation y_0 and pressure P_0 is described by Bernoullis' equation:

$$u^2 = u_0^2 + \frac{2(P_0 - P)}{\rho_0} g_c - 2g(y - y_0) \tag{24}$$

where ρ_0 is the density of the gas and g is the acceleration of gravity (9.807 m/s^2). This equation has utility in estimating the buoyant acceleration of gases rising from the grate of an incinerator.

Case 1. A well-sealed incinerator with a zone of hot gases arising from the grate and a cold zone of stagnant or slowly moving gases above the residue quench tank. The change in static pressure ($P_0 - P$) for the cold gas, as the flow exits through a vertical outlet flue, is also experienced by the hot gas. Noting that the ratio of absolute temperatures is the inverse of the ratio of densities, writing Eq. (24) for both flows and combining yields:

$$(u^2)_{\text{hot}} = (u_0^2)_{\text{hot}} + 2g(y - y_0)\left(\frac{T_{\text{hot}}}{T_{\text{cold}}} - 1\right) \qquad (25)$$

Case 2. For many older furnaces, the furnace is "leaky," so that there is an interaction between the hot and cold gases and the ambient atmosphere, with both gas streams accelerating, but with the acceleration of the hot zone being more pronounced. In this case

$$(u^2)_{\text{hot}} = (u_0^2)_{\text{hot}} + 2g(y - y_0)_{\text{hot}}\left(\frac{T_{\text{hot}}}{T_a} - 1\right) \qquad (26a)$$

$$(u^2)_{\text{cold}} = (u_0^2)_{\text{cold}} + 2g(y - y_0)_{\text{cold}}\left(\frac{T_{\text{cold}}}{T_a} - 1\right) \qquad (26b)$$

Example 6. In a large, well-sealed furnace, 6000 m^3/min of gases leave the burning refuse bed at a temperature of 1100 °C (1373 K) at an elevation of 12.5 m and at a velocity of 1.2 m/s. At the end of the furnace, 25 m^3/min of quench tank vapors leave the furnace at 300 °C (573 K) at 0.1 m/s.

The two gas flows leave through a vertical outlet flue at the top of the chamber (elevation 17.5 m, area of 65 m^2). Estimate the average velocity through the flue as well as the possible peak velocity due to buoyancy effects. Neglect the flow area for the cold gases.

(*a*) Mean Velocity.

$$\bar{v} = \frac{\text{Volumetric flow rate}}{\text{flue area}} = \frac{6000}{65} = 92.31 \quad \text{m/min or 1.54 m/s}$$

(*b*) Buoyancy Effects [Eq. (25)].

$$(u^2)_{\text{hot}} = (1.2)^2 + 2(9.807)(17.5 - 12.5)\left(\frac{1373}{573} - 1\right) = 11.76 \text{ m/s}$$

At this velocity, the gas needs only 6000/(11.76 × 60) or 8.5 m^2 of duct area. Thus, the flow cross-section shrinks as the gas accelerates, exiting as a high velocity stream on the side of the flue nearest the grate with a slowly moving mass of gas filling the remainder of the flue.

III. INCINERATION SYSTEMS FOR MUNICIPAL SOLID WASTE (>100 ton/d)

Incinerators handling large quantities of solid waste exhibit wide variation in design. These variations reflect local conditions, scale of operation, state of technology and, not unimportantly, the personal experiences and prejudices of the design engineer. In a study of incineration practice in the United States [5], over twenty major engineering firms and equipment vendors were asked to identify those design parameters which they believed to be broadly reflected in U.S. practice. Only one such parameter (the burning rate per unit area of grate) was identified.

Thus an incinerator can be and is many things. The paragraphs that follow outline very briefly the principal options in incinerator design. Any one topic would justify a chapter or a book in its own right if the topic was explored at a level of detail fully supporting design decision-making.

A. Receipt and Storage

The system used for the receipt and storage of raw refuse is of major concern to the design engineer. Since this portion of the plant interfaces with the entire refuse collection system and, in some cases with the individual citizen, the traffic patterns and dumping areas should be carefully thought out. Consideration must be given to the rapid processing of incoming vehicles (especially since refuse deliveries are seldom equally spread through the day) and to the safety of all parties concerned.

In most larger plants, refuse is received and stored in a pit below ground level. A traveling crane with a clamshell, "orange-peel," or grapple-type bucket is used to pile and mix the refuse as well as to feed the incinerator furnace.

The crane operator is more than a cog in the materials handling system. The operator should be concerned with: mixing of refuse to even out variations in refuse character, setting aside problem wastes (mattresses, engine blocks, refrigerators, and other "white goods"), moving refuse away from the dumping wall to allow uninterrupted dumping during peak receipt periods, keeping the feed hoppers filled, and planning the refuse withdrawals to effect a systematic cleaning of the pit. The latter responsibility is important for sanitation, good housekeeping, minimization of housefly nuisance (by processing refuse in a shorter time period than the larval–pupal cycle of the housefly) and the elimination of odors from decomposing refuse.

In some cases, a paved floor serviced by a front-end loader or dozer blade-equipped vehicle is used to receive and charge refuse. In such cases particular attention must be given to frequent cleaning to avoid the hazard of a slippery and unsanitary floor area. Special concern should also be devoted to the selection of rugged tires for tipping floor vehicles (to avoid rapid wastage or blow-outs) and in using heavy duty radiators (to avoid dust clogging and overheating).

B. Charging

In almost all larger continuous furnaces (>100 ton/d, T/D), charging is effected by a rectangular chute discharging to a feeding grate (Fig. 7). These chutes are often water-cooled and are best designed with a slight taper opening toward the discharge and with well-ground butt welds to avoid bridging.

In smaller, batch-fed units, a direct feed through a charging hopper or a ram feeder is used. It is recommended that any batch feed system should be designed to (*1*) minimize the inrush of cold air accompanying the entry of a refuse charge (chilling the combustion processes and disrupting the furnace draft) and (*2*) minimize the quantity of fresh refuse introduced at any one time (reducing the magnitude of the load on the combustion air supply system as the fresh feed flashes into flame).

If suspension burning is to be used, the refuse should be properly prepared by shredding and, for many applications, subsequent removal of large metal, stone and glass materials. Feeding is most often by pneumatic means, although mechanical conveyors and "flinger-type" spreader stokers can be used if careful attention is given to avoiding bridging or other congestion of the transport system. Some workers have recommended the use of trommel screening to remove fines and wet wastes which appear, predominantly in the $-2\frac{1}{2}$ cm fraction.

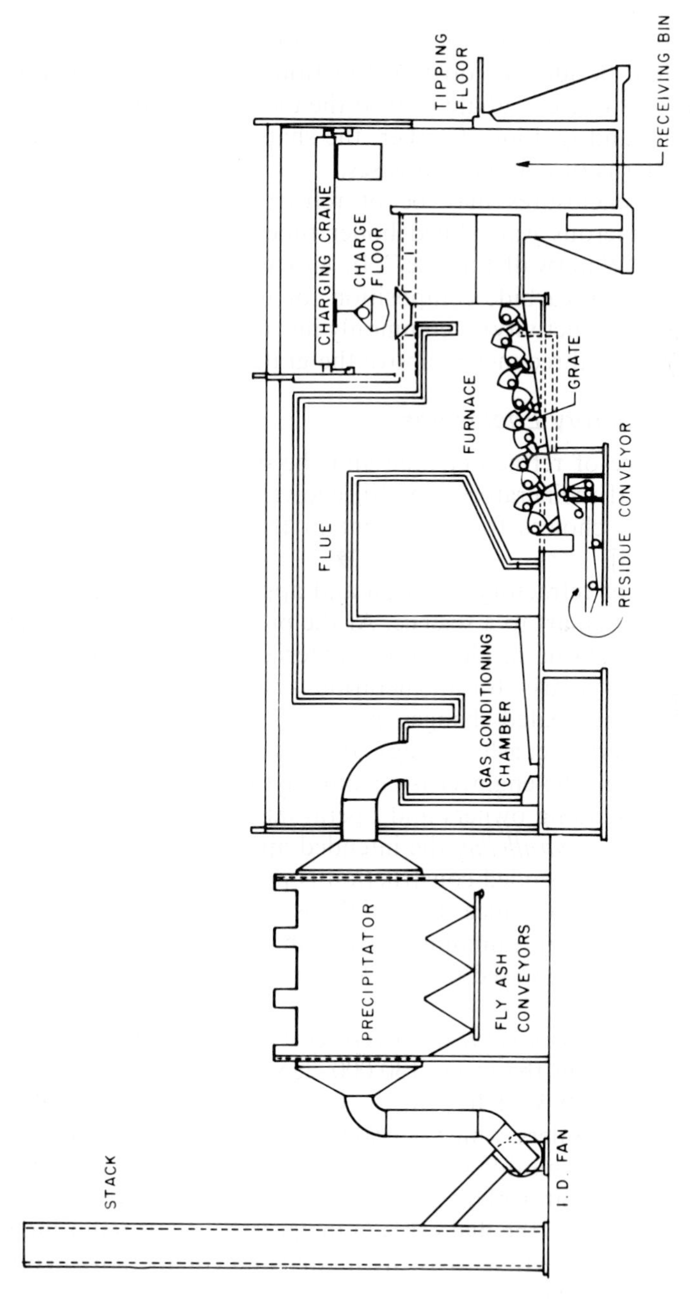

FIG. 7 Continuous feed incinerator.

C. Enclosures

The enclosure surrounding the grate system is a critical element in the system design. Besides the obvious function of containing the fire, the hot enclosure surfaces radiate heat to the incoming feed, thus speeding refuse drying and ignition. Further, the shape of the enclosure defines the flow patterns of combustion gases.

Furnace enclosures may be of refractory construction or may consist of numerous water-filled boiler tubes joined by a short metal bridge—the "Waterwall furnace." In either case, refractory material (usually of silicon carbide composition for its abrasion resistance and high thermal conductivity) is commonly installed at the grate line, often extending one to three meters above the grate.

1. *Refractory Enclosures*

The choice of refractory type and the details of construction for large municipal incinerators are still evolving. Table 11 indicates the more common choices for placement in the different regions of the incinerator [25].

The field of refractory selection and application is just now in the transition from an art to a science. At the present, both art and science must be combined in developing the criteria for proper application of refractory, particularly for incineration systems where frequent and sometimes unanticipated changes in waste character can result in wide swings in temperature, ash (or slag), and gas composition. In the course of system design, the engineer is encouraged to consult with firms and individuals (including owners and operators of existing facilities) with experience *closely paralleling* the intended application.

In general, suspended construction is preferred in incinerator construction in order to reduce the initial cost and to reduce maintenance expense. Castable or ramming mixes (shipped dry or ready-to-use, respectively) are often used for quick repairs during short maintenance outages.

Although designs vary, most refractory furnaces are designed for heat release rates in the range 130,000–225,000 kcal h^{-1} m^{-3} with an average of 180,000 kcal h^{-1} m^{-3}. This compares unfavorably with stoker-fired coal boilers, which operate at 260,000–310,000 kcal h^{-1} m^{-3} heat release rates. This difference reflects, in part, the great conservatism common in incinerator plant design, the lesser mixing energy provided by overfire air jets to the flue gases in incinerators (thus requiring more volume and residence time to assure mixing and completion of combustion), and to the higher excess-air level found in incinerators as compared with coal-fired furnaces (150–300% vs 35–60% excess air, respectively). The latter operating condition is related to the maintenance of low wall temperatures in refractory units. Modern waterwall incinerators usually operate at 75–100% excess air.

The enclosure (refractory or waterwall) also includes the "secondary combustion chamber," which may or may not represent a second and discrete chamber. What is needed, however, is a combustion space discrete relative to the grate area, the primary source of unburned combustible gases. Thus, in the secondary combustion volume, sufficient turbulence and residence time is provided at elevated temperatures such that the complete combustion of CO, soot, pyrolysis products, and other combustibles may be effected.

2. *Waterwall Enclosures*

The containment of combustion processes in refractory-lined chambers limits the peak temperatures. This usually means that refractory incinerators are operated at high levels of excess air with consequent penalties in the cost of air-pollution control devices, fans, and stacks. An alternative enclosure concept, permitting a substantial reduction in excess air, uses water-cooled metal tubes to protect the wall surfaces. A second and often more important benefit is the withdrawal of heat as steam or hot water for use in the incinerator plant or for sale.

When steam generation is the desired objective, the incinerator enclosure design changes to maximize the fraction of the refuse heat content transferred to the water. Often, the metal tubes comprising the wall panels are welded together with a narrow steel strip between the individual tubes to form a continuous, gas-tight membrane or waterwall. Also, banks of tubes are immersed in the hot gas flow to recover heat convectively.

To better understand the system, let us follow the course of the water entering the boiler plant.

a. *Water Treatment.* Raw water, containing dissolved minerals and suspended matter would be an unsatisfactory feed to a boiler. To avoid scale build-up and/or corrosion from these contaminants, the water must be treated. The level of treatment increases with the severity of the water side environment, as characterized by the temperature and pressure of the product steam. Treatment methods include filtration, softening, distillation, and/or ion exchange.

b. *De-aeration.* Water also contains dissolved gases (air components, CO_2, etc.) which would accumulate in the boiler. To remove

Table 11
Suggested Refractory Selection for Incinerators[a]

Incinerator part	Temperature range, °C	Abrasion	Slagging	Mechanical shock	Spalling	Fly ash adherence	Recommended refractory
Charging gate	20–1425	Severe, very important	Slight	Severe	Severe	None	Superduty
Furnace walls, grate to 48 in. above	20–1425	Severe	Severe, very important	Severe	Severe	None	Silicon carbide[b] or superduty
Furnace walls, upper portion	20–1425	Slight	Severe	Moderate	Severe	None	Superduty
Stoking doors	20–1425	Severe, very important	Severe	Severe	Severe	None	Superduty
Furnace ceiling	370–1425	Slight	Moderate	Slight	Severe	Moderate	Superduty
Flue to combustion chamber	650–1425	Slight	Severe, very important	None	Moderate	Moderate	Silicon carbide or superduty

Combustion chamber walls	650–1425	Slight	Moderate	None	Moderate	Moderate	Superduty or 1st quality
Combustion chamber ceiling	650–1425	Slight	Moderate	None	Moderate	Moderate	Superduty or 1st quality
Breeching walls	650–2400	Slight	Slight	None	Moderate	Moderate	Superduty or 1st quality
Breeching wall	650–2400	Slight	Slight	None	Moderate	Moderate	Superduty or 1st quality
Subsidence chamber walls	650–870	Slight	Slight	None	Slight	Moderate	Medium duty or 1st quality
Subsidence chamber ceiling	650–870	Slight	Slight	None	Slight	Moderate	Medium duty or 1st quality
Stack	260–540	Slight	None	None	Slight	Slight	Medium duty or 1st quality

[a] From reference [25].
[b] Air cooling is often used with subsequent discharge of the warmed air as overfire air.

most of the dissolved gases, the treated water is heated with steam or electricity to the atmospheric boiling point in a de-aerator. The water leaving the de-aerator is ready for introduction into the boiler using the feedwater pumps to raise the pressure to the boiler's working level.

c. *Boiler.* At the point of introduction into the boiler, the feedwater is treated, de-aerated, and, perhaps, somewhat preheated such that its temperature is in the range 100–200 °C. In passing the water through the boiler, it is desirable to optimize the temperature difference between the water and the hot combustion gases (maximum heat transfer rate) to minimize the required amount of heat transfer area (capital cost) while still extracting the maximum amount of heat from the combustion gases. In larger boilers, this will include such components as:

- *Economizer.* This component consists of one or more banks of tubes between which the hot flue gases flow and convectively transfer heat to the feedwater. However, the feedwater is usually not heated to the point where evaporation occurs. The economizer is located in the part of the boiler where the flue gas temperature is the lowest.
- *Convection Boiler.* This consists of one or more banks of tubes ("or passes") between which the hot flue gases flow, and where the water (from the economizer) is evaporated. Flow through the tubes may result from buoyancy effects (natural convection) or pumps (forced convection) and is two-phase, containing both liquid water and steam.
- *Radiant Boiler.* These components are water walls or, in some cases, banks of tubes exposed to the combustion zone between which the hot flue gases flow. Heat transfer rates are very high, and radiant energy transport from the incandescent refuse bed, flame, and/or hot gases is the predominant means of heat transfer.
- *Steam Drum.* This device consists of one or more large accumulators with disengagement space and mechanical devices to separate the gaseous steam from the liquid water. The latter is recirculated to the convection or radiant boiler sections. The product steam, in thermodynamic equilibrium with liquid water, is "saturated" at the temperature and pressure of the steam drum contents.
- *Superheater.* This device is a radiantly and/or convectively heated tube bank where, at (roughly) constant pressure, the steam is further heated to produce dry steam with a heat content greater than that of the saturated steam. Such super-

heated steam conditions are often advantageous, since they will produce more mechanical energy in a turbine than will saturated steam and/or will tolerate moderate heat losses without condensation in, say, a steam-distribution system pipeline.

D. Grates and Hearths

Nearly all incineration furnaces employ either a refractory hearth to support the burning refuse or a variety of grate types which stoke or mix the refuse during the combustion process in various ways depending upon the type of grate or stoker. There are many different types of hearth or grate designs, each of which has its own special features. The more common concepts are described below.

1. *Stationary Hearth*

These incinerator furnace systems that operate without grates include the stationary hearth and rotary kiln types. The stationary hearth is usually a refractory floor to the furnace, and may have openings for the admission of air under a slight pressure below the burning material on the hearth. In the absence of underfire air ports, air is admitted along the sides or from the top of the furnace. It is usually necessary to provide manual stoking in order to achieve a reasonable degree of burnout.

Stationary hearth furnaces are used for many commercial and smaller industrial incinerators. They are also used in crematories and for hospital wastes, when assisted with auxiliary gas or oil burners to maintain the furnace temperature above 650–900 °C.

2. *Rotary Kiln*

A rotary kiln consists of a steel, cylindrical shell lined with abrasion-resistant refractory to prevent overheating of the metal. The kiln is inclined toward the discharge end and the movement of the solids being processed is controlled by the speed of rotation. There has been little use of the rotary kiln for municipal incinerator furnaces except to provide additional residence time for improved residue burn-out *after* the burning of refuse on a multiple-grate system.

3. *Stationary Grates*

Stationary grates have been used in incinerator furnaces for a longer time then any grate system except the stationary hearth. The stationary grate is comprised of cast metal or fabricated metal grates with, perhaps, provision for rotating the grate sections to permit dumping of the ash residue. Although some stoking action can be obtained by shaking the grate, stationary grates normally require manual stoking.

4. *Mechanical Grates: Batch Operations*

Mechanically operated grates installed in batch-type furnaces were a natural evolution from the stationary grate furnaces. Although batch-type incineration furnaces have given way to continuous furnaces for new, large installations, many of the new small-capacity incinerators still utilize batch-fed furnaces, either with stationary or intermittently operated grates or without grates, the latter in small commercial and industrial installations.

a. *Cylindrical Furnace Grates.* In the circular batch furnace, the grates form annuli inside the vertical cylindrical walls of the furnace. A solid grate or "dead plate" covers the central area of the annulus. A hollow rotating hub with extended rabble arms rotates slowly above the circular dead plate to provide mechanical stoking or mixing. The rotating hub is covered with a hemispherical "cone," and one or more consecutively smaller "cones" are stacked on top of the first one. Forced air (called "cone air") for combustion is supplied through the hub to the hollow rabble arms, and thence through openings in the arms to the space just above the dead plate. Additional cone air is supplied to each of the cones in order to cool the metal. The annular grate area is divided into pairs of keystone-shaped segments; each pair is arranged to open downward for dumping the ash residue into the ash hopper below. These segmental grates are either hand-operated or hydraulically operated.

b. *Rectangular Batch Furnace Grates.* Mechanically operated grates in rectangular batch-operated incinerator furnaces include reciprocating (pusher) grates, and rocking grates. The grates are installed in a slightly inclined position from the horizontal, with the lower end of the incline at the ash discharge point. With these grates, the furnace is fed intermittently through an opening in the top and at the higher end of the grate, and the fresh refuse is deposited over the bonfire of previously ignited refuse.

As the burning continues, the grates are operated under manual control to move the burning bed of refuse toward the discharge, with manual control; ideally, to prevent the discharge of residue that has not been completely burned. In some instances, a dump grate is installed at the ash discharge point to hold back ash residue that is still burning, with manual operation of the dump grate after the accumulated ash has been completely burned.

5. *Mechanical Grates: Continuous Operations*

Mechanical constant-flow grates have been and are being used in most of the newer continuous-burning incinerators. The constant-flow grate feeds the refuse continuously from the refuse feed chute to the incinerator furnace, provides movement of the refuse bed and ash residue toward the discharge end of the grate, and does some stoking and mixing of the burning material on the grates. Underfire air passes upward through the grate to provide oxygen for the combustion processes, while at the same time cooling the metal portions of the grate to protect them from oxidation and heat damage. Typical grate designs correspond to an average heat release rate of 13,500 kcal/m^2/min. Clearly, the actual rate in different portions of the grate differs widely from this average.

a. *Reciprocating Grate.* The reciprocating or pusher grate is installed stepwise in rows on a slight downward incline toward the discharge. The rows move alternately to convey the refuse from the feed chute through the combustion area to the ash hopper.

Additional stoking and mixing (breaking open packed refuse masses) is obtained by providing a drop-off for tumbling from one grate section (step) to the next. Up to four grate sections are commonly included in this type of grate for a continuous flow incinerator. The reverse-acting reciprocating grate is also inclined downward, though at a steeper angle, toward the discharge end.

b. *Rocking Grate.* The rocking grate also slopes downward from the feed toward the discharge end, with up to four or more grate sections installed in series, and with or without a drop-off or spill-off between grate sections. The rocking grate includes a multiplicity of grate sections or segments that are approximately quarter-cylindrical and include openings for undergrate air. These grates are arranged as successive stair treads with risers of less than 3 cm. Alternate rows of grate sections are rotated approximately 90° about the edge toward the discharge of the grates, with the grate face rising up into the burning mass and thus breaking it up and thrusting it forward toward the discharge.

4. *Traveling Grate*

The traveling grate is widely used in continuous flow incinerator furnaces. There are two types of traveling grate stokers: the chain grate and the bar grate. Both convey the refuse from the gravity feed chute through the incinerator furnace to the ash residue discharge; much as a conveyor belt.

Because the traveling grate stoker does not stoke or mix the fuel bed as it conveys, incinerator traveling-grate stokers are often cascaded in two, three, and even four or more units with spill-offs of a meter or more between grate sections.

E. Combustion Air

For incineration, the term combustion air usually includes underfire air, overfire air, and secondary air. *Underfire air* for grate systems is defined as the air supplied upward through the grates and beneath the burning refuse. Underfire air is required to cool the grates and to furnish oxygen for the combustion reaction. It may be insufficient for complete combustion, yet sufficient to release enough heat to pyrolyze and/or gasify the refuse and remove the volatile components.

It is common practice to supply a quantity of underfire air theoretically sufficient for complete combustion of the refuse on the grates. However, poor air distribution vis-a-vis the distribution of air demand, the high relative rate of gasification and pyrolysis reactions, and poor gas mixing over the bed requires additional air to be supplied as overfire or secondary air for combustion, for mixing, and for dilution of the gases to maintain temperatures below 980–1100 °C. The latter function (temperature control) is of prime importance for refractory lined furnaces.

Overfire air may be defined as that air admitted above the burning bed of refuse on the grate. It is usually admitted either in low- or high-velocity jets to provide the energy to mix the combustible gases rising from the burning refuse with combustion air. *Secondary air*, added for temperature control, may be admitted through high-velocity jets in the side walls and roof of the furnace enclosure, either near the upstream end of the primary furnace or at the transition between the primary and secondary furnace. Also, secondary air can be added at low velocity through a slot or small openings in a bridge wall separating the primary chamber from the secondary chamber. In the latter case, mixing is dependent more on the shape of the chamber and changes in direction of the main gas stream than on the energy carried by the air jets.

Low velocity combustion air can also be admitted through furnace openings as a result of the negative pressure or draft within the furnace chamber. The quantity of combustion air admitted through such openings can be controlled by dampers or by the door opening.

F. Flue Gas Conditioning

Flue gas conditioning is defined as the cooling of the flue gas after it has left the combustion zone to permit discharge to mechanical equipment such as dry air-pollution control devices and fans or a stack. In general, cooling to 230–370 °C is necessary if the gas is discharged to mechanical equipment, although cooling to 480–590 °C is adequate for discharge to a refractory-lined stack.

Both wet and dry methods are used for cooling (or tempering) incinerator flue-gas streams. The paragraphs below discuss the technical and economic features of several embodiments of these methods.

1. *Cooling by Water Evaporation*

In wet methods, water is introduced into the hot gas stream and evaporation occurs. The degree of cooling is controlled by (*1*) the amount and droplet size of the water which is added to the gas and (*2*) the residence time of the gas in the water atmosphere. Presently, two types of wet cooling are used, the wet-bottom method and the dry-bottom method.

The *wet-bottom method*, which is most commonly used, involves the flow of large quantities of water (much more than is required for cooling the flue gas). The water is supplied by coarse sprays operated at relatively low pressures. The excess water falls to the bottom of the cooling zone and is rejected or recycled.

The equipment used in this system consists of several banks of sprays, each with several nozzles with relatively large openings (over 0.5 cm), located in the flue leading to the stack or air pollution control equipment. Line water pressure is adequate for satisfactory operation. The system is generally controlled by measuring the gas temperature downstream of the sprays and modulating water flow, either manually or automatically.

The advantages of the wet-bottom system are that it is relatively simple, reliable, and inexpensive to design and install. Also, there is a reduction in total gas volume during cooling. A disadvantage of the wet-bottom system is that much more water (greater than 100% excess) is used than is necessary for cooling the gas. The excess water is acidified in use and is contaminated with particulate and dissolved solids. Also, the flue gas leaving the spray chamber may carry entrained water droplets or wet particulate matter. These moist particles can cause operating problems with the APC and fan equipment owing to fly ash adherence and accumulation.

In the *dry-bottom method*, only enough water is added to cool the gas to the desired temperature, and the system is designed and operated to assure complete evaporation. In this system, a conditioning tower 10–30 m high is required and fine, high-pressure spray nozzles are used. Booster pumps are necessary to raise the water pressure to assure fine atomization; water pressures from 6 to 36 atm are common in such systems. Alternatively, atomization is effected using compressed air or steam. Control is usually accomplished with a temperature controller measuring the outlet flue gas temperatures and modulating the flow of water to the conditioning tower sprays.

The advantages of the dry-bottom system are that it minimizes water consumption, eliminates water pollution problems, produces a "dry" effluent gas (free from entrained water), and reduces the volume of the flue gas. The disadvantages of the system are that it is expensive to design and install, power consumption is high, control is somewhat complex, and the small orifices of the atomizing nozzles make them susceptible to plugging. Also, the dry-bottom system is more costly than the wet-bottom system because of the need for water filtration and the maintenance required for the high-pressure pumps, nozzles, and control systems.

2. *Cooling by Heat Withdrawal*

The second method of gas conditioning uses a convection boiler in which heat is removed from the flue gas by the generation of steam. The equipment consists of a convection-tube waste-heat boiler, an economizer, and all of the auxiliary equipment required, such as boiler-feed water pumps, steam drums, and water treatment facilities. These boilers have controls similar to conventional boilers and require an additional full-time operator—a licensed stationary engineer.

The advantages of this system are that heat is recovered and that the shrinkage in flue gas is greater than with any other method discussed. No water is added to the system during cooling, which may or may not be desirable. The disadvantages of the method are that the system is expensive to design and install, the boiler installation is complex to operate and requires an experienced licensed operator, corrosion and erosion problems with the boiler tubes have not been completely resolved, sticking of fly ash on the boiler tubes can occur, and reliable markets for the steam must be found or still greater investment is required for air- or water-cooled condensers.

G. *Air Pollution Control*

An incinerator is probably of greatest concern to a municipality because of the fear of the pollution of the contiguous environment. The principal concern is air pollution. The most noticeable forms of air pollution are fly ash, smoke, odors (from the stack as well as other areas), noxious gases, and dust. All emanate from an incinerator at times.

Although the flue gases from incinerators contain a number of pollutants, air pollution control (APC) equipment installed on these units are primarily directed at the problem of particulate removal. For this purpose, a number of devices are in use, ranging in particulate removal efficiency from 5 to 15% to upward of 95%. In light of current (1978) federal particulate emission standards (0.08 grains/dry ft^3 at 50% excess air), control efficiencies in excess of 95% are generally required.

Settling chambers or expansion chambers have been used in the breeching and flue gas ducts, and many of the older installations have employed refractory baffles across the breechings extending downward from the roof or upward from the bottom of the breeching to require the flue gases to pass under and over such baffles. In some instances, a coarse spray of water is directed into the flue gases and toward the baffles with most of the water falling to the floor of the chamber without vaporization. The wet floor and baffles improve particulate removal by preventing reentrainment of settled ash into the flue gas stream. At best, however, such systems only attain a control efficiency of 20–35%, far below modern requirements.

Mechanical collectors are usually "cyclones" in which the flue gas is rotated within the confines of a cylinder after entering tangentially at the periphery. The flue gas then leaves through an axial outlet. Solid particulate concentrates near the cylindrical wall (as a result of centrifugal force) and solids are discharged at the lower end and opposite to the cleaned gas outlet. Listed below in order of decreasing APC effectiveness (and pressure drop) are three general types of such cyclones. The maximum efficiency to be expected with such units is 60–80%.

- A multiple cyclone with many small-diameter (less than 30 cm) cyclone units installed in a tube sheet.
- A multiple cyclone system of larger diameter (over 45 cm) installed in clusters with flue gas manifolded to the inlets of the

Table 12
Average Control Efficiency of APC Systems [a]

APC Type	APC System Removal Efficiency (weight percent)								
	Mineral particulate	Combustible particulate [b]	Carbon monoxide	Nitrogen oxides	Hydro-carbons	Sulfur oxides	Hydrogen chloride	Polynuclear Hydro-carbons [c]	Volatile metals [d]
None (flue settling only)	20	2	0	0	0	0	0	10	2
Dry expansion chamber	20	2	0	0	0	0	0	10	0
Wet-bottom expansion chamber	33	4	0	7	0	0	10	22	4
Spray chamber	40	5	0	25	0	0.1	40	40	5
Wetted wall chamber	35	7	0	25	0	0.1	40	40	7
Wetted, close-spaced baffles	50	10	0	30	0	0.5	50	85	10
Mechanical cyclone (dry)	70	30	0	0	0	0	0	35	0
Medium-energy wet scrubber	90	80	0	65	0	1.5	95	95	80
Electrostatic precipitator	99	90	0	0	0	0	0	60	90
Fabric filter	99.9	99	0	0	0	0	0	67	99

[a] From reference [5].
[b] Assumed primarily $< 5\ \mu m$.
[c] Assumed two-thirds condensed on particulate, one-third as vapor.
[d] Assumed primarily a fume $< 5\ \mu m$.

individual cyclones and the outlet manifolded into a single duct.

- Single or double cyclone units of larger diameter (over 1 m) with a single or split flue duct at the inlet and outlet.

Other devices used for particulate removal from flue gases include scrubbers, which may be open spray chambers, packed chambers, and most importantly, high-pressure drop (40 to more than 100 cm of water) Venturi scrubbers. Fabric filters can, conceptually, be used in incinerator applications, but the results to date have been disappointing; *viz.*, although collection efficiency is high (>99%), maintenance and operating problems have plagued the units. The filter materials are mainly high temperature fabrics, such as silicone-treated glass fiber cloth, arranged in bags or tubes.

The electrostatic precipitator is currently receiving the greatest attention for particulate removal from incinerator flue gases. At the time of this writing, there are over nine operating electrostatic precipitators in the United States associated with refuse incinerators. Electrostatic precipitators used for cleaning incinerator flue gases are the vertical, multiple-plate type. Reported efficiencies exceed federal regulations.

A summary of expected average control efficiencies of various air pollution control (APC) systems is given in Table 12.

H. Special Topics

1. *Heat Recovery*

During the years after 1968, the capital investment and operating costs for conventional refractory incinerators escalated rapidly. These cost increases severely retarded the rate of installation of incineration systems. The energy crisis of the mid-1970s, however, gave incineration economics a major assist through the revenue potential of heat recovered from refuse combustion.

With oil selling for $2.40 per 42-gallon barrel (in 1970), a competitive steam price for refuse-derived steam would be about $1.15/ton of steam. Since a ton of refuse yields about 2.85 tons of steam, the gross energy credit (neglecting any costs associated with heat recovery) would be about $3.30, or about one-fourth of the total incineration cost. Increasing the oil price to $12.00/barrel (in 1976) raises the refuse credit to over $16.00/ton, essentially full coverage of the incineration cost for a large (say, >800–1000 ton/d) plant.

Why then is the heat recovery incinerator not a panacea?

- *Steam Marketing.* Finding a suitable customer who can take all or most of your steam is not easy. Such a customer must be located near to open land (to site the incinerator), near to good transportation routes and areas of refuse generation, must need steam 24 h/d, 7 d/week, 52 week/yr, may have to accept whatever unreliability in steam supply the incinerator exhibits (usually greater than for a fossil-fueled boiler), and is limited to the steam pressures and temperatures that can be developed in refuse fired boilers. These conditions are not easily fulfilled.
- *Capital Requirements.* Even though net incineration costs may be substantially reduced by steam sales, the municipality must pledge tens of millions of dollars in capital investment for large plants.
- *Public Reaction.* Although air pollution and other environmental problems can be solved using modern abatement technology, public reaction to incineration is often still hostile. Also, though heat recovery may, in truth, represent a highly developed form of "recycling," many citizens believe incineration to be a waste of natural resources.
- *Competition.* Although incineration with heat recovery is often an attractive alternative, landfills, RDF (refuse-derived fuel) or other techniques may be economically, technically, environmentally, or politically more appropriate in a particular location.

a. *Energy Recovery Potential.* The energy recovery potential of an incinerator boiler is dependent upon three parameters: (a) the refuse heat content, (b) the excess air used for combustion, and (c) the minimum gas temperature leaving the boiler or air heater. Refuse heat content (see Appendix A) often averages 2500 kcal/kg, but can be much lower, or slightly higher, depending upon the composition, moisture level, and local practices. Excess air levels as low as 50% have been reported but 75–100% is more common for grate-fired units. Excess air above these values probably reflects operator inattention or poor practice. Back-end temperatures can be below 175 °C, although this may require installation of uneconomical air preheater equipment.

b. *Corrosion.* Corrosion of waterwall and tube metal surfaces is, perhaps, the most serious technical problem in the design of refuse-fired boilers.

In the operation of a boiler using wastes as fuels (and for conventional liquid or solid fuels as well), metal wastage owing to corrosion and erosion and tube fouling owing to the build-up of deposits have presented serious problems to the system designer and operator. Detailing the nature and cures for such problems is beyond the scope of this chapter and, at this writing, is still a matter of intense study and speculation. Several basic concepts, however, merit qualitative description:

- Low-Temperature Corrosion. In regions of the boiler, especially in the economizer or on any surfaces used to transfer heat to incoming combustion air (the air heater), slow-flow zones (low-heat transfer rates) and/or feed water or air too cold, can lower metal temperatures to the point at which condensation of moisture in the flue gases will occur. The presence of mineral acids (e.g., sulfuric or hydrochloric acid) in the flue gas (especially sulfur trioxide) leads to condensation at temperatures considerably above 100 °C. The resulting metal wastage rate, accelerated by the presence of soluble chlorides or acids, can become unacceptably high. Clearly, such corrosive mechanisms are always operative during boiler start-up and shut-down unless fossil fuels are fired to warm up the unit.

  The "cure" for this type of corrosion is straightforward; namely, a design to avoid dead zones and to maintain metal temperatures safely above the dew point of the flue gases. Furthermore, frequency and duration of cool-downs should be minimized.
- High-Temperature Corrosion. In regions of the furnace above, say, 400 °C, a variety of mechanisms for chemical attack of the metal tube surfaces can become operative.

  Chlorine, appearing in the flue gases as hydrochloric acid (e.g., from the combustion of chlorinated hydrocarbon wastes or PVC), or in salts such as sodium or potassium chloride, has been shown to participate in corrosive attack of metal tubes. Sulfur, appearing as the dioxide or trioxide or as sulfates, appears to slow the rate of attack of the metal by chlorides [26]. Although the exact mechanism of attack is still in question, it is clear that fireside metal temperature is the single most important parameter in evaluating the potential for rapid metal wastage. The flue gas temperature, however, is a second, though less important variable at gas temperatures of interest.

  Data reported by Battelle [26–31] indicate that a maximum fireside metal temperature of 205 °C (400 °F) should give long (say, 15 yr) carbon-steel boiler-tube service for systems burning

100% municipal refuse. Allowing for a 25 °C temperature drop across the tube wall (a reasonable average for boiler tubes containing liquid water and thus experiencing a high heat transfer rate on the inner wall, but far too little for superheater tubes containing gaseous steam), this corresponds to a maximum (saturated) steam pressure of 9.85 atm (130 psig). For higher pressures and temperatures, increased wastage must be accepted and/or more costly tube metal alloys must be used.

The problems of metal wastage are of special concern for superheater surfaces. The lower heat-transfer rates of the steam (compared to liquid water) results in higher fireside tube temperatures and greatly accelerated corrosion. Some relief from this problem has been found by the use of rammed silicon carbide type refractory coatings on the tubes [32]. This "solution" comes at the cost of lowered heat transfer rates and increased investment and maintenance expense.

Oxidizing and Reducing Conditions. In the incineration of highly nonideal fuels such as raw municipal refuse and especially for mass-burning systems, it is common to produce flue gases which fluctuate in composition between oxidizing (having an excess of oxygen) and reducing (having an absence of oxygen and the showing presence of carbon monoxide, hydrogen, hydrocarbon gases, etc.). Metal surfaces exposed to such changing gas compositions are subject to rapid wastage. In large part, this wastage results from the effects of repeated cycles of surficial metal-oxide formation and then reduction. Flaking of the weak reduced-metal structure is accelerated by the "shot blasting" effect of entrained particulate.

In mass-burning systems, the bed processes always produce reducing gases and, consequently, sidewalls and radiant tube banks are particularly prone to this type of attack. Protection of the sidewalls with refractory up to a distance 3–10 m above the grate line has been used successfully to cure the sidewall corrosion problem. Introduction of sufficient secondary air above the fire and stimulation of high levels of turbulence can greatly assist the burnout of reducing gases prior to their entry of the tube banks.

- Abrasion (Erosion) Wastage. The mechanical erosion of tube surfaces by fast-moving fly ash particles can rapidly lead to tube failure. The fly ash from municipal refuse combustion has been shown to be particularly abrasive (more so than, for example, most coal ash). This problem can be mitigated by reducing the velocity of the flue gases between the tubes (say, to 3.5–4.5 m/s) by coating the tubes with refractory (or letting slag build up to a degree), and by careful design of tube bank geometry and flow patterns. In general, however, these remedies lead to larger, more costly boiler facilities.

- Slagging. Slag build-up is not directly responsible for tube wastage, and as mentioned above can even reduce erosive metal losses. Slag accumulation will, however, reduce heat transfer rates and increase the pressure drop and gas velocity (erosion rate) through the boiler flues. It should be recognized that important corrosion reactions have been shown to occur within the slag layer so that slag build-up cannot be used to infer a lack of chemical attack.

  Slagging can be avoided by designs that maintain fireside metal temperatures below the range where the slag becomes tacky. For municipal refuse, this range is approximately 600–700 °C. Alternatively, the boiler passes can be equipped with "soot blowers" that use steam, compressed air jets, or metal shot periodically to dislodge adherent slag. Note, however, that cleaning the tube surface can also result in the removal of coatings which were performing a protective role with respect to tube attack. Thus, following soot blower activation, corrosive wastage will be initiated, typically at very high rates, until a protective coating of slag and/or corrosion products is reestablished.

2. *Burning in Suspension*

Mass-burning incinerators not equipped for heat recovery offer several advantages, for example, unprocessed, raw refuse may be fired, capital costs represent the norm for large-scale units, and operating controls and/or operator-skill requirements are relatively modest. Disadvantages of mass burning when heat recovery is practical include sluggish response to changes in steam demand and the severe problem encountered during emergency shutdown when the large mass of burning material is difficult or impossible to put out. Also, the degree of burnout of incinerated raw refuse is good in the overall, but somewhat irregular depending on the dimensions and/or moisture content of the refuse fired. A much greater degree of responsiveness to variations in heat release requirement, more even burnout and rapid shutdown is possible when the refuse is first subdivided and then thrust into the incinerator mechanically or pneumatically to burn partially or entirely while it is airborne. In practice, this is accomplished with either a spreader stoker-fired system (sometimes called "semisuspension" burning) or in a true suspension-fired system.

In either case, the refuse is received and stored and then subdivided using a hammermill, rasp, or other device (see Chapter 2). If it is desired to reduce the quantity of residue, to minimize problems owing to the build-up of slag deposits or to recover unincinerated ferrous metal, the shredded refuse may be processed to (a) reduce massive residue content, (b) reduce (somewhat) fine residue content, (c) increase mean heat content, and (d) simplify and reduce problems with transport of the processed refuse. The processed refuse is then stored for later retrieval or fed directly to the furnace using either belt or other mechanical conveyors, or a pneumatic conveyor.

a. *Spreader Stoker.* In a spreader stoker-fired system, the refuse is projected into the incinerator enclosure using a rotating, vaned "flinger," or is pneumatically blown in. The trajectory of an entering refuse particle carries it over a grate moving the bed of burning refuse back toward the firing face, discharging burned-out residue just below the firing chutes. The hot gases rising from the burning refuse bed and radiation from airborne burning particles rapidly heat the incoming refuse, drying and igniting it. The refuse particle may be entirely burned out while suspended in the gas and/or be swept out of the furnace. If the particle is large and heavy, it falls to the grate to aid in the drying and ignition of incoming refuse. Both burning patterns are desirable to provide ignition and to protect the grate. Thus, extra-fine shredding is to be avoided, a benefit since shredding costs rise rapidly as product top-size decreases [5].

b. *Suspension Burning.* In a suspension-fired furnace, the refuse is pneumatically blown into the system and almost all burning occurs while the refuse particle is suspended in the hot furnace gases. To assure that this occurs, tall furnaces (similar in dimensions to anthracite coal fired boilers) are used or a high degree of refuse subdivision is required. Often, a small burnout grate is provided in the floor of the furnace to increase the retention time for larger wood fragments or exceedingly wet refuse materials which survived the subdivision processing steps.

Suspension burning is well adapted to burning several fuels and, indeed, most suspension burning of refuse (up to 20% of the total heat release rate) has been in cofiring applications with coal in large (>100 MW) steam electric plants. Although long-term experience is lacking, extended (several days to several weeks) tests in St. Louis [33] have shown:

- No significant reduction in the performance of the electrostatic precipitators used for air pollution control.
- No marked increase in boiler corrosion or tube fouling.
- Rapid burnout of most refuse particles, but a clear requirement for a burnout grate.
- Significant problems with refuse pneumatic-conveying systems owing to the high abrasiveness of the refuse.

Combustion air for suspension burning, without main grates, requires a different consideration from combustion air for a grate system. In suspension burning, the air that conveys the shredded refuse into the chamber may be half or all of the theoretical air required for combustion; however, it must be sufficient to convey and inject the shredded refuse into the furnace. It is also desirable to add air at points in the furnace chamber in a tangential pattern to create a cyclonic action, with the burning mass in the center of the rotating cyclone and the air injection surrounding the cyclonic flame. If the suspension burning system includes an auxiliary grate at the bottom of the furnace chamber for completing the burnout of the ash residue, a small amount of underfire air is desirable through this grate.

3. *Residue Processing and Disposal*

The residue from the incineration of municipal refuse contains partially oxidized metal, glass, inert mineral matter, unburned organic material, and char. Ideally, the latter combustible fractions are small. Practically, however, wet refuse, incompletely burned heavy wood pieces, unbroken compacted masses, and the like will be found in the residue conveyor.

The residue is, most commonly, trucked to a landfill for disposal. The weight of residue usually comprises about 20% of the refuse fired (dry basis), but its volume (assuming reasonably efficient burning) is only 5–10% of the original refuse volume. The requirements for landfills receiving incinerator residue do not differ greatly from those for raw refuse, although gas formation is unlikely and leachate characteristics are predominantly related to the metal and salt constant rather than to the organic matter. This latter characteristic is important since without the acid formed by biodegradation of organic matter in the fill, the solution rate of the metals in the residue is very slow.

Incinerator residue may be readily processed by use of, say, a 2-cm-opening trommel screen and magnetic separator to produce three products: (a) ferrous metal, (b) mixed glass and (primarily) inorganic fines, and (c) oversize. Such processing is now underway at Baltimore, Maryland's Incinerator number 4 with good results.

The ferrous metal product, though inferior to the ferrous metal recovered from raw refuse, can be sold on the scrap metal market. The primary reasons for the lower quality of this material include its partially oxidized state, the mechanical trapping of impurities by the collapse of the can in processing, and the alloying of copper, tin and other metals that occurs in the incineration environment.

The mixed glass and fines product is useful as a fill and has been used by Baltimore (with further size separation) as an aggregate for road topping ("Glasphalt").

The oversize fraction, importantly containing the unburned combustible, can be reburned or landfilled.

4. *Pyrolysis Systems*

Earlier in this chapter, the basic concept of pyrolysis was described. This section describes the hardware concepts that have been employed to carry out refuse pyrolysis on a commercial scale.

The problem facing the designer of a system operating in the pyrolysis mode is to find a way to apply heat to refuse sufficient to raise its temperature to 500 °C or more without adding oxygen, which will burn the pyrolysis products. This has been accomplished in three ways: (a) indirect heating, (b) zoned partial combustion, and (c) fluid-bed flash pyrolysis.

a. *Indirect Heating.* Using this approach, refuse is placed inside a metal cylinder that is heated on the outside surface by burning the pyrolysis gases. Such a processing method has been used by Pan American Resources of Albuquerque, NM. Their 40-ton/d unit (the Lantz Converter) uses a rotating kiln design with an airlock seal for the feed. Pyrolysis temperatures range between 480 and 820 °C depending on the refuse characteristics. Test work, initiated in 1968 at a Ford Motor Company assembly plant in San Jose, California, was not promising and the operation was discontinued. The Destrugas system, developed in Copenhagen, Denmark, is also reported to use a retort-type, vertically-oriented cylindrical system.

b. *Zoned Partial Combustion.* In general, pyrolysis systems seek to produce a gaseous product suitable for transmission over some (limited) distance and useful as a fuel gas in existing boiler equipment. Thus, the char fraction represents a waste product. The zoned partial combustion approach makes use of the heat of combustion of the char to produce a hot gas which, when passed over or through the incoming refuse, produces the thermal environment necessary for pyrolysis. Careful control of the addition of oxidizing gas (air or pure oxygen) limits the combustion to the char fraction, thus maximizing the product gas yield.

The Monsanto Landgard system, constructed in Baltimore, Maryland, with a design capacity of 910 ton/d is an embodiment of zoned partial combustion using a rotary kiln as the pyrolysis vessel. Fuel oil was burned in the discharge zone to add additional pyrolysis heat. Excess air (relative to the fuel oil) acted to burn the char. The hot, oxygen-deficient combustion products then passed up the kiln effecting pyrolysis and feed drying.

The Monsanto system has encountered severe problems in control and with air pollution. The latter may be associated with an inadequate design of the scrubber system although it has been suggested that volatilization of salts in the extremely hot char-burning zone may have produced such an abundance of submicron particulate that almost any air pollution control device would have been overwhelmed.

A second type of zoned partial combustion system uses a vertical shaft furnace configuration. Combustion occurs in the lower region of the furnace and the hot gases pass upward through the incoming feed plug. The Union Carbide Purox system effects combustion with pure oxygen thus avoiding dilution of the product gas with the nitrogen associated with air-derived oxygen. The Andco Torrax system uses air for combustion, but preheats the air using a portion of the product gas. Both systems have been successfully demonstrated at the 180-ton/d scale.

c. *Flash Pyrolysis.* The third approach to commercialized pyrolysis processes refuse which has been thoroughly pre-prepared (coarse shredding, drying, air classification, and final shredding of the combustibles to a very fine size), in a fluid bed. Fluidizing gases, preheated by burning the pyrolysis char, rapidly heat the incoming refuse. The rapid heating rate maximizes the yield of liquid products. A 180-ton/d demonstration plant was being tested in San Diego, California by its developer, the Garrett Division of Occidental Petroleum, but operating problems have required shutdown of the facility pending the resolution of the problems through additional pilot-scale studies.

In each of these methods to process refuse by pyrolysis, the objective is the same, that is, to convert a heterogeneous, hard-to-handle material into gaseous and liquid fuel products suitable for firing in conventional boilers. Effecting such a fuel conversion step resolves in part the critical step in exploiting refuse energy content to reduce solid waste disposal costs, and thus lead to the marketing of the energy resource.

IV. INCINERATION SYSTEMS FOR MUNICIPAL AND COMMERCIAL WASTES (>100 ton/d)

At the present, two types of incinerators have emerged as the principal competitors for the "on-site" incinerators used in large shopping centers or, possibly, for the smaller municipalities; they are the Los Angeles type and the modular combustion units.

A. Los Angeles Type

For commercial or light-industrial waste disposal in the capacity range from 20 to 1000 kg/h, a well-established design rational has been developed. The design, based on an extensive empirical investigation undertaken by the Air Pollution Control District of Los Angeles County over the years 1949–1956, has been reduced to a series of nomographs, tables, and graphs relating construction parameters to burning rate [34–36].

The Los Angeles designs are meant to be applied to wastes with a heating value from 2400 to 3600 kcal/kg (typically paper, rags, foliage, some garbage, etc.). The unit designs are predicated upon the use of natural draft to draw in air under the fire (about 10%), over the fire (about 70%), and into the secondary (mixing) chamber (about 20%). Typically, the total air ranges from 100 to 300% excess.

The Los Angeles design methods are applied to two configurations, the retort and the in-line. The retort style derives its name from the return flow of flue gases through the secondary combustion chambers, which sit in a side-by-side arrangement with the ignition (primary) chamber. The in-line style places the ignition (furnace), mixing, and combustion (secondary combustion) chambers in a line. The retort design has the advantages of compactness and structural economy, and is more efficient than the in-line design at capacities to 350 kg/h. Above 450 kg/h, the in-line design is superior.

B. Modular Combustion Units

In the 1960s, a number of manufacturers, recognizing the increasing emphasis on smoke abatement, began producing incinerators that limited (or starved) the combustion air supply. This holds flue gas temperatures high and minimizes the fuel consumption associated with any afterburner devices. Refuse is charged repeatedly in batches, and after 10–12 h of charging, the unit is allowed to burn down. Residue is raked out after the burn-down (8–12 h) and the process is repeated. The starved air units consist of a cylindrical or elliptical cross-section primary chamber incorporating underfire air slots in the hearth-type floor. Overfire air is also supplied. All combustion air is provided by forced draft fans. In many such incinerators, the proportion of underfire to overfire air is regulated from a temperature sensor (thermocouple) in the exhaust flue (higher exit temperatures increase overfire air and decrease underfire air).

Most units incorporate a gas- or oil-fired afterburner that is energized whenever the exit gas temperature falls below the set point temperature (usually 650–750 °C). The afterburner is either mounted in a separate, secondary combustion chamber or in a refractory-lined stack.

Starved air incinerators are available in capacities from 100 to 1000 kg/h and, if properly operated and maintained, can meet most federal, state, and local air pollution codes when fed with typical office and plant trash (principally paper, cardboard, and wood).

The modular combustion units can be equipped with a boiler (usually of the firetube type) for the recovery of heat. The larger units can also be equipped with automatic feed and residue removal systems thus increasing the daily throughput and somewhat stabilizing the steam generation rate.

V. SPECIAL PURPOSE INCINERATORS

The disposal of bulky combustible wastes (tree stumps, overstuffed furniture, brush, etc.) cannot readily be accomplished using a conventional incinerator. Although these materials are often innocuous in a landfill, their low bulk density makes direct disposal wasteful of the landfill volumetric capacity. In larger cities, the hauling costs to remote landfills is often prohibitive.

To respond to this special incineration problem, two incineration system configurations have been developed, namely, the bulky waste incinerator and the pit incinerator.

A. Bulky Waste Incinerator

In the late 1960s, Elmer Kaiser, an engineer particularly active in the incineration field, developed a concept for the disposal of bulky wastes. Working under an EPA grant with the city of Stamford, Connecticut, a successful prototype was built and tested. The Stamford unit discharged its off-gases into the flues of the Stamford refuse incinerator and thence through an electrostatic precipitator.

The design of large units, suitable for the combustion of 1000–2500 kg/h of tree stumps, demolition waste, sofas, chairs, mattresses, etc. (heat of combustion approximately 4000 kcal/kg) can be based on the following [37]):

- Furnace-floor hearth area corresponds to 75 kg/h/m^2 (300,000 kcal/h/m^2).
- Furnace height, 3.75 m (to assure clearance during charging).
- Furnace length, 6–8 m.
- Furnace width, 2.5–5 m.
- Secondary chamber volume, 65% of furnace.
- Forced air supply corresponds to 150% excess, supplied 25% through 7-cm-diameter floor ports at 30 m/s; 75% through 8-cm-diameter side-wall ports. Side-wall-port nozzle velocity selected to give 10 m/s at furnace midplane. Assume 500-m/min inflow velocity through charging door clearances.
- Area of flue ("flame port") between furnace and secondary combustion chamber designed to flow at 500 m/min for 175% excess air at mean firing rate; 650 m/min between secondary chamber and outlet breeching to the air pollution control system.

Such furnaces have been constructed and tested [37–39] and been found satisfactory in this service for an operating cycle of 7-h charging at 15-min intervals, and a burn-down through the next 17 h. Similar designs can be developed for car bodies, drum-cleaning, and other bulky waste incineration, although consideration should be given to the heat release rate relative to the above. Auxiliary burners may be added to the furnace to hold temperatures at or above 550 °C to assure acceptably high burning rates. The heat input requirements may be estimated using the basic heat and material balances, heat loss, and infiltration estimates presented above.

B. Pit-type Incinerators

The disposal of scrap wood, pallets, brush, and other readily combustible, low-ash wastes is often a problem. Open burning is prohibited in many communities and, because of its slowness and inefficiency is both costly (labor costs) and presents safety hazards.

Monroe [40] developed a simple refractory-lined pit incinerator which has shown utility in these applications. His concept uses perforated pipes in the floor of a rectangular in-ground or above-ground pit to supply forced underfire air. Jet nozzles, directed to "bounce" air off a side-wall onto the fire, are mounted along one side of the pit.

Similar incinerator designs have been used for the disposal of heavy pitch and tar residues, but with variable success and, generally, great concern on the part of air-pollution regulatory agencies.

VI. ECONOMICS OF INCINERATION

The costs of incineration have increased rapidly over the period 1965–1978. The reasons underlying the cost increases parallel those which have gripped the overall U.S. economy during this period, such as rising interest rates, construction costs, and labor and energy rates. For refractory incinerators (and particularly for upgraded older units) where operation at high excess-air levels yields large flue-gas flow rates, the requirement for sophisticated air-pollution control systems was also a major factor.

The cost for municipal incinerators varies widely. The estimates shown in Tables 13 and 14 were developed for 1975 in the Richmond, Virginia area. The estimates are based on a throughput of 920 and 1395 ton/d, respectively. The air pollution control system which was assumed here was a four-stage electrostatic precipitator yielding a design effluent loading of 0.03 grains/dscf (dry standard cubic foot) corrected to 50% excess air. This is, very likely, the most stringent incinerator emission code in the United States. Amortization was charged at 7% for all capital assets. Land, site development, by-product distribution costs and indirect costs (engineering, construction expense, working capital, taxes) are not included.

No economics are presented here for the pyrolysis systems, although some data have been made available based on the preliminary results of the full scale plants. Since these systems are proprietary in nature, a large component of the systems' cost is associated with the technology

Table 13

Capital and Operating Cost Estimates for Refractory Incineration Systems in 1975

	Plant I	Plant II
Plant throughput, ton/week	6,425	9,760
Furnace size, ton/d	360	410
Number of furnaces	3	4
Landfill area, hectares	15.4	22
Fill depth, including cover, m	6.1	6.1
Capital cost, $		
Incinerator		
Furnaces	2,650,000	3,830,000
Stack	560,000	810,000
Fans and ducts	760,000	1,100,000
Piping	850,000	1,230,000
Electrical	1,240,000	1,780,000
Conveyors	1,290,000	1,860,000
Air pollution control	3,730,000	5,370,000
Building and cranes	10,550,000	13,860,000
Total incinerator cost	21,630,000	29,840,000
Residue landfill		
Site preparation	40,000	55,000
Leachate collection	622,100	874,800
Equipment	80,000	118,500
Total landfill cost	742,100	1,048,300
Total system cost	22,372,100	30,888,300
Cost/ton/d	20,715	18,835
Operating cost (annual), $		
Labor	932,000	1,130,000
Electricity	498,000	747,000
Water and sewer	134,000	201,000
Fuel	13,000	18,000
Maintenance	1,082,000	1,490,000
Residue disposal	78,000	106,000
Amortization (7%, 20 yr)		
Incinerator	2,042,000	2,817,000
Landfill	97,000	131,000
Gross operating cost	4,876,000	6,640,000
Ferrous metal salvage[a]	(258,000)	(392,000)
Net operating cost	4,618,000	6,248,000
Cost/ton	13.82	12.31

Basis: Incinerator operation 7 d/week, 50 week/yr, 24 h/d operation requires four shifts overall; plant availability, 85 percent; associated residue landfill handling residue comprising 20% of the incoming refuse. Residue density is 1900 kg/m^3.

[a] Ferrous metal estimated at 7.8% of incoming refuse, 90% recovery, sale at $11.00/ton (net).

Table 14

Capital and Operating Cost Estimates for Refractory Incineration Systems in 1975

	Plant III	Plant IV
Plant throughput, ton/week	6,425	9,760
Furnace size, ton/d	360	410
Number of furnaces	3	4
Landfill area, hectares	15.4	22
Fill depth, including clover, m	6.1	6.1
Capital cost, $		
Incinerator		
Refuse receiving and handling	3,980,000	5,730,000
Boiler system	14,870,000	21,440,000
Air pollution control	2,420,000	3,490,000
Ash handling	2,170,000	3,130,000
Utilities and miscellaneous	850,000	1,270,000
Feedwater supply and treatment	530,000	790,000
Standby boilers	650,000	750,000
Total incinerator cost	25,470,000	36,600,000
Residue landfill		
Site preparation	40,000	55,000
Leachate collection	622,100	929,800
Equipment	80,000	118,500
Total landfill cost	742,100	1,048,300
Total system cost	26,212,000	37,648,300
Cost/ton/d	24,270	22,960
Operating cost (annual), $		
Labor	980,000	1,190,000
Electricity	456,000	648,000
Water and sewer	110,000	166,000
Fuel	11,000	16,000
Maintenance	1,274,000	1,830,000
Overhead and insurance	255,000	366,000
Residue disposal	78,000	106,000
Amortization (7%, 20 yr)		
Incinerator	2,404,000	3,455,000
Landfill	97,000	131,000
Gross operating cost	5,665,000	7,575,000
Ferrous metal salvage[a]	(258,000)	(392,000)
Steam credit[b]	(5,569,500)	(8,460,400)
Net operating cost (credit)	(162,500)	(1,277,400)
Cost (credit)/ton	(0.48)	(2.52)

Basis: Incinerator operation, 7 d/week, 50 week/yr, 24 h/d operation requires four shifts overall; plant availability 85%; associated residue landfill handling residue comprising 20% of the incoming refuse. Refuse density is 1900 kg/m^3.

[a] Ferrous metal estimated at 7.8% of incoming refuse, 90% recovery, sale at $11.00/ton (net).

[b] Steam credits of 100% of the generation rate (70% efficiency) of 2.85 ton steam/ton refuse at a price competitive with oil, at $12/barrel fired at 80% efficiency, *viz.* $16.67/ton refuse.

fee paid to the developer. Also, the costs for labor, utilities, and maintenance, as well as the magnitude of the energy credit, are still unknown. Preliminary indications, however, suggest that the net costs for pyrolysis will be slightly higher than for a comparable heat-recovery incinerator.

VII. AN APPROACH TO DESIGN

How then do you design an incinerator? The author wishes the answer were straightforward; the underlying principles uncluttered with contradiction and free of the need to apply both technical and value judgments. This section, however, can only scratch the surface of the challenge of system design; we will attempt generally to structure, if not guide in detail, the design process.

A. Characterize the Waste

Obtain the best practical characterization of the quantity and composition of the waste. Keep in mind future growth and the impact of changes in technology and economics on operational patterns and decision-making.

B. Lay Out the System in Blocks

Too often, incineration facilities are developed in pieces with insufficient attention being given to the mating of interfaces between various elements. Remember the concept of system.

C. Establish Performance Objectives

Review present and prospective regulatory requirements for effluent quality. Evaluate the needs for volume reduction, residue burnout, or detoxification. Apply these to appropriate points in the facility layout.

D. Develop Heat and Material Balances

Using the techniques developed early in this chapter, determine the flows of material and energy in the waste, combustion air, and flue gases. Take into consideration probable materials of construction and establish reasonable limits on temperatures. Explore the impact of variations from the "average" waste feed composition and quantity. In practice, these off-average characteristics will generally better characterize the day-to-day operating conditions.

E. Develop Incinerator Envelope

Using heat release rates per unit area and per unit volume, the overall size of the system can be established. Using burning rate, flame length and shape, kinetic expressions, and other analysis tools (see Ref. 1), establish the basic incinerator envelope. The final shape will depend on judgment, as well as on these calculations. Draw on the literature and the personal experience of others. Interact with other engineers, vendors, operators, and designers of other combustion systems with similar operating goals or physical arrangements. Attempt to find the balance between being overly conservative at high cost and the unfortunate fact that few of the answers are tractable to definitive analysis and computation. Particularly, talk to operators of systems. Too often the designers speak only to one another, and the valuable insights of direct personal experience go unheard and, worse, unasked for.

F. Evaluate Incinerator Dynamics

Apply the jet evaluation methodology, buoyancy calculations, empirical relationships, and conventional furnace draft and pressure drop evaluation tools to grasp, however inadequately, the dynamics of the system.

G. Develop the Designs of Auxiliary Equipment

Determine the sizes and requirements of burners, fans, grates, materials handling systems, pumps, air compressors, air pollution control systems and the many other auxiliary equipments comprising the system. Here, again, the caution is to be generous, protective, rugged. The cardinal rule is to prepare for *when* "it" happens, not to argue about *whether* "it" will happen.

H. Review Heat and Material Balances

This self-explanatory step will help to reinforce the systems perspective by tracking the flows through one after another component part.

I. Build and Operate

In many cases, fortunately, nature is kind—reasonable engineering designs will function, though perhaps not to expectations. At times, plants built using the most detailed analysis and care result in failure. Such is the lot of workers in the complex but fascinating field of combustion and incineration.

APPENDIX

WASTE THERMOCHEMICAL DATA

The chemical composition and heat of combustion of waste materials is a matter of concern to the design engineer, to the operator of an incinerator, and to the engineer involved in incinerator troubleshooting.

Waste composition and heat content affect combustion air requirements, flame temperatures and bed burning processes, and impact on corrosion, air pollution, and other important incinerator operating characteristics. Ideally, refuse composition is determined empirically in a comprehensive and scientifically planned and executed program of waste sampling and analysis. Often, however, an engineer requires compositional data for use in preliminary studies. In such cases, data are needed which might be called "typical" or "average." The data presented below may serve this need. A guideline, however, is offered: There is no "typical" municipal refuse.

For a more comprehensive assembly of data on the properties and generation rates of municipal refuse and other wastes, the reader is directed to Refs. 1 and 2.

A. Refuse Composition

The seasonal and annual average compositions shown in Table A-1 were derived from an analysis of over 30 data sets from municipalities throughout the United States. Using these data as defining a base year (1970), composition forecasts were prepared (Table A-2) based on estimates of future consumption rates of paper, metal cans, and other consumer products. In the forecasts, consideration was given to the mean useful life of the products.

The notation "as-discarded" refers to the composition as if measured at the time the waste component was thrown away and before any exchange of moisture between components takes place. Food wastes, for example, will transfer significant quantities of moisture to paper and textiles. The as-discarded basis is useful in indicating the true relative magnitude of waste generation for the various categories, as the appropriate basis for estimating salvage potential, and as the basis for forecasting refuse generation rates and chemical and physical properties.

B. Solid Waste Properties

The categorical composition is the starting point for the development of parameters of interest to the incinerator designer. Although the manipulation of gross categorical data to establish average chemical composition, heat content, and the like requires assumptions of questionable accuracy, it is a necessary compromise. Typically, between 1 and 3 tons of waste from a waste flow of 200–1000 ton/d are analyzed to produce a categorical composition. Then a still smaller sample is hammermilled, mixed, and a 500-mg sample is taken. Clearly, a calorific value determination on this sample is, at best, a rough reflection of the energy content of the original waste.

1. *Thermochemical Analysis*

Stepping from the categorical analysis to a mean thermochemical analysis provides the basis for stoichiometric calculations and energy balances.

Mixed Municipal Refuse. Chemical data for average municipal refuse components and for the mixed refuse shown in Table A-1 are presented in Tables A-3, A-4, and A-5. These data were developed with the average refuse of 1970 in mind, but may be used to develop the composition of "future" refuse using the category projections in Table A-2.

Specific Waste Components. Data for specific waste components are given in Table A-6. These data may be used when detailed categorical analyses are available or to explore the impact of refuse compositional changes. Once a weight-basis generation rate is established, data on the heating value of waste components are of great interest to the combustion engineer.

Table A-1
Estimated Average Municipal Refuse Composition, 1970,[a] Weight Percent as Discarded

					Annual Average	
Category	Summer	Fall	Winter[b]	Spring	As-Discarded	As-Fired
Paper	31.0	39.9	42.2	36.5	37.4	44.0
Yard wastes	27.1	6.2	0.4	14.4	13.9	9.4
Food wastes	17.7	22.7	24.1	20.8	20.0	17.1
Glass	7.5	9.6	10.2	8.8	9.8	8.8
Metal	7.0	9.1	9.7	8.2	8.4	8.6
Wood	2.6	3.4	3.6	3.1	3.1	3.0
Textiles	1.8	2.5	2.7	2.2	2.2	2.6
Leather and rubber	1.1	1.4	1.5	1.2	1.2	1.5
Plastics	1.1	1.2	1.4	1.1	1.4	1.4
Miscellaneous	3.1	4.0	4.2	3.7	3.4	3.6
Total	100.0	100.0	100.0	100.0	100.0	100.0

[a] From Ref. 5.
[b] For southern states, the refuse composition in winter is similar to that shown here for fall.

Table A-2
Projected Average Generated Refuse Composition, 1970–2000, Weight Percent as Discarded[a]

Category	1970	1975	1980	1990	2000
Paper	37.4	39.2	40.1	43.4	48.0
Yard wastes	13.9	13.3	12.9	12.3	11.9
Food wastes	20.0	17.8	16.1	14.0	12.1
Glass	9.0	9.9	10.2	9.5	8.1
Metal	8.4	8.6	8.9	8.6	7.1
Wood	3.1	2.7	2.4	2.0	1.6
Textiles	2.2	2.3	2.3	2.7	3.1
Leather and rubber	1.2	1.2	1.2	1.2	1.3
Plastics	1.4	2.1	3.0	3.9	4.7
Miscellaneous	3.4	3.0	2.7	2.4	2.1
Total	100.0	100.0	100.0	100.0	100.0

[a] From Ref. 5.

Table A-3

Estimated Final Analysis of Refuse Categories, Percent, Dry Basis[a]

Category	C	H	O	N	Ash	S	Fe	Al	Cu	Zn	Pb	Sn	P[b]	Cl	Se	Fixed carbon
Metal	4.5	0.6	4.3	0.05	90.5	0.01	77.3	20.1	2.0	—	0.02	0.6	0.03	—	—	0.5
Paper	45.4	6.1	42.1	0.3	6.0	0.12	—	—	—	—	—	—	—	—	trace	11.3
Plastics	59.8	8.3	19.0	1.0	11.6	0.3	—	—	—	—	—	—	0.01	6.0	—	5.1
Leather and rubber							—	—	—	2.0	—	—	—	—	—	6.4
Textiles	46.2	6.4	41.8	2.2	3.2	0.2	—	—	—	—	—	—	0.03	—	—	3.9
Wood	48.3	6.0	42.4	0.3	2.9	0.11	—	—	—	—	—	—	0.05	—	—	14.1
Food wastes	41.7	5.8	27.6	2.8	21.9	0.25	—	—	—	—	—	—	0.24	—	—	5.3
Yard wastes	49.2	6.5	36.1	2.9	5.0	0.35	—	—	—	—	—	—	0.04	—	—	19.3
Glass	0.52	0.07	0.36	0.03	99.02	—	—	—	—	—	—	—	—	—	—	0.4
Miscellaneous	13.0[a]	2.0[a]	12.0[a]	3.0[a]	70.0	—	—	—	—	—	—	—	—	—	—	7.5

[a] From Ref. 5.
[b] Estimated (varies widely).
[c] Excludes phosphorus in $CaPO_4$.

Table A-4
Ultimate Analysis of Annual Average 1970 Mixed Municipal Refuse[a]

Category	Wt%		% Moisture	Composition of average refuse, kg/100 kg dry solids						
	As fired	As discarded	As discarded	Ash	C	H_2	O_2	S	N_2	TotalWt, kg
Metal	8.7	8.2	2.0	10.13	0.50	0.067	0.481	0.0011	0.0056	11.19
Paper	44.2	35.6	8.0	2.74	20.70	2.781	19.193	0.0547	0.1368	45.59
Plastics	1.2	1.1	2.0	0.17	0.90	0.125	0.285	0.0045	0.0150	1.50
Leather and rubber	1.7	1.5	2.0	0.24	1.23	0.170	0.390	0.0062	0.0205	2.05
Textiles	2.3	1.9	10.0	0.08	1.10	0.152	0.995	0.0048	0.0523	2.38
Wood	2.5	2.5	15.0	0.09	1.43	0.178	1.260	0.0033	0.0089	2.96
Food waste	16.6	23.7	70.0	2.17	4.13	0.574	2.730	0.0248	0.2772	9.90
Yard waste	12.6	15.5	50.0	0.54	5.31	0.701	3.890	0.0378	0.3129	10.79
Glass	8.5	8.3	2.0	11.21	0.06	0.008	0.041	—	0.0034	11.32
Miscellaneous	1.7	1.7	2.0	1.62	0.30	0.046	0.278	—	0.0696	2.32
	100.0	100.0		28.99	35.66	4.802	29.543	0.1372	0.9022	100.00

[a] From Ref. 5.

Table A-5
Average Refuse Summary, As-Fired Basis: 100 kg Average Refuse

Component	Wt %	Moles
Moisture, H_2O	28.16	1.564
Carbon, C	25.62	2.135
Hydrogen, H_2-bound	2.65	1.326
Oxygen, O-bound	21.21	1.326
Hydrogen, H_2	0.80	0.399
Sulfur, S	0.10	0.003
Nitrogen, N_2	0.64	0.023
Ash	20.82	—
Total	100.00	

Note: Higher heating value (water condensed): 2472 kcal/kg.
Lower heating value (water as vapor): 2167 kcal/kg.

Table A-6
Ultimate Analyses and Heating Value of Waste Components[a]

Waste component	Weight %, dry						HHV, kcal/kg	
	C	H	O	N	S	Non-combustible	As received	Dry
Paper and paper products								
Paper, mixed	43.41	5.82	44.32	0.25	0.20	6.00	3778	4207
Newsprint	49.14	6.10	43.03	0.05	0.16	1.52	4430	4711
Brown paper	44.90	6.08	47.34	0.00	0.11	1.07	4031	4281
Trade magazines	32.91	4.95	38.55	0.07	0.09	23.43	2919	3044
Corrugated boxes	43.73	5.70	44.93	0.09	0.21	5.34	3913	4127
Plastic-coated paper	45.30	6.17	45.50	0.18	0.08	2.77	4078	4279
Paper food cartons	44.74	6.10	41.92	0.15	0.16	6.93	4032	4294
Food and food waste								
Vegetable food waste	49.06	6.62	37.55	1.68	0.20	4.89	997	4594
Fried fats	73.14	11.54	14.82	0.43	0.07	0.00	9148	9148
Mixed garbage I	44.99	6.43	28.76	3.30	0.52	16.00	1317	4713
Mixed garbage II	41.72	5.75	27.62	2.97	0.25	21.87	—	4026

Trees, wood, brush, plants								
Green logs	50.12	6.40	42.26	0.14	0.08	1.00	1168	2336
Demolition softwood	51.0	6.2	41.8	0.1	<0.1	0.8	4056	4398
Waste hardwood	49.4	6.1	43.7	0.1	<0.1	0.6	3572	4056
Lawn grass	46.18	5.96	36.43	4.46	0.42	6.55	1143	4618
Ripe Leaves	52.15	6.11	30.34	6.99	0.16	4.25	4436	4927
Wood and bark	50.46	5.97	42.37	0.15	0.05	1.00	3833	4785
Brush	42.52	5.90	41.20	2.00	0.05	8.33	2636	4389
Grass, dirt, leaves	36.20	4.75	26.61	2.10	0.26	30.08	—	3491
Domestic wastes								
Upholstery	47.1	6.1	43.6	0.3	0.1	2.8	3867	4155
Tires	79.1	6.8	5.9	0.1	1.5	6.6	7667	7726
Leather	60.00	8.00	11.50	10.00	0.40	10.10	4422	4917
Leather shoes	42.01	5.32	22.83	5.98	1.00	22.86	4024	4348
Rubber	77.65	10.35	—	—	2.00	10.00	6222	6294
Mixed plastics	66.00	7.20	22.60	—	—	10.20	7833	7982
Polyethylene	84.54	14.18	0.00	0.06	0.03	1.19	10,932	10,961
Polystyrene	87.10	8.45	3.96	0.21	0.02	0.45	9122	9139
Polyvinyl chloride	45.14	5.61	1.56	0.08	0.14	2.06[b]	5419	5431
Linoleum	48.06	5.34	18.70	0.10	0.40	27.40	4528	4617
Rags	55.00	6.60	31.20	4.12	0.13	2.45	3833	4251
Municipal wastes								
Street sweepings	34.70	4.76	35.20	0.14	0.20	25.00	2667	3333

[a] From Ref. 2.

[b] Remaining 45.41% is chlorine.

REFERENCES

1. W. R. Niessen, *Combustion and Incineration Processes*, Dekker, New York, 1978.
2. D. G. Wilson, *Handbook of Solid Waste Management*, Van Nostrand-Reinhold, New York, 1977.
3. H. C. Hottel, G. C. Williams, N. M. Nerheim, and G. Schneider, Combustion of Carbon Monoxide and Propane in *10th International Symposium on Combustion*, pp. 111–121. Combustion Institute, Pittsburgh, Pa. 1965.
4. M. A. Field, D. W. Gill, B. B. Morgan, and P. G. W. Hawksley, *Combustion of Pulverized Coal*, The British Coal Utilization Research Assn., Leatherhead, Surrey, England, 1967.
5. W. R. Niessen, S. H. Chansky, E. L. Field, A. N. Dimetriou, C. P. La Mantia, R. E. Zinn, T. J. Lamb, and A. S. Sarofim, *Systems Study of Air Pollution from Municipal Incineration*, NAPCA, U.S. DHEW, Contract CPA-22-69-23, March, 1970.
6. E. R. Kaiser and S. B. Friedman, paper presented at 60th Annual Meeting, AIChE, November, 1968.
7. D. A. Hoffman and R. A. Fritz, *Env. Sci. Technol.* **2** (11), 1023 (1968).
8. R. S. Burto, III and R. C. Bailie, *Combustion*, 13–18 (February 1974).
9. S. B. Alpert and F. A. Ferguson, et al., *Pyrolysis of Solid Waste: A Technical and Economic Assessment*, NTIS Pb. 218–231, September, 1972.
10. M. A. Kanury, *Combustion and Flame* **18**, 75–83 (1972).
11. U. K. Shivadev and H. W. Emmons, *Combustion and Flame* **22**, 223–236 (1974).
12. P. Nicholls, *Underfeed Combustion, Effect of Preheat, and Distribution of Ash in Fuel Beds*, U.S. Bureau of Mines, Bulletin 378 (1934).
13. R. H. Stevens et al., *Incinerator Overfire Mixing Study—Demonstration of Overfire Jet Mixing*, OAP, U.S. Contract 68020204 (1974).
14. E. R. Kaiser, personal communication to W. R. Niessen, C. M. Mohr, and A. F. Sarofim (1970).
15. H. Eherhardt and W. Mayer, "Experiences with Refuse Incinerators in Europe," in *Proceedings of the 1968 National Incineration Conference*, pp. 142–153, ASME, New York, 1968.
16. F. Nowak, *Brennst.-Wärme-Kraft* **19** (2), 71–76 (1967).
17. R. L. Stenburg, R. R. Horsley, R. A. Herrick, and A. H. Rose, Jr., *J. Air Poll. Cont. Assn.* **10**, 114–120 (1966).
18. A. B. Walker and F. W. Schmitz, "Characteristics of Furnace Emissions from Large, Mechanically Staked Municipal Incinerators," in *Proceedings of the 1964 National Incineration Conference*, pp. 64–73, ASME, New York, 1964.
19. F. R. Rehm, *J. Air Poll. Cont. Assn.* **6** (4), 199–204 (1957).
20. H. R. Johnstone, *Univ. Illinois Eng. Exper. Sta. Bull.* **228**, 221 (1931).
21. A. H. Rose, Jr. and H. R. Crabaugh, Research Findings in Standards of Incinerator Design, in *Air Pollution*, Reinhold, New York, 1955.
22. Y. V. Ivanov, *Effective Combustion of Overfire Fuel Gases in Furnaces*, Estonian State Publishing House, Tallin, 1959.
23. G. N. Abramovich, *The Theory of Turbulent Jets*, M.I.T. Press, Cambridge, Mass., 1963.

24. "Layout and Application of Overfire Jets for Smoke Control in Coal Fired Furnaces," Section F-3, Fuel Engineering Data, National Coal Association, Washington, DC, December 1962.
25. "Municipal Refuse Disposal," p. 179, Institute for Solid Wastes, American Public Works Association, 1970.
26. H. H. Krause, D. A. Vaughan, and W. K. Boyd, "Corrosion and Deposits from Combustion of Solid Waste. Part III. Effects of Sulfur on Boiler Tube Metals," in *Proceedings of the ASME Winter Annual Meeting*, 1974.
27. H. H. Krause, D. A. Vaughan, and W. K. Boyd, "Corrosion and Deposits from Combustion of Solid Waste. Part IV. Combined Firing of Refuse and Coal," in *Proceedings of the ASME Winter Annual Meeting*, Houston, Texas, 1975.
28. H. H. Krause, D. A. Vaughan, and P. D. Miller, *J. of Eng. Power, Trans. ASME, A.* **95**, 45–52 (1973).
29. H. H. Krause, D. A. Vaughan, and P. D. Miller, *J. Eng. Power, Trans. ASME, A.* **96**, 216–222 (1974).
30. P. D. Miller and H. H. Krause, "Corrosion of Carbon and Stainless Steels in Flue Gases from Municipal Incinerators," in *Proceedings of the 1972 ASME National Incineration Conference*, New York, 1972.
31. P. D. Miller and H. H. Krause, *Corrosion* **27**, 31–45 (1971).
32. F. Nowak, "Considerations in the Construction of Large Refuse Incinerators," in *Proceedings of the 1970 ASME National Incineration Conference*, pp. 86–92, New York (1970).
33. D. L. Klumb, "Union Electric Facilities for Burning Municipal Refuse at the Meramec Power Plant," paper presented at the Union Electric Co. Solid Waste Seminar, St. Louis, Mo., 26 October 1972.
34. J. E. Williamson, R. J. MacKnight, and R. L. Chass, *Multiple-Chamber Incinerator Design Standards*, Los Angeles APCD, October, 1960.
35. J. A. Danielson, ed. *Air Pollution Engineering Manual*, 2nd Edition, pp. 434ff., US EPA, 1973.
36. R. C. Corey, ed. *Principles and Practices of Incineration*, p. 83. Wiley-Interscience, New York, 1969.
37. E. R. Kaiser, "Incineration of Bulky Refuse. III," in *Proceedings of the 1966 ASME Incinerator Conference*, ASME, New York, 1966.
38. E. R. Kaiser, "Incineration of Bulky Refuse," in *Proceedings of the 1966 ASME Incinerator Conference*, ASME, New York, 1966.
39. E. R. Kaiser, "The Incineration of Bulky Refuse. II," in *Proceedings of the 1968 ASME Incinerator Conference*, p. 129, ASME, New York, 1968.
40. E. S. Monroe, "New Developments in Industrial Incineration," in *Proceedings of the 1966 ASME Incinerator Conference*, pp. 226–230, ASME, New York, 1966.

4

Sanitary Landfill

P. Michael Terlecky, Jr.
Frontier Technical Associates, Inc., Buffalo, New York

I. INTRODUCTION

Planned disposal of refuse on land has been practiced for centuries, but the concept of sanitary landfilling has been used for only a short time. The continued increase in the amounts as well as variety of solid waste produced in the United States has necessitated incorporation of controlled, planned, and environmentally compatible methods of refuse disposal. The elimination of open dumping and the practice of sanitary landfilling has brought a significant reduction in air and water pollution, unsightliness, health hazards, and infestation formerly associated with disposal sites.

Increases in the standard of living and population as well as the development of a "disposable"-oriented generation has resulted in the annual production of over 227 million metric tons (250 million short tons) of solid wastes from residential, commercial, and institutional sources [1]. It has become apparent that former methods employed for disposal of solid wastes such as uncontrolled incineration, open dumping, and collection and disposal practices are inadequate today. It has been estimated that more than 90% of our nation's solid waste has been directly disposed of on land [1], and because of this a great many adverse effects have resulted that have long range consequences.

Communities have discovered that not only can sanitary landfills be administered in an environmentally compatible and economic manner, but also that otherwise marginal land areas have become more valuable for other uses such as parks, recreation, and industrial sites.

The objectives of this chapter are to outline what sanitary landfills are (and are not!), the methods of choosing potential sites, engineering and construction methods and options, environmental considerations, and the outlook for the future with respect to sanitary landfill technology. Only by sound, careful, and advance engineering planning will the practice of sanitary landfilling and solid waste disposal improve to a point that it can be an asset rather than a liability to municipalities.

A. Definition

A sanitary landfill is a land disposal site that employs the principle of spreading solid wastes in thin layers, compacting the material to the smallest possible volume, and applying cover material at the conclusion of each operating day [1–3]. The method of sanitary landfilling is often confused with open dumps that have smoke, odor, insect, and rodent problems, but this is a gross misconception. A true sanitary landfill minimizes and eliminates these problems because of sound engineering and construction practices, detailed planning, and efficient operation.

Open dumps are areas where uncontrolled disposal of garbage, trash, and other solid wastes takes place usually without detailed planning and without regard to the consequences of its presence such as smoke, odor, rodents, and other pests, and health and water pollution hazards.

II. SITE SELECTION

Because a sanitary landfill is an engineering undertaking requiring detailed planning and proper site selection, perhaps the most important work associated with this solid waste disposal method takes place before the project is in actual construction or operation. Besides selection of a competent design engineer with knowledge and experience in site selection, design, and operation of a sanitary landfill, it is also important to consider a public information program, a survey of the types and amounts of wastes to be disposed of, possible zoning restrictions, and potential end use(s) of the completed site prior to embarking

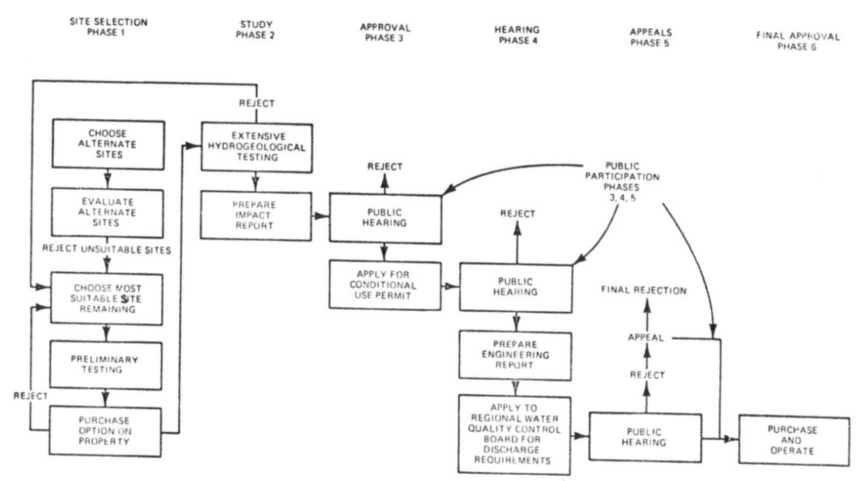

FIG. 1 The steps usually required in the development of a sanitary landfill from site selection to final approval. From Ref. 4.

on an actual program. Recently Hekimian summarized the major steps required for acquisition of a sanitary landfill site [4]. Figure 1 outlines the necessary stages from site selection to final approval of a proposed project. At several important steps, public participation is essential to ensure that the operation of the site will be conducted in an environmentally sound manner and will be compatible with its surroundings. The following discussion emphasizes the site selection criteria and legal requirements that dictate the location of a sanitary landfill operation.

A. Site Characteristics

Besides location of sanitary landfill disposal sites with consideration of the transportation costs, many special site constraints affect the success or failure of a particular operation. These factors include soils, geology, location, topography, and climate.

1. *Soils*

According to Ref. 5 desirable soil properties for solid waste disposal by sanitary landfill include those which "(1) prevent leachate from polluting the ground or surface water; (2) hold and absorb any undesirable gases generated by decomposition of wastes; (3) prevent insect or rodent infestation; (4) provide needed aeration and moisture to properly decompose organic wastes; (5) lack stones or rock which would hinder operations; and (6) have the qualities which would not hinder the movement of hauling vehicles."

Consideration of these properties and physical characteristics of sites such as depth to seasonal water table, soil drainage, flooding probability, permeability, and slope has resulted in the development by the Soil Conservation Service (SCS) of a series of limitation classes. These classes generally indicate the suitability or degree of difficulty of use of a soil for a sanitary landfill operation.

SCS soil limitation ratings for area- and trench-type sanitary landfills are listed in Tables 1 and 2.

The acceptance or rejection of potential sanitary landfill sites, therefore, must strongly depend upon location in areas with a reasonable chance of finding satisfactory soil conditions. One source [5] recommends that areas of soil associations with a 75% or greater chance of severe soil limitations be excluded categorically.

It has been recommended that any potential landfill sites be located on a map of the soil associations of the areas usually published by the Soil Conservation Service. Thus, it is possible to limit potential landfill sites to areas that meet recommended characteristics.

Table 1

Soil Limitation Ratings for Area-Type Sanitary Landfills[a]

	Degree of Soil Limitation		
Item affecting use	Slight	Moderate	Severe
Depth to seasonal[b] water table	More than 60 in.	40–60 in.	Less than 60 in.
Soil drainage[b] class	Excessively drained, somewhat excessively drained, well drained, and moderately well drained	Somewhat poorly drained	Poorly drained and very poorly drained
Flooding	None	Rare	Occasional or frequent
Permeability[c]	Not class determining if less than 2 in./h		More than 2 in./h
Slope	0–8%	8–15%	More than 15%

[a] From Reference 6.

[b] Reflects influence of wetness on operation of equipment.

[c] Reflects ability of the soil to retard movement of leachate from landfills; may not reflect a limitation in arid and semiarid areas.

Table 2
Soil Limitation Ratings for Trench-Type Sanitary Landfills[a,b]

Item affecting use	Degree of soil limitation		
	Slight[b]	Moderate[c]	Severe
Depth to seasonal high water table	Not class determining if more than 72 in.		Less than 72 in.
Soil drainage class	Excessively drained, somewhat excessively drained, well drained, and some[d] moderately well drained	Somewhat poorly drained and some[c] moderately well drained	Poorly drained very poorly drained
Flooding	None	Rare	Occasional or frequent
Permeability[e]	Less than 2.0 in./h	Less than 2.0 in./h	More than 2.0 in./h
Slope	0–15%	15–25%	More than 25%
Soil texture[f] dominant to a depth of 60 in.	Sandy loam, loam, silt loam, sandy clay loam	Silty clay loam,[g] clay loam, sandy clay, loamy sand	Silty clay, clay, muck, peat, gravel, sand
Depth to bedrock			
Hard	More than 72 in.	More than 72 in.	Less than 72 in.
Rippable	More than 60 in.	Less than 60 in.	Less than 60 in.

[a] From Reference 6.

[b] Based on soil depth (5–6 ft), commonly investigated in making soil surveys.

[c] If probability is high that the soil material to a depth of 10–15 ft will not alter a rating of slight or moderate, indicate this by an appropriate footnote, such as "probably slight to a depth of 12 ft," or "probably moderate to a depth of 12 ft".

[d] Soil drainage classes do not correlate exactly with depth to seasonal water table. The overlap of moderately well-drained soils into two limitation classes allows some of the wetter moderately well-drained soils to be given a limitation rating of moderate.

[e] Reflects ability of soil to retard movement of leachate from the landfills; may not reflect a limitation in arid and semiarid areas.

[f] Reflects ease of digging and moving (workability) and trafficability in the immediate area of the trench where there may not be surfaced roads.

[g] Soils high in expansive clays may need to be given a limitation rating of severe.

Cover material used at a landfill is intended to perform functions such as insect, rodent, and bird control, prevention of litter in the local area, esthetic appearance, and control of leachate production and gas movement. Table 3 presents information on the suitability of general soil types as cover material and their different functions.

Table 3
Suitability of General Soil Types as Cover Material[a]

Function	Clean gravel	Clayey-silty gravel	Clean sand	Clayey-silty sand	Silt	Clay
Prevent rodents from burrowing or tunneling	G	F–G	G	P	P	P
Keep flies from emerging	P	F	P	G	G	E[b]
Minimize moisture entering fill	P	F–G	P	G–E	G–E	E[b]
Minimize landfill gas venting through cover	P	F–G	P	G–E	G–E	E[b]
Provide pleasing appearance and control blowing paper	E	E	E	E	E	E
Grow vegetation	P	G	P–F	E	G–E	F–G
Permeability for venting decomposition gas[c]	E	P	G	P	P	P

[a] From Brunner and Keller [6].
Key: E, excellent; G, good; F, fair; P, poor.
[b] Except when cracks extend through the entire cover.
[c] Only if well drained.

2. *Geology*

Since solid waste after disposal in a sanitary landfill becomes a part of the geological environment, normal geological processes such as weathering, erosion, and percolation of groundwater through the site must be considered. Health and other hazards associated with surface- and groundwater pollution are too great to overlook this important aspect of site selection. Site selection should include a detailed geological investigation of the site subsurface, possibly in conjunction with a survey of surface soil materials for use as cover. Location of the groundwater table and detailed data on its seasonal fluctuations are necessary to determine historical high-water levels and general movement of the groundwater [1, 7].

A detailed topographic survey of the site, subsurface geological structure, depth to bedrock, and potential aquifers should be performed prior to final approval and selection.

Sites located near rivers, streams, and lakes present special problems. Generally, a landfill should not be located in a flood plain because of potential flooding conditions and water pollution hazard, and because the site may be unusable during and after floods or periods of heavy rains and snow melts [1, 7].

The most important geological requirements have been summarized in Refs. 5 and 7 as follows:

1. The base of a proposed landfill should be in relatively fine-grained material and more than 6–9 m above the shallowest aquifer.
2. The base of a proposed landfill should be above the highest seasonal level of the water table.
3. A proposed site should not be subject to flooding.
4. Adequate medium-texture cover material must be available near a proposed site.

These requirements apply mainly to the area method of sanitary landfilling; items 1 and 4 must be compromised in terms of soil textures for the trench or ramp methods of sanitary landfilling to be used [5, 8].

A recent survey of sanitary landfill practices in Canada [9] indicated that depth to groundwater was less than 6.1 m (20 ft) in 44% of the towns and greater than 6.1 m (20 ft) in 19% of the total (38% of towns not responding). Some towns reported depth to groundwater as less than 3.05 m (10 ft). The possible use of liners or liner materials should be considered under such circumstances.

3. *Location and Accessibility*

The U.S. Environmental Protection Agency (USEPA) has recommended the following site location requirements [10]: (a) easily accessible in any type of weather to all vehicles expected to use it; (b) safeguard against uncontrolled gas movement originating from the disposed solid waste; (c) have an adequate quantity of earth cover material that is easily workable, compactible, free of large objects (usually rocks) that would hinder compaction, and does not contain organic matter of sufficient quantity and distribution conducive to the harborage and breeding of vectors (disease carrying insects); and (d) conform with land use planning of the area.

If a site is remotely situated from the area(s) that it services, the haul costs may be high and total costs for disposal will consequently increase. The normal economical hauling distance for collection vehicles is 16–24 km [11], varying with factors such as volume of refuse to be hauled, type of refuse, road conditions, transport routes, and use of transfer stations. The sanitary landfill sites should be selected so as to minimize the costs of transportation. In some cases, single-disposal sites will be optimum for a county, but at others, multiple sites may be necessary. The evaluation of a regional system of solid waste disposal requires that a least-cost system should be developed for a region (either county or 3–4 county areas) as well as for each of the individual counties. Examples of this approach are given in Ref. 5 for several counties in Indiana that illustrate the utility of this approach to the spatial distribution of sites.

B. Land Volume (Area) Required

In order to accomplish proper design and maximum utility for a sanitary landfill, the amount of refuse that will be produced by the communities served by the disposal area must be estimated in order to determine the land area or volume of space required. The area or volume required is primarily dependent upon the character and quantity of the solid wastes generated in the service area, the compaction of the wastes, the depth of the landfill, the thickness of the daily cover used, and the service life of the landfill. It is generally recommended that the land area and volume available should be capable of supporting the area serviced for a 20–40 year period [3]. Normally, economics favor large-capacity sites that can be used for at least 15 years. Smaller sites are often easier to obtain and may create less public resistance, however.

Detailed data on the quantity and types of commercial, industrial, and residential solid wastes to be landfilled should be obtained by means of a survey of the service area and existing methods of disposal. Municipal refuse generation usually averages from approximately 1.6 kg (3.5 lb) to 2.5 kg (5.5 lb) per person per day (Fig. 2). Thus, a site would have to provide 12,328–24,655 m^3 (16,125–32,249 yd^3) per year per 10,000 population.

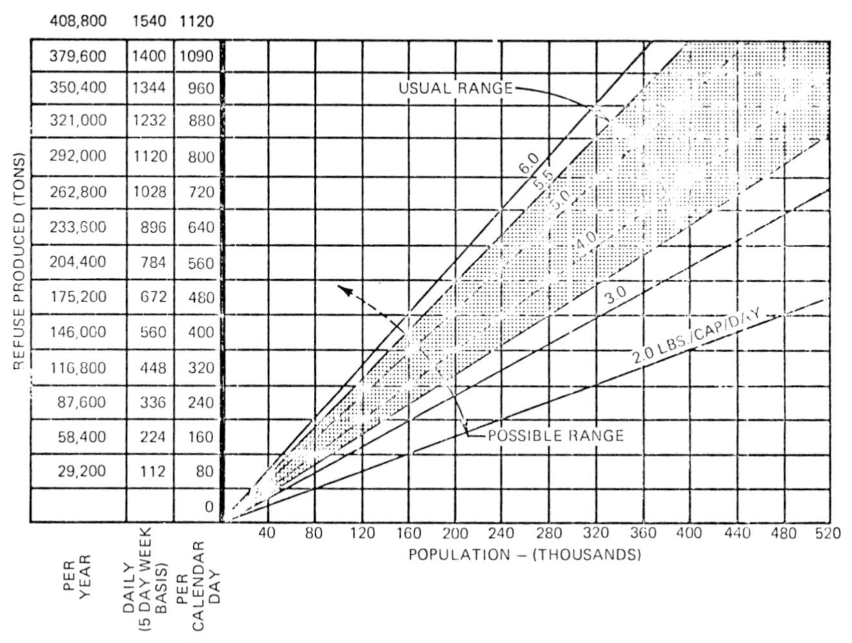

FIG. 2 Refuse produced: varying per capita figures. From Ref. 11.

The space needed for refuse disposal is a function of population served, per capita refuse contribution, density of the refuse in place, total amount of earth cover used, and time in use. This may be expressed as [3]:

$$Q = \frac{peck}{d}$$

where Q is the space or volume required in m^3 (yd^3) per year, p is the population served, e is the ratio of earth to compacted fill. (Use 1.25 if one part earth is used to four parts fill; use 1.20 if one part earth is used to five parts fill; use 1.0 if no earth is used.) c is the kilograms (pounds) collected per capita per day, k is 365 days per year, and d is the density of compacted fill. A density of 475 to 593 kg/m³ (800 to 1000 lb/yd³) is readily achieved with proper operation; 356 kg/m³ (600 lb/yd³) is poor; 712 kg/m³ (1200 lb/yd³) or more is very good.

An example problem will serve to illustrate the above equation for space requirements.

> What is the yearly volume required to accommodate a community with a population of 10,000 people, where a ratio of earth to fill is one to four (1.25); solid waste contributed is 2 kilograms per day and the density of compacted fill is 593 kg/m³ (~1000 lb/yd³)?

$$Q = \frac{peck}{d}$$

$$Q = \frac{(10{,}000 \text{ pop.})(1.25)(2 \text{ kg/capita/day})(365 \text{ days})}{(593 \text{ kg/m}^3)}$$

$$Q = \cong 15{,}388 \text{ m}^3 \ (20{,}128 \text{ yd}^3)$$

It has been estimated [3] that the average amount of refuse increases at the rate of 4% per year. This factor plus the solid waste contributions from industrial and agricultural sources must be included in estimates for future land and volume requirements. The factors in the above calculation will vary for each community; therefore, each project should be based upon individual estimates.

C. Legal Requirements

Most state and local sanitary and zoning codes or laws usually require that a new disposal or landfill area have an operating permit prior to establishment of the operation. In the application for this permit to the proper agency having jurisdiction, the site characteristics and method of operation must be specifically approved.

A survey conducted by the American Public Works Association [1] indicated that a significant number of cities have restrictions in their zoning ordinances concerning the acquisition of disposal sites. As a result, all city, town, and county zoning ordinances should be reviewed and appropriate agencies coordinated to avoid acquisition of land that cannot be used for the intended purpose. With the current emphasis on municipal, county, and state formulation of master plans for future development, these documents should be reviewed to circumvent any potential conflicts in designated and potential usage.

In 1974, the Office of Solid Waste Management Programs, United States Environmental Protection Agency (USEPA-OSWMP) published guidelines in the Federal Register for the land disposal of solid wastes [12]. The requirements contained in the guidelines delineate minimum levels of performance required of any solid-waste land-disposal operation. The recommended procedures are based on the practice of sanitary landfilling municipal solid waste that is normally residential and commercial solid waste generated within a community. The authority for the recommended guidelines is contained in Section 209(a) of the Solid Waste Disposal Act of 1965 (P.L. 89-272) as amended by the Resource Recovery Act of 1970 (P.L. 91-512).

Recommended procedures with respect to design, operations, and requirements are summarized below as excerpted from Ref. [12]:

Solid Wastes Accepted. Only wastes for which the facility has been specifically designed shall be accepted, in general. Plans should specify the procedures to be employed for special handling. Routine sanitary landfill techniques of spreading, compacting and placing cover material at the end of each operating day is required. Water treatment plant sludges containing no free moisture should be placed on the working face along with municipal solid wastes and covered with soil or municipal solid wastes. Incinerator and air pollution control procedures containing no free moisture should be incorporated into the working face and covered at such intervals as necessary to prevent them from becoming airborne.

Solid Wastes Excluded. The waste generator, responsible agency and the disposal site owner/operator shall jointly determine wastes to be excluded and identify them in the plans. The criteria used in considering whether a waste is unacceptable shall include the hydrogeology of the site, the chemical and biological characteristics of the waste, alternative methods available, environmental and health effects, and the safety of personnel. Disposal of pesticides and pesticide containers shall be consistent with the Federal Environmental Pesticides Control Act of 1972 (P.L. 92-516).

Site Selection. Site selection shall be consistent with public health and welfare, air and water quality standards, and adaptable to the appropriate land use plan.

(a) The hydrogeology of the site should be evaluated in order to protect or minimize the impact on groundwater resources. Precipitation, evapotranspiration, and other climatological conditions should be considered in site selection and design.
(b) Characteristics of on site soil should be evaluated with respect to their effect on site operations, such as vehicle maneuverability and use as cover material.
(c) Environmental factors, climatological conditions and socioeconomic factors should be given full consideration as selection criteria.
(d) The site should be accessible to vehicles by all-weather roads leading from the public road system.
(e) The site should not be located in an area where the attraction of birds would pose a hazard to low flying aircraft.

Design
(1) The types and quantities of all solid wastes expected should be determined by a survey and analysis.
(2) Site development plans should be prepared or approved by a professional engineer and should include:

(a) Initial and final topography at a contour interval of 1.5 m or less.
(b) Land use and zoning within one-quarter mile of the site including locations of all residences, buildings, wells, watercourses, arroyos, rock outcrops, roads, and soil or rock borings. Airports in the vicinity of the site should be identified to aid in assessing the bird strike hazard.
(c) Location of all utilities within 152 m (500 ft) of the site.
(d) Employee convenience and equipment maintenance facilities.
(e) Narrative descriptions, indicating site development and operation procedures.

(3) Plans should describe the projected use of the completed land disposal site. In addition to maintenance programs and provisions, where necessary, for monitoring and controlling decomposition gases and leachate, the plans should address the following ultimate use criteria:

(a) *Cultivated area.* If cultivation is to be practiced, a sufficient depth of cover material should be present to allow cultivation and support vegetation.
(b) *Structures.* Major structure construction is not recommended on a completed land disposal site.

Water Quality. The location, design, construction, and operation of the land disposal site shall conform to the most stringent of applicable water quality standards established under the provisions of the Federal Water Pollution Control Act, as amended. In the absence of such standards, the site shall be located, designed, constructed, and operated to provide protection to ground and surface waters used as drinking water supplies. Plans should include: (1) Current and projected use of water resources in the zone of influence of the land disposal site; (2) Groundwater elevation and movement and proposed separation between the lowest point of the lowest cell and the predicted maximum water table elevation; (3) Potential interrelationships of the land disposal site, local aquifers, and surface waters; (4) Background quality of water resources in the potential zone of influence; (5) Proposed location of observation wells, sampling stations, and testing program planned, when appropriate; (6) Description of soil and other geologic materials to a depth adequate to allow evaluation of the water quality protection provided by the soil and other geologic material; (7) Provision for surface water runoff control to minimize infiltration and erosion of cover material; (8) Potential of leachate generation and proposed control; (9) If a land disposal site is located in a flood plain, it should be protected against at least the 50 year design flood by impervious dikes and other appropriate means to prevent the floodwaters from contacting municipal solid waste.

Surface water courses and runoff should be diverted from the land disposal site (especially from the working face) by trenches, conduits and proper grading. Construction of the site should promote rapid runoff without excessive erosion. Regrading should be done to avoid ponding of precipitation and to maintain cover material integrity.

Air Quality. The design, construction, and operation of the land disposal site shall conform to applicable ambient air quality standards and source control regulations established under authority of the Clean Air Act, as amended, or State or local standards effective under that Act, if the latter are more stringent. Plans should include an effective dust control program. Open burning should be prohibited.

Gas Control. Decomposition gases generated within the land disposal site shall be controlled on site, as necessary, to avoid posing a hazard to occupants of adjacent property. Plans should assess the need for gas control and indicate the location and design of any vents, barriers, or other control measures to be provided.

Vectors. Conditions should be maintained that are unfavorable for the harboring, feeding, and breeding of vectors.

Aesthetics. The land disposal site shall be designed and operated at all times in an aesthetically acceptable manner. Plans should include an effective litter control program. Portable litter fences or other devices should be used in the immediate vicinity of the working face and at other appropriate locations to control blowing litter. Wastes easily moved by wind should be covered, as necessary to prevent their becoming airborne. On-site vegetation should be cleared only as necessary. Natural windbreaks should be maintained where they will improve the appearance of the land disposal site. Buffer strips should be planted and/or berms constructed to screen the working force from nearby residences or major roadways.

Cover Material. Cover material should be applied as necessary to minimize fire hazards, infiltration of precipitation, odors, and blowing litter; control gas venting and vectors; discourage scavenging; and provide a pleasing appearance. Plans should specify: (a) Cover material sources and soil classifications; (b) Surface grades and side slopes necessary for maximum runoff, without excessive erosion, to minimize infiltration; (c) Procedures to promote vegetative growth to combat erosion and improve appearance of idle and completed areas; (d) Procedures to maintain cover material integrity, regrading and recovering.

Recommended procedures include application of daily cover of at least 15.2 centimeters (6 inches) after compaction regardless of weather. Sources of cover should be accessible on all operating days. Intermediate cover of at least 30.5 centimeters (1 foot) should be applied on areas where additional cells are not to be constructed for extended periods of time. Final cover should be applied on each area as it is completed or if the area is to remain idle over one year. The thickness of the compacted final cover should not be less than 61 centimeters (2 feet).

Compaction. Municipal solid waste and cover material should be compacted to the smallest practicable volume. The solid waste should be spread in layers no more than 61 centimeters (2 feet) while confining it to the smallest practicable area. Several such compacted layers will form a cell. The cover material should be placed and spread over the cell at least by the end of each day's operation.

Safety. The land disposal site should be operated in such a manner as to protect the health and safety of personnel associated with the operation and also to conform to the pertinent provisions of the Occupational Safety and Health Act of 1970 (P.L. 91-596).

Other procedures, requirements, and design features are outlined in the *Federal Register* [12] in a more complete manner.

On October 21, 1976, Congress enacted the Resource Conservation and Recovery Act of 1976 (RCRA), Public Law 94-580. The primary thrust of this law is to provide technical and financial assistance for plans, facilities, and projects involving the recovery of energy and resources from municipal and other wastes. Additionally, the law provides for the safe disposal and management of hazardous waste and requires standards for generators, transporters, and disposers of hazardous wastes. According to the USEPA, enactment of this law marked the first major commitment by the United States to bring the health-related aspects of solid wastes under control [13].

A survey conducted in 1975 indicated that there were approximately 19,000 land disposal sites accepting municipal wastes. Fewer than 6000 met existing state standards. Most of these sites also accepted and received industrial wastes [13]. Subtitle C of RCRA places management of hazardous wastes under federal and state regulatory control. The USEPA was required to identify the hazardous wastes, set standards for their management, and issue guidelines for state programs in 1978. Subtitle D provides for grants to the states for development of comprehensive programs of environmentally sound disposal, resource recovery, and conservation. In order to be eligible, a states' solid waste plan must meet minimum criteria which include the requirement that all solid waste be utilized for resource recovery, disposed of in a sanitary landfill, or disposed of in some other environmentally sound manner. The plan must also provide for the closing or upgrading of all existing open dumps. A national inventory of all open dumps is to be published and the Act subsequently requires that all open dumps be closed or upgraded by 1983. Section 1008 requires the USEPA to develop suggested guidelines on solid-waste management practices, with emphasis on methods for protecting the public health and environmental quality. Guidelines are required under this provision for land disposal of municipal solid waste and on-land disposal of sewage sludge.

Pursuant to the requirements of Sections 1008(a) (3) and 4004(a) of RCRA, the USEPA published its "proposed classification criteria for solid waste disposal facilities" in the Federal Register on February 6, 1978 (40CFR Pt. 257). The purpose of the proposed regulations was to "provide minimum criteria to be used by the States to define those solid waste management practices which constitute the open dumping of solid wastes" and to promulgate regulations containing such criteria so as to determine which solid waste disposal facilities pose "no reasonable probability of adverse effects upon health or the environment from disposal at such facility." The criteria require that the solid-waste disposal facility may pose no reasonable probability of adverse effects upon health, safety, or the environment if it is located, designed, constructed, operated, completed and maintained so that it meets the following criteria:

1. Will not be located in a wetland, floodplain, permafrost area, critical habitat, or in the recharge zone of an aquifer (all subject to certain exclusions).
2. Does not adversely affect surface water quality.
3. Does not adversely affect ground water quality.
4. Controls air emissions (prohibits open burning of certain solid waste and permits burning of agricultural and silicultural wastes if in compliance with state and local regulations).
5. Sets the maximum application or criteria for cadmium, pathogens, pesticides, and other potentially harmful substances for land application of solid waste.
6. Protects the public health by controlling disease vectors.
7. Does not pose a safety hazard (explosives, fires, gases, birds).

III. ENGINEERING, CONSTRUCTION, AND OPERATION OF SANITARY LANDFILL SITES

A. Sanitary Landfilling Methods

1. *General*

The two basic methods of sanitary landfilling are the area and trench methods. Other methods, such as the ramp or slope method, or valley and low-area fill methods, are only modifications of the two basic methods. In general, the area method can be utilized on flat or gently rolling terrain, is adaptable to most topographies, and is often used if large quantities of solid wastes must be disposed of [1]. The trench method is used primarily on level ground, although it is also suitable for moderately sloping terrain. The groundwater table is generally low and the soil must be more than 6 ft deep for the trench method to be used effectively. Scattering of refuse by wind is minimized and therefore more direct dumping control can be achieved. At many sites a combination of the two methods is used.

The basic mode of emplacement for all methods is the construction of a cell (Fig. 3). In this technique, all solid wastes received are spread and compacted in layers within a confined area. Either on a daily or on a more frequent basis, a thin continuous layer of cover is applied that is then compacted. A series of cells, all of the same height, constitutes a lift with the completed landfill consisting of one or more lifts [1].

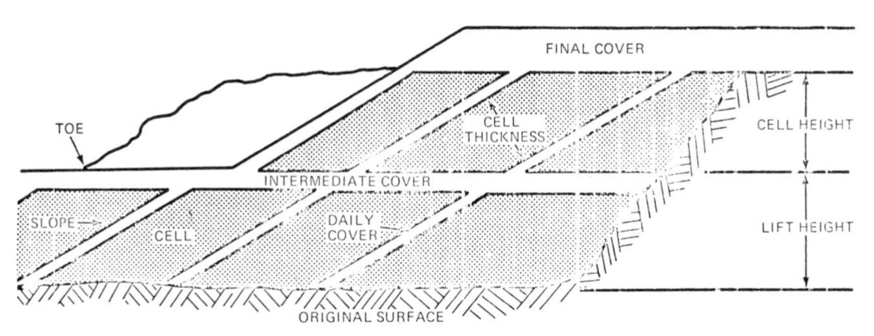

FIG. 3 Cross section of cell construction. From Ref. 6.

Cell dimensions are determined by the volume and density of the compacted solid waste. Density of most compacted solid waste in a cell should be at least 475 kg/m^3 (800 lb/yd^2). If the material to be spread is of low field density, 0.6 m thick layers should be covered with at least 15.3 cm (6 in.) of soil. Over this, mixed solid waste should be spread and compacted to keep elastic or low density materials compressed [1].

There are no rules concerning the ideal or proper height of a cell, although some designers recommend 2.44 m (8 ft) or less. If land and cover material are readily available, the height restriction may be appropriate; however, large operations using cell heights up to 9.14 m (30 ft) are common.

For proper utilization of a site, the designer and operator should keep cover material at a minimum while disposing of as much waste as possible [1].

In general, it is recommended that the cell should be square with steep slopes. Side slopes of 20–30° will keep the surface area and thus the cover material at a minimum.

2. *Area Method*

The area method utilizes the natural surface and slope of the land. The width and length of the fill slope are dependent upon the slope of the terrain, the volume of refuse delivered daily, and the number of vehicles that must be unloading at one time [3].

In an area sanitary landfill, the solid wastes are placed on the land, a bulldozer or similar equipment spreads and compacts the wastes, and earth cover is applied and compacted (Fig. 4). The working face should be kept as small as practical so that truck compaction can be used, and dumping and scattering of debris can be kept to a confined area. This method is also used in quarries, ravines, valleys, swamps, or where other suitable land depressions exist. Earth cover is usually hauled in or obtained from adjacent areas.

In the ramp or slope method (a variation of either the area or trench landfill), the solid wastes are dumped on the side of an existing slope and earth cover is scraped from the base of the ramp (Fig. 5). In the area method, the cover material is usually hauled in from some nearby source.

3. *Trench Method*

The trench method is used primarily on level ground, although it is also suitable for moderately sloping ground. In this method a trench is excavated and refuse is placed within the excavated area, compacted, and covered at the end of each day (Fig. 6). Cover material is readily avail-

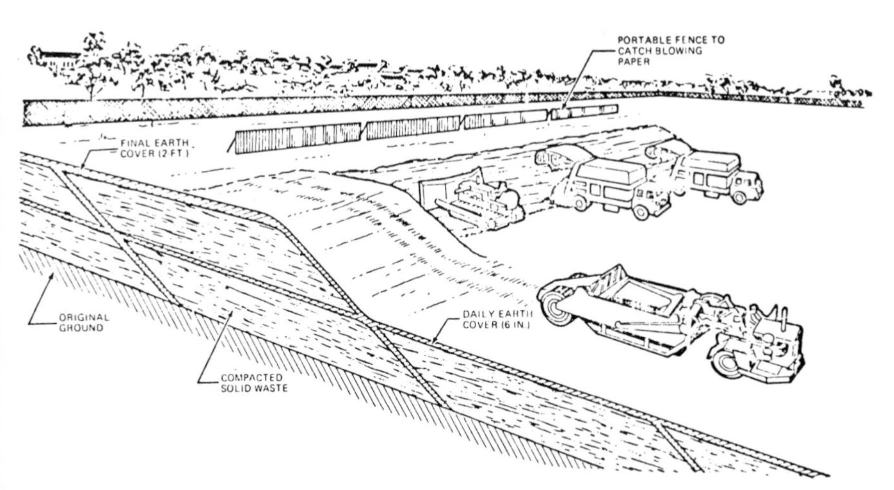

FIG. 4 Area method of sanitary landfill. From Ref. 6.

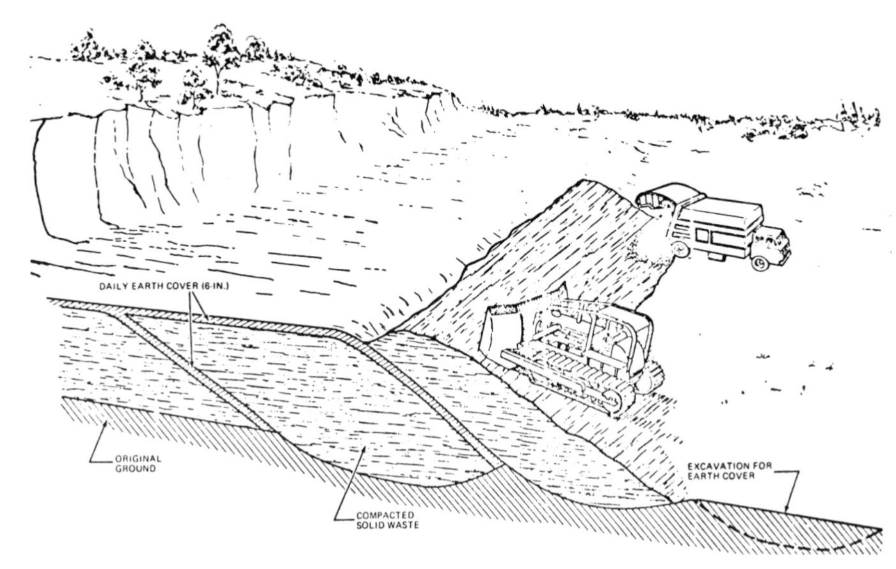

FIG. 5 Ramp variation combined method. From Ref. 6.

able as a result of the excavation and placed so that it may be used for daily cover or as a cover for the area at the completion of the operation. Earth for cover is obtained from the area where the next day's refuse is to be placed, and thus a trench for the next day's refuse is completed as cover material is being excavated [3]. The trenches are generally 6.1–7.6 m (20–25 ft) wide, although they should be at least twice as wide as any compacting machinery that may work in it. The depth of fill is determined by the finished grade desired and the depth to groundwater or rock. Deep trenches make most efficient use of the available

FIG. 6 Trench method of sanitary landfill. From Ref. 6.

land area. Cohesive soils, such as glacial till or silty clay, are desirable for use in a trench operation because the walls between the trenches can be thin and steep [1].

4. *Combination Methods*

Combinations of the area and trench methods are possible, and if applied by the designer properly, provide a great amount of flexibility for a specific sanitary landfill operation. Figure 5 is an example of a combined method.

Both methods may be used at the same site if an extremely large amount of solid waste is to be disposed of. For example, a site may have a varying thickness of soil over the entire area. The landfill designer may use the excess soil obtained to provide cover for the area method over the rest of the site [1].

B. Sanitary Landfill Equipment

1. *General*

The size, type, and amount of equipment necessary for efficient operation at a site depends upon the size, method of operation, and soil conditions. A wide variety of equipment is generally available from which to select the proper size and type needed at an operation. Standby equipment may also be obtained for emergencies, breakdowns, or periodic maintenance.

2. *Types*

The most common equipment used on sanitary landfills is the crawler or rubber-tired tractor. The tractor can be used with a dozer blade, trash blade, or front-end loader. Tractors are versatile and can normally perform all of the operations such as spreading, compacting, covering, trenching, and hauling the cover material. Selection of a rubber-tired or a crawler-type tractor plus front-end accessories must be based on the conditions at each individual site. Other equipment used are scrapers, compactors, draglines, and graders (Fig. 7), in addition to a sheepsfoot roller and a water tank truck equipped with a sprinkler to keep dust down, or a power sprayer to wet down refuse for better compaction.

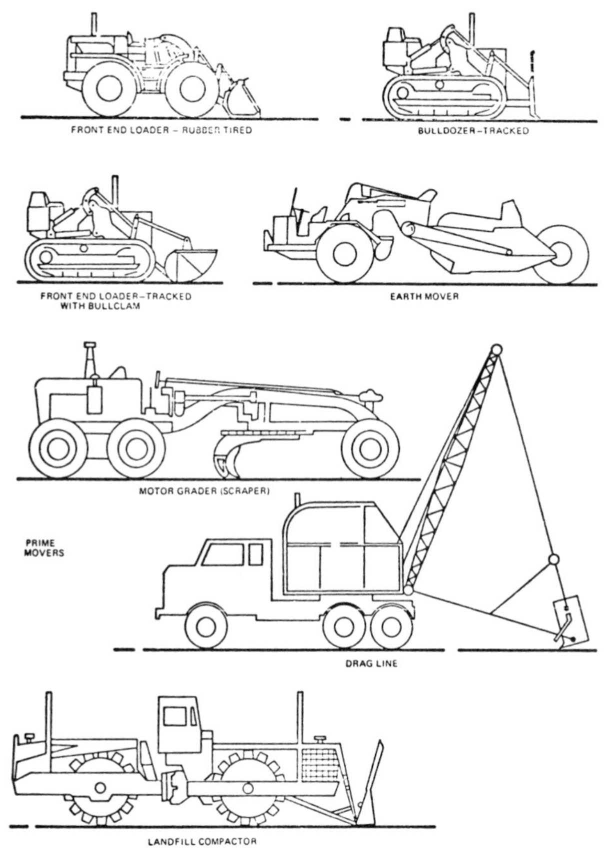

FIG. 7 Sanitary landfill equipment. From Ref. 3.

The average equipment requirements for different size operations are given in Table 4. Performance characteristics of landfill equipment are presented in Table 5.

C. Seepage Prevention and Runoff Control

1. *General*

Experience indicates that mixed refuse placed no deeper than approximately 1–2 m above the groundwater table and bedrock does not present any serious hazard of groundwater pollution, provided surface water is drained off the site [14]. However, if the site is used for the disposal of liquid or sludge-type wastes or wastes that are hazardous,

Table 4
Average Equipment Requirements[a]

Population	Daily tonnage	Number	Type	Equipment size, lb	Accessory
0–15,000	0–46	1	Tractor, crawler or rubber-tired	10,000–30,000	Dozer blade, landfill blade, front-end loader (1–2 yd)
15,000–50,000	46–155	1	Tractor, crawler or rubber-tired	30,000–60,000	Dozer blade, landfill blade, front-end loader (2–4 yd), multipurpose bucket
		*	Scraper, dragline, water truck		
50,000–100,000	155–310	1–2	Tractor, crawler or rubber-tired	30,000 or more	Dozer blade, landfill blade, front-end loader (2–5 yd), multipurpose bucket
		*	Scraper, dragline, water truck		
100,000 or more	310 or more	2 or more	Tractor, crawler, or rubber-tired	45,000 or more	Dozer blade, landfill blade, front-end loader, multi-purpose bucket
		*	Scraper, dragline, steel-wheel compactor, road grader, water truck		

[a] From Sorg and Hickman [7].
* Variable, depends upon local conditions and nature of operation.

Table 5
Performance Characteristics of Landfill Equipment[a,b]

Equipment	Solid waste			Cover material		
	Spreading	Compacting	Excavating	Spreading	Compacting	Hauling
Crawler dozer	E	G	E	E	G	NA
Crawler loader	G	G	E	G	G	NA
Rubber-tired dozer	E	G	F	G	G	NA
Rubber-tired loader	G	G	F	G	G	NA
Landfill compactor	E	E	P	G	E	NA
Scraper	NA	NA	G	E	NA	E
Dragline	NA	NA	E	F	NA	NA

Key: E, excellent; G, good; P, poor; NA, not applicable.
[a] From Weiss [1].
[b] Basis of evaluation: haul distance greater than 305 meters (1000 ft). For easily workable soil and cover material.

a greater separation or the use of natural or synthetic liners is recommended. When issuing permits or certificates, many states require that groundwater and deposited solid waste be 0.6–9.1 m (2–30 ft) apart [1]. Generally, however, a 2-m separation removes enough coliform bacteria to ensure safety from a bacteriological standpoint. Typical sanitary landfill leachate composition is given in Tables 6 and 7.

Mineral or heavy metal pollutants can travel for long distances laterally through soil or rock formations. In addition to other considerations, the sanitary landfill designer must evaluate [1]:

1. Current and projected water use of the area water resources.
2. Effect of potential leachate on groundwater quality.
3. Direction of groundwater movement.
4. Interrelationship of this aquifer with other aquifers and surface waters.

If refuse is placed below the water table or directly on bedrock, groundwater pollution may result. This pollution may be the result of

Table 6

Typical Sanitary Landfill Leachate Composition[a]

	Range of values[b]	
Parameter	Low	High
pH	3.7	8.5
Hardness ($CaCO_3$)	35	8,120
Alkalinity ($CaCO_3$)	310	9,500
Calcium	240	2,570
Magnesium	64	410
Sodium	85	3,800
Potassium	28	1,860
Iron (total)	6	1,640
Chloride	96	2,350
Sulfate	40	1,220
Phosphate	1.5	130
Organic nitrogen	2.4	550
Ammonia nitrogen	0.2	845
Specific Conductance	100	1,200
BOD	7,050	32,400
COD	800	50,700
Suspended solids	13	26,500

[a] From Wegman [15].

[b] Values are given in milligrams per liter, except pH (pH units) and conductivity (micromhos per centimeter).

Table 7
Chemical Composition of Typical Sanitary Landfill Leachate[a]

Parameter	Concentration, mg/L
pH	6.9
Hardness ($CaCO_3$)	1100
Alkalinity ($CaCO_3$)	4220
COD	1340
TDS	5910
NH_4-nitrogen	862
As	0.11
B	29.9
Ca	46.8
Cl	3484
Na	748
K	501
Mg	233
Fe	4.2
Cr	<0.1
Hg	0.87
Ni	0.3
Si	14.9
Zn	18.8
Cu	<0.1
Cd	1.95
Pb	4.46

[a] From Griffen et al. [16].

leaching of the contents of the refuse (e.g., heavy metals, decomposed organic material, nutrients) and the transfer of gases, such as carbon dioxide, methane, nitrogen and nitrogen oxides, ammonia, and hydrogen sulfide, that are produced by decomposition of solid waste components. The presence of these gases and subsequent reactions with the leachate may further solubilize refuse components such as heavy metals.

It has been recommended that sites be at least 61 m (200 ft) from streams, lakes, or other bodies of water and well above groundwater [3]. A geological and hydrological study should be made to determine if the distance from a water body or the groundwater table is adequate to contain and prevent leachate entry into surface- or groundwater.

2. *Seepage Prevention*

If a large quantity of liquid infiltration may be expected or a hazardous or deleterious effect may be caused by leachate, preventative methods may be sought or required, such as selection of site with impervious subsoils, or use of "natural" or synthetic liner materials.

The California State Department of Public Health recommends the following criteria for impermeable soils [20]:

(1) Fine-grained soils generally falling into group classification CL (clay, low liquid limit), CH (clay, high liquid limit), or OH (organic clays, medium to high plasticity) soils according to the Unified Soil Classification System.
(2) Permeability of 10^{-8} cm/s or less.
(3) Not less than 30% by weight passing a 200 sieve (U.S. Standard).
(4) Liquid limit of not less than 30 (ASTM test D424).
(5) Plasticity Index of not less than 15 (ASTM test D424).

An impermeable clay liner may be employed to control the movement of leachate. Well-compacted natural clay soil either from a local source or brought in from outside areas may be used. Usually a layer of clay approximately ~0.5–~1 m (1–3 ft) thick is constructed. If sufficient clay soil is not available locally, additives such as bentonite (montmorillonite) may be disked into it to form an effective liner. Excellent results with this technique have been reported by commercial suppliers [18]. In this technique, bentonite is used to line the bottom and sides of a pit with a material that has the ability to swell to many times original volume. Often an application of bentonite of 9.8 kg/m^2 (18 lb/yd^2) is made, followed by disking to a depth of 15 cm (6 in.), three or four passes are usually employed. Slurry trenches can be constructed (Fig. 8) around the perimeter of a landfill that isolate the leachate from within the landfill and prevent its escape.

The materials being used or proposed as synthetic liners for land disposal sites include conventional paving asphalts, hot-sprayed asphalt, asphalt-sprayed fabric, polyethylene, polyvinyl chloride, butyl rubber, Hypalon®, ethylene propylene diene monomer, chlorinated polyethylene, and soil-cement mixtures [16]. Leaks may develop after the barrier is exposed to leachate because of leachate–liner reactions [16]. However, no current data are available.

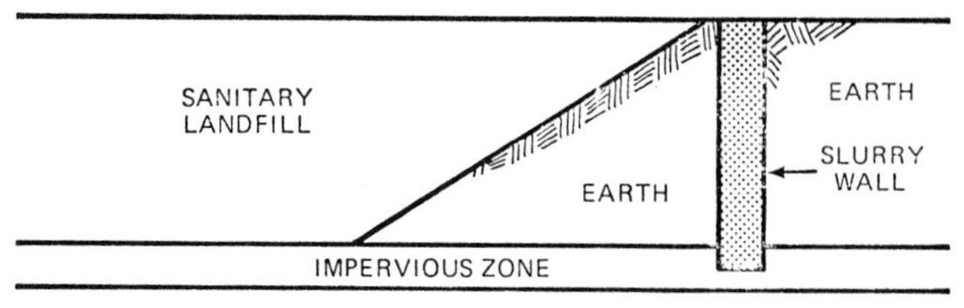

FIG. 8 Slurry trench technique in a sanitary landfill site. From Ref. 18.

Even though groundwater may be prevented from coming into direct contact with the solid waste, surface water may enter the site, leach materials from the refuse, and then percolate through the underlying porous soil. Highly permeable material may thus create considerable difficulty and should be examined very closely in preliminary site investigations. Use of liners may necessitate removal of contained fluids and subsequent waste water treatment. Use of runoff and infiltration control may thus be necessary at many sites.

3. *Runoff or Drainage Control*

Disposal sites should be isolated from surface flows and impoundments to prevent or minimize pollution potential and provide erosion control. In order to accomplish this, diversion ditches are excavated surrounding the refuse disposal site to exclude surface runoff from the area (Fig. 9).

Besides diverting surface runoff or water courses, pipes may be used in canyons, valleys, or gullies to aid transmittal of drainage through the landfill site. Ditches, flumes, pipes, trench drains, and dikes may be used to assist in preventing runoff to the sites from snow melt, rainfall, or streams. Sanitary landfill sites or construction methods that drastically alter natural drainage should be avoided where possible. Landfills located in flood plains should be protected by impervious dikes and liners [1].

Establishment of a vegetative cover of grasses in areas of potential sheet wash and erosion assist in stabilization of surface materials, con-

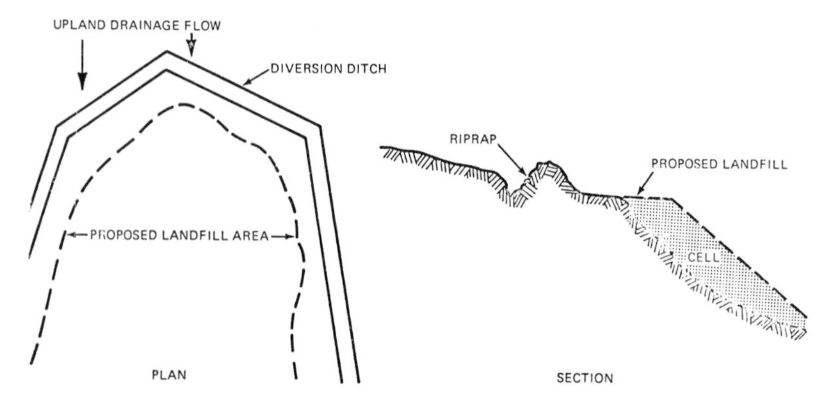

FIG. 9 Plan and section views of the use of a diversion ditch to transmit upland drainage around a sanitary landfill. From Ref. 6.

trol of erosion and sedimentation, and improvement of the esthetic aspects of the area.

The top cover material of a landfill should be graded to allow runoff of rainfall and prevention of impoundment. The grade of the completed cover depends upon the cohesiveness of the soil and future uses of the site. Recommended grades vary from 1 to 4%. Temporary or permanent drainage channels may assist in the interception and conduction of runoff water from the site.

D. Development and Operating Costs

1. *Initial Investment*

The amount of the initial investment for a sanitary landfill depends upon the size, period of expected operation, amount of equipment to be used, sophistication of the landfill, as well as other factors. A typical breakdown of items is listed below.

1. Land (adequate land area for the desired period of operation).
2. Planning and designing.
   a. Consultant (engineering, geology, soils, etc.).
   b. Solid waste survey (types, amounts, volume, collection-transport-distance models).
   c. Site investigation (soil survey, geology, drilling, depth to water table, etc.).
   d. Design, plans, specifications.
   e. Coordination with federal, state, and local agencies, and planning and zoning boards, etc.
3. Site development.
   a. Land development (clearing, landscaping, drainage diversion, etc.).
   b. Access roads.
   c. Utilities (water, electricity, telephone).
   d. Fencing, signs.
4. Facilities.
   a. Administration.
   b. Equipment maintenance.
   c. Sanitation.
   d. Weight scales.

5. Equipment (as required).
   a. Tractor and blades.
   b. Scraper.
   c. Water truck, etc.
6. Liner and liner installation (if required).

In general, the purchase of the land and equipment constitute the major initial purchases. However, when required, liners [19] and their installation costs can be significant (Table 8). A sizable portion of the initial investment for land and equipment can often be recovered through the subsequent development or use of the site and salvage value of the equipment.

Table 8
Costs of Sanitary Landfill Liner Materials[a]

Material	Installed cost[b] ($/yd^2)
Polyethylene (10–20[c] mils[d])	0.90–1.44
Polyvinyl chloride (10–30[b] mils)	1.17–2.16
Butyl rubber (31.3–62.5[b] mils)	3.25–4.00
Hypalon (20–45[c] mils)	2.88–3.06
Ethylene propylene diene monomer (31.3–62.5 mils)	2.43–3.42
Chlorinated polyethylene (20–30[c] mils)	2.43–3.24
Paving asphalt with sealer coat (2 in.)	1.20–1.70
Paving asphalt with sealer coat (4 in.)	2.35–3.25
Hot-sprayed asphalt (1 gal/yd^2)	1.50–2.00[e]
Asphalt sprayed on Polypropylene fabric (100 mils)	1.26–1.87
Soil bentonite (9.1 lb/yd^2)	0.72
Soil bentonite (18.1 lb/yd^2)	1.17
Soil cement with sealer coat (6 in.)	1.25

[a] From Haxo [20].
[b] Cost does not include construction of subgrade or the cost of earth cover. These can range from $0.10 to $0.50/yd^2 per foot of depth.
[c] Material costs are the same for this range of thickness.
[d] One mil = 0.001 in.
[e] Includes earth cover.

2. *Operating Costs*

The operating costs of a sanitary landfill are strongly influenced by the costs of labor and equipment, maintenance and repair of facilities and equipment, energy costs, and the method(s) and efficiency of the operation. The principal items to be considered in the evaluation of operating costs include:

1. Wages, fringe benefits, etc.
2. Equipment required.
   a. Number and types of equipment and accessories.
   b. Operating expenses (gasoline, oil, etc.).
   c. Maintenance and repair.
   d. Rental, depreciation, or amortization.
3. Cover material (including haul costs).
4. Administration and overhead.
5. Miscellaneous (tools, utilities, insurance, road maintenance, fences, facilities, drainage control, revegetation, etc.).

Wages generally comprise 40–50% of the total operating costs, equipment usually 30–40%. Cover material, administration, overhead, and miscellaneous expenses amount to about 20%.

The operating costs per ton vs. the amount of solid waste handled and the population served are plotted in Fig. 10. The cost of a small operation handling less than 50,000 tons/yr varies from $1.25 to approximately $5.00 per ton [7]. This wide range is due primarily to the low efficiency of the smaller operations that are often operated on a part-time basis.

Full-time personnel, full-time use of equipment, more use of specialized equipment, full-time and better management, and other

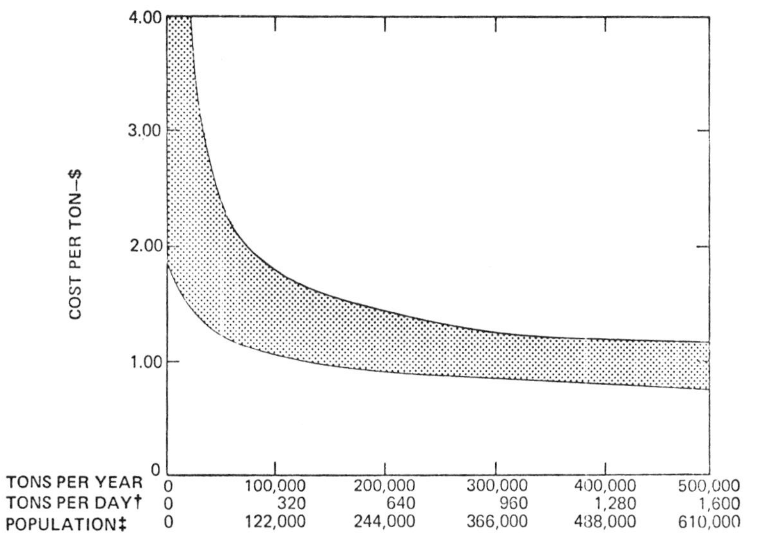

FIG. 10 Sanitary landfill operating costs. From Ref. 7.
† Based on 6-day work week. ‡ Based on national average of 2 kg (4.5 lb) per person per calendar day.

factors may lead to higher efficiency at large sanitary landfills. The increased efficiency results in a lower unit cost of disposal. The unit cost of a large landfill handling more than 45,350 metric tons/yr (50,000 short tons/yr) generally is between \$0.83 and \$2.20 per metric ton (\$0.75 and \$2.00 per short ton) [7].

Even with rising costs resulting from increased environmental control, landfilling is generally less expensive than other final disposal procedures. For sludge disposal, an important consideration can be the dewatering requirements. Figure 11 presents a range of capital and operating and maintenance (O/M) costs for landfills excluding land costs. Although this illustration was derived for wet sludge disposal, other investigators have reported landfill costs of \$1.02 to \$4.41 per metric ton (\$1–\$4 per short ton) of dry solids. It should be noted that in any presentation of cost data several variables must be taken into

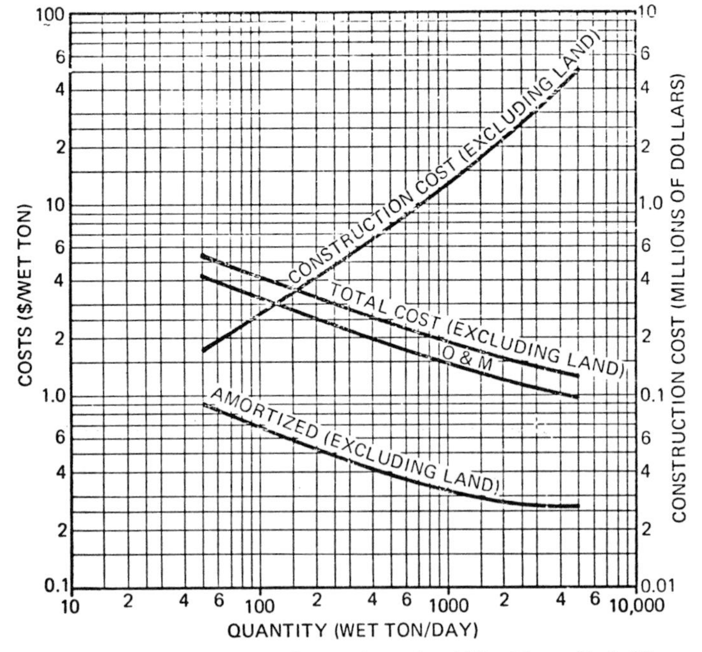

FIG. 11 Capital and O/M costs for sanitary landfills. From Ref. 23.
Key: 1. Minneapolis, March 1972. ENR construction cost index (1827).
2. Amortization of 7% for 20 yr.
3. Labor rate of \$6.25/h.
4. Quantity assumes 6-day work week.
5. Wet sludge must be considered for cost per ton.
6. Source: USPHS and Stanley Consultants.

account, such as the location where the landfill is operated, labor rates and interest rates which may vary with time and location, and collection, handling, and transfer costs which may not be included.

3. *Landfill Costs Compared to Other Methods of Disposal*

In 1975, the USEPA-OSWMP sponsored a study of landfill, shredder, and transfer station costs [21, 22]. This study indicated that the cost of land was less than 4% of the solid-waste disposal costs for 17 sites surveyed. In addition, it was found that in most instances a high price for land close to the center of waste generation (or the addition of liners and leachate control) adds less to the overall costs than the installation and operation of shredding or transfer stations.

The table compares the costs of collection, shredding, transfer, and landfilling of solid waste.

Table 9
Breakdown of Costs of Solid Waste Disposal

Operation	No. of communities surveyed	Av. cost/metric ton[a] ($)
Collection	102	23.15 (21.00)
Shredding (16.1 km haul)	7	8.22 (7.46)
Shredding only	10	6.43 (5.83)
Transfer (27.4 km haul)	11	5.74 (5.21)
Landfilling (mixed solid waste)	14	3.34 (3.33)
Landfilling (shredded solid waste)	3	2.03 (1.84)

[a] $ per short ton in parenthesis.

The land disposal sites surveyed above disposed of an average of 345 metric tons (380 short tons) per day. The figures included land and development costs but omitted interest.

To compare the costs of sanitary landfills with the costs of incineration, composting or other methods, it is important to consider (a) costs and returns on initial investments, (b) hauling costs, (c) disposal of incinerator residue, (d) possibility of operating in conjunction with resource recovery or other methods, and (e) return of the site to other uses.

IV. ENVIRONMENTAL CONSIDERATIONS

A. General

The degradation of solid wastes in a sanitary landfill begins immediately upon refuse deposition. Some materials such as rubber, plastics, glass, and wastes from building demolition degrade very slowly, whereas organic materials such as food wastes and ferrous metals decompose or are oxidized at a relatively rapid rate. Liquid waste products of microbial degradation, such as organic acids, increase chemical activity within the fill. Some factors affecting degradation include the physical, chemical, and biological properties of the wastes, oxygen and water content within the landfill, temperature, microbial populations, and type of synthesis [6]. Since the solid wastes are very heterogeneous and nonuniform with respect to composition and size, it is difficult to predict the quantity and/or production rate of contaminants or waste gases.

With respect to biological activity in a landfill, solid waste initially decomposes aerobically, but as the oxygen supply diminishes, anaerobic microrganisms predominate and produce methane and other gases [6]. Temperatures are known to rise to the range of 15–65 °C (60–150 °F) because of microbial activity. Data have been presented in Table 6 on typical sanitary landfill leachate composition. Characteristic products of aerobic decomposition are carbon dioxide, water, and nitrate. Typical products of anaerobic decomposition are methane, carbon dioxide, water, organic acids, nitrogen, ammonia, and sulfides of iron and hydrogen.

Unfortunately, most of the data available today on the environmental effects of sanitary landfills reflect local conditions prevalent at the site rather than regional characteristics. Recently, model sanitary landfill studies have been sponsored by the U.S. EPA in an effort to develop guidelines that may be applied throughout an entire region. Generally, though, these studies have been undertaken to define design criteria.

In addition to the problems mentioned above of leachate and gas production, the problem of subsidence is one that must be given serious consideration from a short-term viewpoint of stability, erosion, etc., but also from an ultimate end-use view.

B. Leachate and Groundwater Monitoring

The chemical composition of leachate is important in determining potential effects on the quality of nearby surface and groundwaters. The identification of leachate composition has been the object of many laboratory and field lysimeter studies as well as test well-drilling programs in the vicinity of sanitary landfills. A typical design for a field lysimeter is given in Fig. 12. In order to determine the effects of leachate production and movement in the subsurface, test wells and lysimeters should be installed for monitoring not only the quality but also the quantity and rate of movement of leachate produced at a given site.

If disposal of wastes that are highly toxic or degrade over an extended period of time is anticipated, liner materials must be considered in order to protect the quality of surface- and groundwaters.

Leachate percolating through soils underlying and surrounding the landfill site may be subject to many processes capable of attenuating the potentially harmful materials. Ion exchange, filtration, adsorption, complexing, precipitation, and biodegradation are some of the methods by which contaminants may be removed [6]. The rate of flow and attenuation is dependent upon many factors such as the soil particle size,

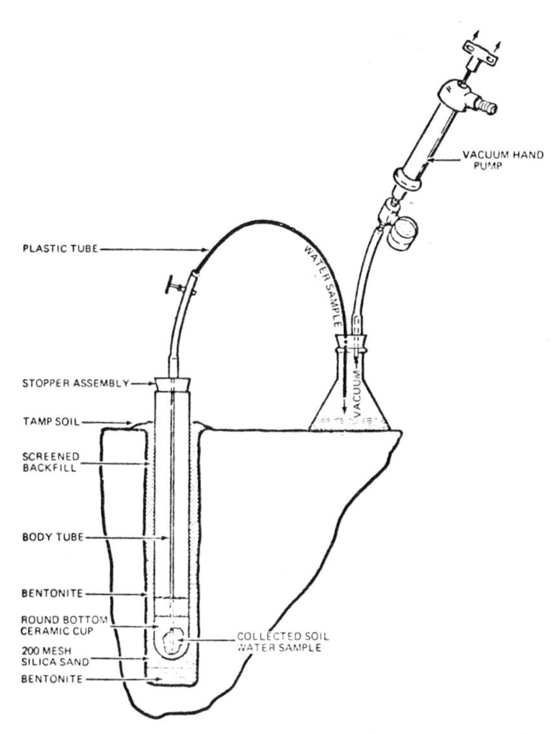

FIG. 12 Typical field lysimeter assembly for soil water sampling. From Soilmoisture Equipment Corp., Santa Barbara, Calif.

shape, and composition, and whether the medium is saturated or unsaturated.

Studies of leachate migration through clay [16, 24] determined that with typical landfill leachate the clay minerals kaolinite, illite, and montmorillonite could be ranked according to their attenuating capacity as follows: montmorillonite > illite > kaolinite. Montmorillonite attenuates pollutants approximately four times better than illite and five times better than kaolinite. Surface area was not considered as important in attenuation as the cation-exchange capacity (CEC). For the 15 substances tested [16, 24], the relative degree of attenuation is:

High: Pb, Zn, Cd, Hg
Moderate: Fe, Si, K, NH_4, Mg
Low: COD, Na, Cl
None: B, Mn, Ca

This listing ranks the chemical constituents in municipal landfill leachate according to their mobility through clay columns [24]. The conclusion drawn from the studies [16, 24] is that the principal factors affecting pollutant attenuation mechanisms by clay minerals in landfill leachates are: (a) microbial reduction of permeability, (b) the cation-exchange capacity of the clay, (c) chemical composition of the leachate, (d) the cations originally present on the exchange complex, and (e) the effect of pH on the formation of hydroxide and carbonate precipitates. Such studies assist in the design of liner systems that effectively protect groundwater from contamination by landfill leachates.

Natural degradation and purification processes have only a limited ability to remove contaminants because the number of adsorption sites and exchangeable ions is finite. In general, the residence time is also shortened by rapid flow rates. In addition, the production of leachate from a landfill tends to decrease over a period of time.

Leachate production and movement may be prevented or minimized to the extent that it will not create a water pollution problem. The most obvious means of controlling leachate production and movement is to prevent water from entering the landfill to the greatest extent possible. This may be accomplished by the use of diversion dikes, site selection, and proper grading. An alternative method of control is the construction of "natural" or artificial liners to prevent leachate movement from the landfill. Monitoring of the sites by groundwater sampling is the best way to determine if potential problems exist in site selection, development, or after closure.

C. Decomposition Gas Production

Gas is produced naturally when solid wastes, particularly organic ones, decompose. The quantity of gas and its composition depend upon the oxygen content of the landfill, the types of materials deposited, and the methods used in isolation of individual cells. A waste with a large percentage of easily degradable organic materials yields more gas than one consisting of inorganic materials such as ash or demolition debris. The rate of gas production is governed by the rate of microbial decomposition, composition of the refuse, and moisture content. Methane and carbon dioxide are the major components of the gas produced, but other gases are present such as hydrogen sulfide, nitrogen, and ammonia [25–27].

Several methane recovery operations have proved the feasibility of extraction of landfill gas as well as demonstrated the commercial possibilities and economics associated with this potential resource. In a project conducted at Mountain View, California, methane comprised approximately 40–45% by volume at a 1.4 m^3/min (50 ft^3/min) withdrawal rate [26]. Gas production at the 61-ha, 12-m (150-acre, 40-ft) deep site has been estimated at 210,000 m^3/day (7.5 million ft^3/day). With optimum spacing of withdrawal wells, the site may be capable of producing approximately 288,400 m^3/day (10.3 million ft^3/day) of 44% methane gas. In the Sheldon-Arleta area of Los Angeles, recovery of 28 million m^3/yr (1 billion ft^3/yr) of 126 kcal (500 BTU) gas over an 8-year period is thought to be possible [28]. A facility that also recovers methane from a sanitary landfill is located in the city of Rolling Hills, California. The facility generates about 28,000 m^3/day (1 million ft^3/day) of methane for distribution by the Southern California Gas Company [29].

Landfill gas is important with respect to environmental considerations, because methane is explosive and carbon dioxide dissolved in water forms a weak acid, carbonic acid. This can result in dissolution of minerals present in the soils and rocks present at a site. Methane is explosive when present in air at concentrations of 5–15%. In the oxygen-poor environment of a landfill this does not present a problem. If methane vents to the atmosphere, it may accumulate in buildings or other enclosed spaces or close to a sanitary landfill [6]. The potential movement of gas is an essential element in site selection and development, particularly if the landfill is located near industrial, commercial, or residential areas.

Landfill gas movement can be controlled. Of the several methods tested, permeable vents and impermeable barriers are the two basic

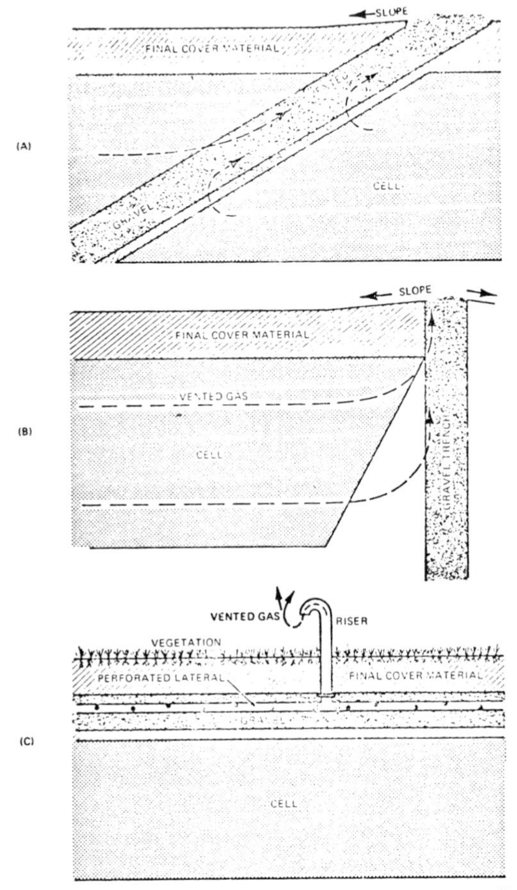

FIG. 13 Method of gas movement control: (A) gravel vent, (B) gravel trench, (C) vent pipes. From Ref. 6.

types. Figure 13 illustrates three methods of gas control in a sanitary landfill.

D. Subsidence

A sanitary landfill settles as a result of waste decomposition, filtering of fine material, superimposed loads, and by its own weight. Consolidation of the fill may continue for many years, and therefore construction of permanent buildings over a landfill should be avoided for a sufficiently long time to allow settling.

A lower waste-to-cover ratio reduces settlement in a landfill, whereas well-compacted waste settles less. Studies of landfills have indicated that approximately 90% of the ultimate settlement occurs in the first 5 years [1].

Most important for settling is the problem of allowing water to infiltrate or be impounded. A periodic inspection of the site and regrading where necessary should be part of the development plan for every sanitary landfill site.

V. TERMINATION OF OPERATIONS

One of the greatest benefits of filling and raising the ground surface is reclaiming land and returning it to an alternative use that may benefit man. There are many ways in which sanitary landfills can be converted to other end uses, all requiring planning prior to actual operation of the landfill. In addition to conversion of the site to alternative uses, it is desirable to conduct periodic surveillance and monitoring of the site both from an environmental and health safety standpoint, but also in order to maintain the closed disposal area.

A. Reuse of Site

The possible ways in which a site may be reused is probably limited only by the imagination of the planners, needs of the community, and physical characteristics of the completed site. Generally, the site can be redesigned for agricultural, recreational, or light construction purposes. Each potential use must be evaluated both from a technical feasibility and economic viewpoint. More suitable land might be available elsewhere that would not require the expensive construction techniques necessary at a sanitary landfill.

1. *Agricultural Uses*

Agricultural use involves few or no structures on the land. A green area is one of the most common uses of a completed landfill, established for the benefit and pleasure of the community. Some maintenance would be required to prevent surface erosion and additional grading is often necessary to prevent water from ponding or infiltrating. The type of plants, grass, trees, etc. planted depend upon the thickness of the final cover and the type of soil present.

A completed sanitary landfill can be made productive by turning it into pasture or crop land. A thick final cover (in excess of 1.23 m (4 ft)) should be applied because of the length of roots commonly encountered in agricultural crops such as corn and wheat. If the landfill is to be cultivated, it is advisable to place a 0.3 to 0.6 meter (1–2 ft) layer of soil such as clay on top of the solid waste to provide an impermeable layer. Agricultural soil of appropriate thickness can then be placed over the clay layer. A means of escape for gas must be provided such as gravel trenches or pipes.

Some of the other types of agricultural uses that have been suggested are tree farms, nurseries, and orchards.

2. *Recreational Uses*

Many factors must be considered in the selection of appropriate uses of a sanitary landfill. These include location and access, size and configuration, vegetation, and topography.

A few examples of the type of recreational uses that could be made available to the public on landfill sites include:

Parks	Nature trails
Wildlife preserve	Hiking and riding trails
Hunting area	Archery or rifle range
Zoo	Court games
Golf course	Field games
Driving range	Bicycle paths

It has been suggested that some recreational uses can coexist with the development or utilization of the landfill [1]. Where this is done, proper screening and good operational practices must be undertaken.

3. *Construction*

Many communities have turned former landfill areas into industrial sites suitable for light building construction. For uses of this type, foundations engineers should be consulted to determine what methods should be utilized during development to ensure safety of the buildings to be constructed at a later time. One of the methods used is to preplan the use of earth islands to avoid settlement, corrosion, and bearing-capacity problems. Building construction should not require extensive excavation. Piles can be used to support buildings, but should penetrate the refuse and be firmly anchored in soil or rock. Angling of the piles is desirable to resist lateral movements that may occur in the fill.

Generally, light, one-story buildings are constructed on landfill sites. Special preparation should be made for foundations, roads, parking lots, etc., because of differential movement in the landfill below. Gas accumulation problems must be recognized and dealt with appropriately.

B. Maintenance and Surveillance

Once a sanitary landfill or lift of a landfill is completed, it is necessary to maintain the surface in order to take care of differential settlement. Settlement may vary, ranging up to approximately 20–30%, depending on the compaction, depth, and character of the refuse. Maintenance of the cover is necessary to prevent ponding and excessive cracking, that permits gases to escape and insects and rodents to enter the fill.

Proper drainage is necessary to prevent the seepage of surface water into the landfill and subsequent contaminated leachate water from entering the groundwater table or the surface. Generally a 1–2% grade with culverts and lined ditches is needed. Proper grading is an essential element, as is the thickness of final cover to be applied to the completed landfill. Gas and water controls and drains require periodic inspection and maintenance.

Test wells and lysimeters as well as gas monitoring stations installed during construction should be periodically sampled until the landfill stabilizes. After approximately 5–7 yr of final cover completion, test-well monitoring should be adequate for the determination of environmental effects as gas production should be at a minimum after this period of time.

ACKNOWLEDGMENTS

The author is indebted to Calspan Corporation, and especially D. Barry Dahm, Roland J. Pilié, Richard P. Leonard, and Richard G. Swan for technical and administrative assistance during the course of preparation of the manuscript. Final preparation of the manuscript took place at Frontier Technical Associates, Inc. Dr. L. K. Wang and N. C. Pereira provided generous editorial assistance throughout the period of preparation. Mr. George Niesyty offered advice with respect to economic considerations. Mr. Ben Warner, editor of *Solid Waste Systems*, kindly granted permission for use of an illustrative chart (Figure 1). Thanks are due also to Mrs. Helen Fabrizzi who typed the manuscript.

REFERENCES

1. S. Weiss, *Sanitary Landfill Technology*, Noyes Data Corporation, Park Ridge, N.J. 1974.
2. Office of Solid Waste Management Programs, "Land Disposal of Solid Waste," *Waste Age*, 20–28, 64 (March/April, vol. 4, No. 2, 1973).
3. J. A. Salvato, Jr., *Environmental Engineering and Sanitation*, 2nd ed., Wiley-Interscience, New York, 1972.
4. K. K. Hekimian, "Landfill Siting Calls for Homework, Engineering, EIR and Good Public Relations," *Solid Waste Systems Magazine* (August 1977), 20–22.
5. K. C. Clayton and J. M. Huie, *Solid Wastes Management—The Regional Approach*, Ballinger Publishing Company, Cambridge, Mass., 1973.
6. D. R. Brunner and D. J. Keller, "Sanitary Landfill Design and Operation," USEPA SWMO,† Report No. EPA-SW-65ts, 1972.
7. T. J. Sorg and H. L. Hickman, Jr., "Sanitary Landfill Facts," U.S. Department of Health, Education and Welfare, Public Health Service Publication No. 1792, 1970.
8. N. K. Bleuer, "Geologic Considerations in Planning Solid Waste Disposal Sites in Indiana," Environmental Study No. 1, Geological Survey Special Report 5, Indiana Department of Natural Resources, Bloomington, Indiana, 1970.
9. T. Viraraghavan, "Sanitary Landfill Practices in Canada" *Public Works*, 92–94 (March 1973).
10. "Recommended Standards for Sanitary Landfill Design, Construction, and Evaluation, and Model Sanitary Landfill Operation Agreement," USEPA, Publication SW-86ts, 1971.
11. "Sanitary Landfill—Planning, Design, Operation, Maintenance," New York Department of Environmental Conservation, Albany, 1971.
12. USEPA-OSWMP, "Guidelines for the Land Disposal of Solid Wastes," Federal Register, Vol. 39, No. 158, Aug. 14, 1974, p. 29328; and as amended on Feb. 4, 1975, p. 5159.
13. "Words Into Deeds"—Implementing the Resource Conservation and Recovery Act of 1976," USEPA, August, 1977, 7 pp.
14. J. A. Salvato, W. G. Wilkie, and B. E. Mead, "Sanitary Landfill Leachate Prevention and Control," *J. Water Poll. Cont. Feder.* **43**, 2084–2100 (1971).
15. Leonard S. Wegman, Co. Inc., "Typical Specifications of an Impermeable Membrane," Lycoming Board of Commissioners, Lycoming, Pa., Unpublished Data, 1974 (see Ref. [22]).
16. R. A. Griffen, N. F. Shimp, J. D. Steele, R. R. Ruch, W. A. White, and G. M. Hughes, "Attenuation of Pollutants in Municipal Landfill Leachates by Passage Through Clay," *Environ. Sci. Technol.* **10**, (3), 1262–1268 (1976).
17. "Guidelines for Hazardous Waste Land Disposal Facilities," California State Department of Public Health, January, 1973.
18. American Colloid Company, *Soil Sealants, Slurry Trenching, and Sanitary Landfill Information Brochures*, 1974. Skokie, Ill.

† U.S. Environmental Protection Agency, Office of Solid Waste Management Programs, Washington.

19. U.S. Environmental Protection Agency, "Liners for Land Disposal Sites—An Assessment," USEPA Report No. EPA/530-SW-137, March, 1975.
20. H. E. Haxo, Jr., "Evaluation of Liner Materials," USEPA Contract No. 68-03-0230, October, 1973 (See Ref. [22]).
21. "Cost Estimate Handbook for Transfer, Shredding, and Sanitary Landfilling of Solid Waste," Booz, Allen, and Hamilton, Washington, August, 1976, 85 pp. (NTIS PB-256-444-1WP).
22. J. W. Thompson, "Economics of Landfill Location," Presented at the National Solid Waste Management Association 5th National Congress on Waste Management Technology and Resource and Energy Recovery, Dallas, Texas, December 7–9, 1976.
23. R. C. Merz and R. Stone, "Special Studies of a Sanitary Landfill" U.S. Department of Health, Education and Welfare, 1970 (Natl. Tech. Info. Service, PB-196 148).
24. R. A. Griffen and N. F. Shimp, "Leachate Migration Through Selected Clays," in Gas and Leachate from Landfills—Formation, Collection and Treatment, USEPA Report No. EPA-600/9-76-004, March, 1976, 92–95.
25. T. J. Borg and H. Lanier-Hickman, Jr., "Sanitary Landfill Facts," U.S. Dept. of Health, Education, and Welfare, Public Health Service, Publication SW-4ts, 1970, 30 pp.
26. J. A. Carlson, "Recovery of Landfill Gas at Mountain View—Engineering Site Study," USEPA, Report EPA/530/SW-587d, May, 1977, 63 pp.
27. M. J. Blauchet, "Treatment and Utilization of Landfill Gas—Mountain View Project Feasibility Study," USEPA Report SW-583, 1977, 115 pp. (Reproduced by the USEPA with permission of the Pacific Gas and Electric Company).
28. Los Angeles Bureau of Sanitation, "Examination of the Quantity and Quality of Landfill Gas from the Sheldon-Arleta Sanitary Landfill," October 17, 1975, 34 pp.
29. "Gas from Garbage a Reality in Southern California," *Pipeline Gas J.*, **202**, (11), 69 (1975).

5

Composting Process

Raul R. Cardenas, Jr.

Department of Civil Engineering, Polytechnic Institute of New York, Brooklyn, New York

Lawrence K. Wang

Department of Mechanical Engineering, Stevens Institute of Technology, Hoboken, New Jersey

I. INTRODUCTION

A. General Description

Composting consists of the biological decomposition of solid or semi-solid materials by microorganisms over a prolonged period of time resulting in the degradation of the organic content of the material and a reduction in volume. During the process the temperature rises and the material is changed in character from one that has an identity such as manures, sludges, leaves, grass, or newspaper, etc., to a stable end product, compost, that resembles a granulated dark brown humus or a soil-like material. Modern composting is an aerobic biological process subject to the constraints of all biological activities.

Utilization of the end product obtained from a composting process represents resource recovery. For this reason it has been viewed favorably by conservation-minded groups. The chief use of the humus-like compost is as a soil conditioner. The finished product, after drying, contains organic nitrogen, carbonaceous matter, microorganisms, important micronutrients, and about 10–15% moisture.

The use of composting in processing of animal manures and sewage sludges is considered to be feasible [1, 2]. The major raw materials today employed in natural organic fertilizers are cow manure and other animal excrement and bedding straw. Processed sewage sludge can also be marketed as an organic fertilizer for gardens and lawns.

Municipal refuse containing much cellulosic material can also be successfully composted. Blending with higher nitrogen sewage sludge hastens the composting process and upgrades the final product [3, 4].

Composting costs are comparable to those for incineration. A primary problem in the United States has been lack of a market for the stable compost which is needed to produce revenue from the product's sale to offset the cost of the process.

B. Historical Development

Natural composting is a very old process that has always been a part of the global ecosystem and which is continuously taking place on the surface of the earth. If natural composting did not take place, leaves, grass, tree bark, and other semisolid debris that falls to the surface of the earth would not be degraded.

As an applied tool used by man to assist him in his agricultural practices, composting has been employed by farmers for centuries as an organic soil supplement. Farmers and gardeners, for example, have used composted night soil, vegetable matter, animal manure, garbage, refuse, etc. In its most primitive form these materials were placed in piles or pits located in some convenient place and allowed to decompose as natural conditions would permit until the material was composted or "ready" for the soil and the farmers could apply it to the land. In general, such natural processes involved little, or no control and required long periods of standing in heaps to provide a good organic humus or soil-like material that would contain the appropriate organic nutrient for addition to the soil. This represented a conservation of nitrogen, phosphorus, and potassium or potash by farmers.

Modern composting had its origin in these agricultural practices. The first modern development of composting can be probably traced to the work of Howard [5] in the early 1920s in India in collaboration with Jackson and Wad [6] as well as others. These investigators took the local traditional composting procedures and organized or systematized them into a process called the Indore process. This was a primitive process that involved the anaerobic degradation of animal wastes, night soil, sewage sludge, leaves or straw, or any other organic material. These materials were stacked and turned twice during the composting period of approximately six months or longer. In the late 1920s and 1930s, this process was modified to include turning the compost during degradation to encourage aerobic digestion to occur [7]. Similar practices were followed in Malaya, East Africa, South Africa, and China.

The basis of composting principles can be traced to the 1950s probably beginning with the work of Gotaas and his associates [9] as well as to a number of European processes that have been patented since the 1920s. Among these are the Beccari process in Italy [8] as well as the Bordas process developed in the 1930s, the Earp–Thomas method of 1939, the Frazer process of 1949, the Hardy digester and others [9, 10]. In Europe the Dano process developed in Denmark has been especially popular, and in Holland the Vam processing procedure has been used [11]. In Europe, especially in Germany, there are a number of municipal compost plants in operation. An excellent review can be found in the work of Gotaas [9], as well as in others [12, 13]. In America, during approximately the same period engineers and researchers also carried out fundamental research on the aerobic decomposition of vegetable residues, stable manure, and agricultural residues. A series of publications resulted from Waksman's investigation which set forth many of the basic principles in composting with respect to temperature, rate of decomposition, the role of microorganisms, and so forth [14, 15]. Since World War II, especially since 1950, a large urban problem with respect to solid waste disposal has developed in the United States. Until recently, the traditional methods of treating solid wastes consisted of either sanitary landfill or incineration. With increased urbanization, sanitary landfill has become less profitable because of the increased cost of land and the scarcity of the land for filling. Incineration costs have also risen, accelerated by the fact that incinerators have had to install air pollution devices to improve the quality of the discharged air. Accordingly, opportunistic entrepreneurs, and environmental conservationists combined their forces to promote the use of composting as a major method of solid waste disposal within the United States [12–17].

A summary of some typical composting processes together with a brief description of the process and location, adapted from a U.S. Environmental Protection Agency review [12], is shown in Table 1. In general, a variety of composting processes have been used throughout the world.

Table 1
Typical Composting Processes

Process name	General description	Location
Bangalore (Indore)	Trench in ground, 2–3 ft deep. Material placed in alternate layers of refuse, night soil, earth, straw, etc. No grinding. Turned by hand as often as possible. Detention time 120–180 days.	Common in India
Caspari (briquetting)	Ground material is compressed into blocks and stacked for 30 to 40 days. Aeration by natural diffusion and air flow through stacks. Curing follows initial composting. Blocks are later ground.	Schweinfurt, Germany
Dano Biostabilizer	Rotating drum, slightly inclined from the horizontal, 9–12 ft in diameter, up to 150 ft long. One to 5 days digestion followed by windrowing. No grinding. Forced aeration into drum.	Predominately in Europe
Earp–Thomas	Silo type with eight decks stacked vertically. Ground refuse is moved downward from deck to deck by ploughs. Air passes upward through the silo. Uses a patented inoculum. Digestion (2–3 days) followed by windrowing.	Heidelberg, Germany: Turgi, Switzerland; Verona & Palermo, Italy; Thessaloniki, Greece
Fairfield–Hardy	Circular tank. Vertical screws, mounted on two rotating radial arms, keep ground material agitated. Forced aeration through tank bottom and holes in screws. Detention time, 5 days.	Altoona, Pa., and San Juan, Puerto Rico
Fermascreen	Hexagonal drum, three sides of which are screens. Refuse is ground and batch loaded. Screens are sealed for initial composting. Aeration occurs when drum is rotated with screens open. Detention time, 4 days.	Epsom, England
Frazer-Eweson	Ground refuse placed in vertical bin having four or five perforated decks and special arms to force composting material through perforations. Air is forced through bin. Detention time, 4–5 days.	None in operation
Jersey (John Thompson system)	Structure with six floors, each equipped to dump ground refuse onto the next lower floor. Aeration effected by dropping from floor to floor. Detention time, 6 days.	Jersey, Channel Islands, Great Britain, and Bangkok, Thailand

Metrowaste	Open tanks, 20 ft wide, 10 ft deep, 200–400 ft long. Refuse ground. Equipped to give one or two turnings during digestion period (7 days). Air is forced through perforations in bottom of tank.	Houston, Tex., and Gainesville, Fla.
Naturizer or International	Five 9-ft wide steel conveyor belts arranged to pass material from belt to belt. Each belt is an insulated cell. Air passes upward through digester. Detention time, 5 days.	St. Petersburg, Fla.
Riker	Four-story bins with clam-shell floors. Ground refuse is dropped from floor to floor. Forced air aeration. Detention time, 20–28 days.	None in operation
T. A. Crane	Two cells consisting of three horizontal decks. Horizontal ribbon screws extending the length of each deck recirculate ground refuse from deck to deck. Air is introduced in bottom of cells. Composting followed by curing in a bin.	Kobe, Japan
Tollemache	Similar to the Metrowaste digesters.	Spain; Rhodesia
Triga	Towers or silos called hygienizators. In sets of four towers. Refuse is ground. Forced air aeration. Detention time, 4 days.	Dinard, Plaisir, and Versailles, France; Moscow, U.S.S.R.; Buenos Aires, Argentina
Windrowing (normal, aerobic process)	Open windrows, with a "haystack" cross-section. Refuse is ground. Aeration by turning windrows. Detention time depends upon number of turnings and other factors.	Mobile, Alabama; Boulder, Colorado; Johnson City, Tennessee; Europe; Israel; and elsewhere
van Maanen process	Unground refuse in open piles, 120–180 days. Turned once by grab crane for aeration.	Wijster and Mierlo, the Netherlands.
Varro conversion	Enclosed digester, eight decks, 160 ft long, 10 ft wide, and approximately 1 ft deep. Continuous flow, mixed by harrows. Forced air aeration. 40 h detention time. Nutrients and water added to the process.	Brooklyn, N.Y., pilot plant in Stuttgart, Germany. None in operation.

II. MICROBIOLOGY AND CLASSIFICATION

A. Microbiology

Composting is essentially a microbiological process and the transformations that are carried out are biochemical in nature. The breakdown of organic waste materials is carried out by microorganisms and accompanied by temperature elevations. In general, the microorganisms that carry out composting originate either from the soil that is intermixed with the refuse or from food or organic waste materials present. At times, special inoculums or preconditioned, "acclimated" microorganisms have been added to compost to provide the appropriate microorganisms. Some such inoculums are described as mixtures of several pure strains of laboratory cultured organisms especially prominent in the decomposition of organic substances and in nitrogen fixation. Others are purported to contain, as well, such constituents as activated factors, biocatalysts, enzyme systems, hormones, etc. The benefits associated with such practices are controversial and have been noted by a number of investigators [18–23]. At least one commercial composting process [22] has been based on the use of special inoculums. In general, the majority of the technical evidence indicates that special inoculums provides no special benefits or advantage to the composting process [23].

The most active microorganisms in the composting process are a wide variety of bacteria, fungi, and actinomycetes. At times, however, yeasts and protozoa are also seen [23]. All of these organisms are active at different times and represent a variety of physiological conditions relative to oxygen and temperature.

With respect to oxygen requirements, microorganisms can be characterized as those requiring oxygen (obligate aerobes), those to which oxygen is toxic (strict anaerobes) and facultative, where the microbes are mostly aerobic, but can grow in the absence or presence of small amounts of oxygen, depending on the environment. Most of the microorganisms found in compost operations are of the facultative type.

Microorganisms that work in the temperature range of 10–40 °C are called mesophilic organisms, and those that work in the temperature range of 40–70 °C are called thermophilic organisms.

Bacteria are single-celled, microscopic organisms, capable of rapid growth that are found in both aerobic and anaerobic environments. They contain about 80% water and 20% dry material, of which about 90% is organic and 10% inorganic. The organic fraction consists of proteins, carbohydrates, and lipids found in different parts of the cell. The inorganic components consists of phosphorus, sodium, calcium, magnesium, potassium, and iron as well as trace minerals. The aforementioned organic and inorganic fractions can also be applied to sludges (e.g., excess activated sludges) generated from biological wastewater treatment plants.

Fungi are filamentous, spore-forming, nonphotosynthetic, heterotrophic (organic-consuming) microorganisms that have the ability to degrade a wide variety of organic compounds, under low moisture and a broad range of pH conditions.

Actinomycetes are a group of microorganisms possessing properties similar to both the fungi and the bacteria. In form they are similar to fungi except that they are less filamentous, spore forming, and adapted to soil growth. The actinomycetes have a distinct role in the degradation of semidry materials in composting.

Yeasts are a stage of fungal growth adapted to unicellular, vegetative growth. In general yeasts prefer soluble carbohydrate or sugar substrates. The protozoa are a group of motile, single-celled microorganisms that often feed on other microorganisms and which may or may not be present in a composting operation.

Although members of each of the aforementioned microorganism groups can be found that are capable of decomposing all the raw materials in solid wastes and residues, as a group they prefer different compounds. Typically, fungi, yeasts, and actinomycetes prefer celluloses and hemicelluloses, whereas bacteria are particularly effective in the decomposition of simple water-soluble sugars.

Aside from metabolic requirements, the predominance of microorganisms varies during the course of the composting process. In the University of California studies [24], bacteria were characteristically predominant at the start of the process; fungi appeared in 7 to 10 days, and actinomycetes became conspicuous only in the final stages of composting. Bacteria were found in all parts of a pile when the windrow method was used, whereas actinomycetes and fungi were confined to a sharply defined outer zone, 5–13 cm thick, beginning just under the surface of the pile. The limitation of these two groups to the outer zone probably was a function of temperature and/or aeration. The microorganisms that have been reported in composting mixtures during different stages can also be found in the work of Fesenstein et al. [25], and Kane and Mullins [26].

Composting involves the biological decomposition of the organic constituents of wastes under controlled conditions, and is therefore subject to the constraints of all biological activities. To continue to grow and function properly, microorganisms must have a favorable environment (e.g., temperature, pH, moisture content, and oxygen), and a source of energy and carbon for the synthesis of their new cellular material [27]. Nutrients, such as nitrogen, phosphorus, and other trace elements are vital for all synthesis. Environmental and nutritional requirements are described briefly in Section II.B, Classification, and fully in Section IV, Process Parameters.

B. Classification

Composting systems can be classified according to oxygen usage, temperature, technological approach, operational modes, raw material, and operating methods.

1. *Aerobic and Anaerobic Systems*

The microbiological, biochemical transformations take place either aerobically or anaerobically. For satisfactory operations it is important to note the differences between aerobic and anaerobic biochemical reactions with respect to composting. Aerobic decomposition of the organic matter in solid wastes by microorganisms results in end products that generally consist of carbon dioxide, water, and a variety of oxidized end products. The amount of oxygen required for aerobic stabilization of municipal solid wastes can be estimated by using Eq. (1):

$$\begin{aligned} C_aH_bO_cN_dS_eP_f + 0.5\,(nw + 2s + r - c + t)O_2 &\longrightarrow \\ nC_uH_vO_wN_xS_yP_z + sCO_2 + rH_2O + (d - nx)NH_3 & \\ + (e - ny)SO_4^{2-} + (f - nv)PO_4^{3-} & \qquad (1) \end{aligned}$$

$$\begin{aligned} \text{where } r &= 0.5\,[b - nv - 3(d - nx)] \\ s &= a - nu, \quad \text{and} \\ t &= 2(e + f - ny - nz) \end{aligned}$$

The terms $C_aH_bO_cN_dS_eP_f$ and $C_uH_vO_wN_xS_yP_z$ represent the empirical mole composition of the organic wastes initially present and at the conclusion of the process, respectively. If complete aerobic conversion is accomplished, the corresponding expression is:

$$\begin{aligned} 4\,C_aH_bO_cN_dS_eP_f + (4a + b - 2c + 5d + 6e + 5f)O_2 &\longrightarrow \\ 4a\,CO_2 + (2b - 2d - 4e - 6f)H_2O + 4d\,NO_3^- & \\ + 4e\,SO_4^{2-} + 4f\,PO_4^{3-} + (4d + 8e + 12f)H^+ & \qquad (2) \end{aligned}$$

If nitrogen in the organic waste is only converted to ammonia, Eq. (2) must be modified:

$$4\,C_aH_bO_cN_dS_eP_f + (4a + b - 2c - 3d + 6e + 5f)O_2 \longrightarrow 4a\,CO_2 + (2b - 6d - 4e - 6f)H_2O + 4dNH_3 + 4e\,SO_4^{2-} + 4f\,PO_4^{3-} + (8e + 12f)H^+ \quad (3)$$

It can be seen from Eq. (2) that the completely oxidized end products represent essentially carbon dioxide, water, nitrates, sulfates, phosphates, and other end products that are considered to be stable, that is, compounds in which biological decomposition no longer takes place. Aerobic decomposition generally is characterized by high temperatures; therefore, it requires less detention time than anaerobic decomposition and results in an end product that is less offensive, less odorous, and more rapidly stabilized. Because of the advantages of aerobic systems most modern composting processes are basically aerobic—or attempt to be.

Other factors that must be considered with respect to aerobic composting are those related to the rate (i.e., detention time), mixing devices, and air circulation. At the practical level, oxygen is provided to the process through contact with air, or air renewals. This is usually achieved by mixing or forced aeration. Generally, if a mixture of solid wastes remain undisturbed, the biological requirements are such that oxygen, especially in a moist environment (e.g., over 70% moisture) is depleted locally and anaerobic conditions follow. Aerobic composting must be achieved by mixing, turning, tumbling, agitating, or otherwise forcing air through or into the composting mass or pulp.

Anaerobic composting occurs in the absence of oxygen, that is, no air renewal, or aeration. The end products, beside carbon dioxide and water, consist of methane and ammonia as well as other reduced compounds such as hydrogen sulfide, organic acids, ketones, aldehydes, etc. The overall conversion can be represented by the following equation:

$$C_aH_bO_cN_dS_e \longrightarrow nC_uH_vO_wN_xS_y + mCH_4 + sCO_2 + rH_2O + (d - nx)NH_3 + tH_2S + 2(ny - e)H^+ \quad (4)$$

where $s = a - nu - m$

$r = c - nw - 2s$, and

$t = e - ny$

Again, the terms $C_aH_bO_cN_dS_e$ and $C_uH_vO_wN_xS_y$ represent the empirical mole composition of the organic wastes at the start and the end of the process. When the organic wastes are stabilized completely under primarily anaerobic conditions, the biological conversion can be represented by Eq. (5) [28]:

$$8C_aH_bO_cN_dS_e + (8a - 2b - 4c + 6d + 4e)H_2O \longrightarrow (4a - b + 2c + 3d + 2e)CO_2 + 8dNH_3 + (4a + b - 2c - 3d - 2e)CH_4 + 8eH_2S. \quad (5)$$

Anaerobic decomposition is characterized by low temperatures (unless heat is applied from an external source), and generally proceeds at a slower rate than does aerobic composting. Besides, the end products from anaerobic composting process are generally more odorous and offensive substances than those formed aerobically. In the 1920s and 1930s, the early agricultural practices related to composting were anaerobic in nature and required prolonged time for completion. Although most accelerated modern municipal compost operations are aerobic, anaerobic composting offers two important advantages: First, the anaerobic process can be carried on with a minimum of attention, and as such it can be sealed from the environment. Second, more cellulose materials can be degraded by anaerobic composting, compared to aerobic composting, because microbes in anaerobic system have a long time to digest, hydrolize, and decompose the waste material.

2. *Temperature*

Composting processes are naturally characterized by temperature elevations in the decomposing pulp material. Classified according to temperature range, composting is either mesophilic or thermophilic. As the term implies, the former is conducted at temperatures from 10 to 40 °C which in most cases is the ambient temperature. The latter is conducted at temperatures from 40 to 70 °C. In actual practice, most composting processes include both mesophilic and thermophilic ranges.

The temperature elevations result from the microbial activities on the substrate and the physical conditions present. Self-heating masses are dynamic with respect to moisture, oxygen, substrate, and other abiotic factors. A number of excellent studies have been made relating to temperature elevations from microorganism in a number of substrates including wastes [29].

3. *Technological Approach*

Classified according to technology, there is windrow (or open) composting and mechanical (or enclosed) composting. A windrow is a long, low heap of organic matter. Windrow composting is carried out entirely in the open, and the materials to be composted are usually stacked in elongated windrows.

In mechanical composting systems, initial composting activity usually takes place in an enclosed digester that permits closer process controls of the important environmental parameters such as temperature, aeration, moisture, and addition of nutrients or supplements. This will be discussed in later sections.

4. *Operational Mode*

There are two general operational modes that can be used in carrying out municipal composting; they are batch and continuous flow, both of which are discussed in more detail later in this chapter. In batch methods a fresh mass of material is processed all at once and no new material is introduced into the process once it is started. After a period of time, the microbiological compost reactions are successfully completed and the process is terminated. Both operations are used to a large extent in Europe where the production of a product that can be used as an agricultural adjunct is required. Batch operations are usually best suited for small communities or rural areas where land is available and costs are small.

In a continuous-flow system material is added continuously to the process and the end product, compost, is continuously removed. Continuous-flow operations are more desirable for larger municipalities or areas where there is a large and continuous input of solid waste material or where a process must meet a time schedule. In general, continuous-flow operations are more highly mechanized and require more complicated designs, engineering, and planning, as well as closer process control and more highly skilled personnel.

5. *Raw Materials*

The composting process can be classified according to the raw materials into municipal refuse composting, sludge composting, combined refuse-sludge composting, manure composting, paper composting, etc. They are self-explanatory. This chapter emphasizes mainly the municipal refuse composting in which the solid wastes collected from municipalities are disposed of by modern aerobic conventional composting system.

In sludge composting, sludge is composted without the inclusion of solid waste fractions. Generally, the sludge must be blended with some bulking material if windrows are to be used. This bulking material can be soil, sawdust, wood chips, etc. If a mechanical aeration system is used, bulking agent requirements are less severe. Various raw, dewatered, digested, or combined sludges can be composted to a different degree of success. Raw primary or secondary sludges are generally preferred to digested sludge, although anaerobically sludge can also be composted with little difficulty during favorable weather. When anaerobically digested sludge is to be composted aerobically, the production of offensive odors and the survival of pathogens are possible. To cope with these anaerobic conditions, the compost mixture (wood chips to raw sludge at a ratio of 1:3) should be windrowed and forced aeration used to draw air through the stack for a period of from 16 to 20 days. In general, either digested or raw sludge can be composted in an aerated pile. Deodorization can be achieved by passing the gas through a pile of screened compost, and destruction of bacteria is much greater than in windrows that do not employ forced aeration.

When combined refuse-sludge composting is practiced for balancing nutrients and moisture, sludge to refuse ratios of roughly 0.50:1.00 by wt may be employed. A moisture content between 45 and 65% by wt is desirable. Either raw or digested sludge can be used. Since sludge has large amounts of water, and since excessive amounts of water adversely affect composting operations, attention must be given to mixing ratios. Raw sludge is preferred to digested sludge because it can be dewatered more readily and has a higher nutrient content. However, other factors such as drying time and percent moisture desired, especially with raw activated sludge can be important. Furthermore, most sludge also contains heavy metals that may or may not affect the composting process. Sludge ranges from 3–6% solids content; and 12% is about the limit for pumping. Raw sludge contains nitrogen levels of approximately 1500 mg/liter of which one third is ammonia. Thus if sludge containing 95% water is mixed equally (by wt) with processed refuse of 25% moisture, the resulting mixture contains 60% moisture and has approximately 0.2% nitrogen (dry wt), of which only about one-third is readily available. Thus, for accelerated composting, nitrogen supplementation may be necessary, since optimal composting usually requires higher concentrations of nitrogen.

6. *Operating Methods*

In general, composting operations consist of the methods previously described, e.g., digestion using moisture supplements which is conventional composting. Any other composting method is nonconventional composting.

A new method used successfully in municipal refuse composting, for example, in the Altoona, Pa., plant, is hydropulping. In this process, large quantities of water or waste waters are added to the solid material which is then mechanically reduced with a pulper to a liquid slurry. The slurry is then dewatered and the pulp is added to a digester for composting. Decomposition usually proceeds quickly and a satisfactory end product is obtained that has been used as a soil supplement. In wet pulpers such as are used in Altoona, cans, bottles and other noncompostable items are not normally part of the incoming refuse [12].

Other nonconventional composting processes are in developmental stages.

III. DESIGN APPROACHES

A. General Approach

The quality of compost depends directly on the raw solid wastes and the type of preprocessing and/or postprocessing carried out on the wastes. Prior to receiving material to be composted some control can be exerted that will affect the quality of the solid wastes. In a well-developed solid-waste disposal program, collection of refuse may be controlled at the consumer level to restrict the materials received as solid wastes at pickup and thus eliminate bulky, unwieldy, or otherwise unsatisfactory material for composting, (construction debris, refrigerators, dead animals, large metal objects, trees, etc.) Grinding or shredding equipment can be damaged by the input of certain objects or materials that are best eliminated at the source.

There are no textbook approaches that can be applied to the design of a compost plant or for operation. In general, adequate background information concerning the community to be served is essential. Solid wastes from a highly urbanized area has a different composition compared to refuse from a suburban or rural area. The difference is in paper content (i.e., cellulose) and other degradable components. Other factors that might appear to be obvious should also be considered such as geographical location and weather (temperature and rainfall), since this affects the solid wastes added to the process. The level of commercial or industrial input to the wastes may also be important since such material affects the wastes composition. Data should be collected with respect to the amount of solid waste generated in the community. In the United States, approximately 3.2 kg (7 lb) of solid wastes are generated per capita per day. However, production rates for different locations vary and must, therefore, be determined by analysis of existing data or surveys.

It is believed [30] that plastics, most textiles, leather, and rubber are essentially noncompostable. The organic fraction of most compostable municipal solid wastes includes, but is not limited to: (a) water-soluble constituents, such as sugars, starches, amino acids, etc.; (b) hemicellulose; (c) cellulose; (d) fats, oils and waxes that are esters of alcohols and higher fatty acids; (e) lignin, a material present in some pulp and paper products or by-products [31]; (f) lignocellulose, a combination of lignin and cellulose; and (g) proteins.

Another valuable criterion affecting solid wastes is the length of the work week. For example, a plant operating on a 5-day work week (or some variation thereof) should plan to accept solid wastes at 1.4 times the rate of a 7-day per week operation. Since the compost plants must often operate according to refuse pick-up schedules, much of the solid wastes pretreatment (e.g., grinding, sorting, etc.) must be geared to this schedule. Grinder outputs and waste materials movement and storage must be designed or planned around this schedule. In a continuous-flow operation, for example, storage facilities must be designed to provide refuse inputs through long weekends or holidays. In this case a consideration would be the prevention of premature decomposition or odor problems with the stored refuse.

It has been mentioned earlier that the quality of compost also depends on the type of preprocessing and/or postprocessing carried out on the solid wastes. In general, municipal solid wastes composting consists of at least five major process operations that must be carried out. These are summarized in Fig. 1, and consist of (a) pretreatment or sample preparation, (b) digestion, (c) curing, (d) finishing or upgrading, and (e) storage of the end product, which will be described in subsequent sections.

B. Pretreatment

An ideal processing system for composting would remove all glass, ferrous, and nonferrous metal, aluminum, as well as plastics, leather, rubber, and most textiles. In practice this is never achieved. These materials are generally considered to be nonbiologically degradable and contribute little to the overall final quality of the composted product. In at least one process (the Varro process, see Fig. 2), municipal refuse is shredded and added to the digester, with the exception of ferrous metals. However, pretreatment or sample preparation prior to composting is an important step.

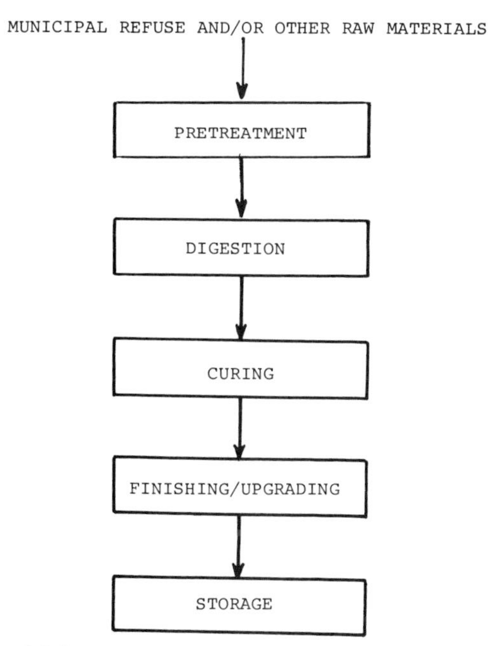

FIG. 1 Major process operation sequence in composting.

Pretreatment involves receiving, sorting, size reduction, ferrous removal (if it is to be carried out), as well as storage and moisture adjustments. Pretreatment measures are generally built around the digestion or the microbiological stabilization phase of the process. A detailed description of modern pretreatment processes is discussed in other chapters. In this chapter, only the pretreatment measures and their proper selection to correspond to the composting process are presented.

Size reduction is achieved mainly by grinding, shredding, rasping, or chopping the solid wastes to a smaller particle size of uniform density. A variety of mills (e.g., hammer mills, chain mills, etc.), as well as wet pulpers have been used. The most commonly used machinery for shredding or grinding is the hammer mill consisting of high-speed swing-type hammers that force the refuse through grates, reducing the size. Usually a considerable amount of noise as well as vibration is associated with hammer mills. In Europe, raspers are used to shred refuse and are favored since they require less power. In at least one compost plant in the United States, wet pulping is carried out. This involves adding moisture to the incoming refuse and shredding the refuse while wet, which facilitates the shredding. However, prior to

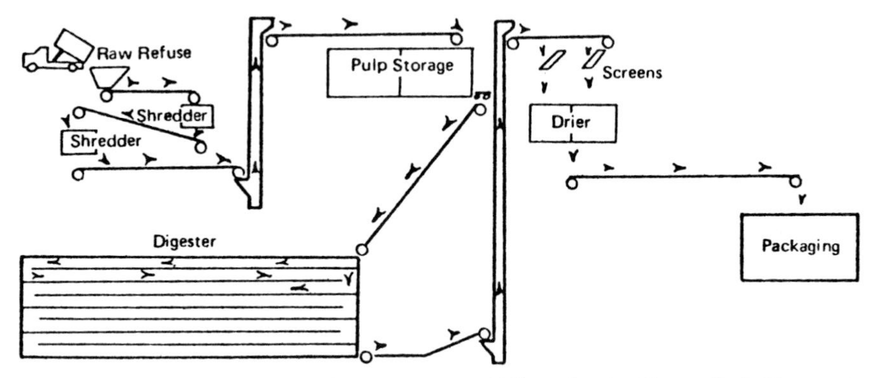

FIG. 2 Varro process flow diagram. *From:* "Completely Controlled Compost Plant Opens," *Eng. News. Rec.*, April 1, 1971, copyright, McGraw-Hill, Inc., all rights reserved.

further processing, the material must be dewatered and the water recovered is now a waste water that must be treated.

In general, grinding or shredding, in one or two stages, is the most important pretreatment operation. This step improves materials handling measures as well as biological decomposition. Grinding, if rapid composting is to be carried out, should reduce the refuse to less than 2.5 cm particle size or smaller, if possible. In general, reduction in particle size increases the exposed surface available for microbiological activity and for moisture and nutrient supplement additions, and thus is directly related to the process. Sorting, including ferrous removal, should be consistent either with the overall solid waste recovery program, if one is used, or sufficient to protect the grinders or shredders.

C. Digestion

Digestion or microbiological stabilization in composting operations as indicated above, is carried out by either batch or continuous-flow methods. In general, the same overall nutrient and moisture considerations apply to either process. However, in continuous-flow operations optimal activity is much more dependent on nutrient, moisture, and environmental control for sustained, optimal production.

Digestion, as will be seen, can also be achieved through the use of windrows, pits, trenches, cells, tanks, multistoried or multidecked towers or building, drums, or bins. Some of these processes combine more than one of these digester types. Usually a special digester is used (e.g., batch or continuous flow) and this is combined with a storage area for curing which is carried out in a windrow or a bin.

1. *Batch Operations*

The most common batch process used in municipal composting is the windrow method. In the windrow process, after pretreatment—which may consist only of grinding—the material is heaped into stacks called windrows. These windrows may be from 1.5 to 5 m high. Nutrients as well as moisture supplements, which will be discussed later, may or may not be added to accelerate the process. Biological activity quickly elevates the temperature in the stack as high as 65–80 °C and persists for a period of time while the volume of the pulp gradually becomes reduced. Periodically, the windrow is mechanically turned, that is, mixed so that aeration is achieved and parts of the stack do not become anaerobic. At the practical level, aeration is achieved by mixing turning, agitating, or forcing air through or into the composting "pulp." Gradually the appearance of the material changes from recognizeable ground municipal refuse into a soil-like material, or humus, that is darker in color and begins to resemble a rich soil. After the available organic material has been degraded, the temperature in the center of the windrow decreases to ambient level and the product is said to be "stabilized" since no further biological degradation is taking place.

When batch preparations are used, sample preparations may be minimal and can consist only of sorting out the large materials, primary grinding, and, possibly, moisture adjustments. Land requirements can be estimated from the quantity of solid wastes, the bulk density, and the detention time of the process selected. If nutrients are to be used, the amounts of nitrogen and phosphorus added should be based upon the amount of available carbon calculated on a ratio of approximately 30–35 to 1 (C:N); phosphorus should be approximately one-fifth of the nitrogen level. The addition of sludges, sewage or industrial wastes has been used in composting operations. Sludges and sewage, providing moisture and nutrients, are also potential sources of pathogenic microorganisms, heavy metals, and other contaminating organic compounds. Thus end product utilization should be evaluated in this light. In general, available nitrogen and phosphorus from such sources should be calculated together with moisture additions. For windrow operations, moisture content should be 35–65%. In practice, however, in windrow operations little effort is made to control the process closely. Batch process operations also requires equipment to be used in recovery operations as well as to turn the material periodically. Screens are often used to sort or classify the material or end product. Usually large areas must also be provided for "curing," which will be discussed in a later section.

2. *Continuous-Flow Operations*

Continuous-flow, semi-continuous, or even intermittent composting process operations can also be used. In general, after pretreatment, the ground or shredded refuse is stored and continuously added to the digester or composting unit. Continuous-flow composting operations, as opposed to batch operations, use considerably shorter detention times to reach stability. Continuous-flow operations can be carried out through the use of open pit or enclosed digesters. Enclosed digesters have the advantage that closer control of the end product is possible, thus detention times are shorter and more control of the end product can be achieved.

a. *Open-Pit Digestion.* Open-pit digestion is carried out in the open in a tank or open pit. The pretreated refuse (e.g., ground, sorted, etc.) is added to the pit on a continuous or semi-continuous schedule. Usually a mechanical device is used to periodically turn or aerate the composting solid wastes. As in the windrow method, temperature rises in the composting pulp and the volume is reduced after a period of time. Quite often additives are introduced to accelerate the process. In the now closed Metro plant located in Houston, Texas, sewage sludge was added [32]. Usually the material is introduced through conveyors to the digester, where a tracked mixer turns or mixes the composting mass approximately once a day. Detention time in the digester, for this type of operation, is about 5 days. In the Houston Metro plant, this was followed by an additional curing period during which reduced biological activity takes place and the material becomes stable and more homogeneous and closely resembles humus or soil in appearance.

b. *Enclosed Digesters.* Enclosed digesters are usually highly mechanized, but offer the advantage of close environmental control. As for the open-pit digester, mixing or agitation of the composting pulp must take place. A wide variety of configurations has been used in closed-digester construction.

The more common types of configurations are shown in Fig. 3. Agitation or mixing to achieve aeration, is accomplished by mechanical means such as screws, plows, rakes, bucketlifts, etc.

As shown in Fig. 3, the composting or digesting pulp can be agitated through the use of moving conveyors as occurs in the Naturizer process [33]. In this process the conveyors are stacked in such a way that material from the preceding conveyor or from the incoming pretreated refuse is discharged to the conveyor below, which in turn discharges the pulp to the conveyor belt below it. Mixing of the digesting pulp can be assisted through other mechanical devices (projecting bars,

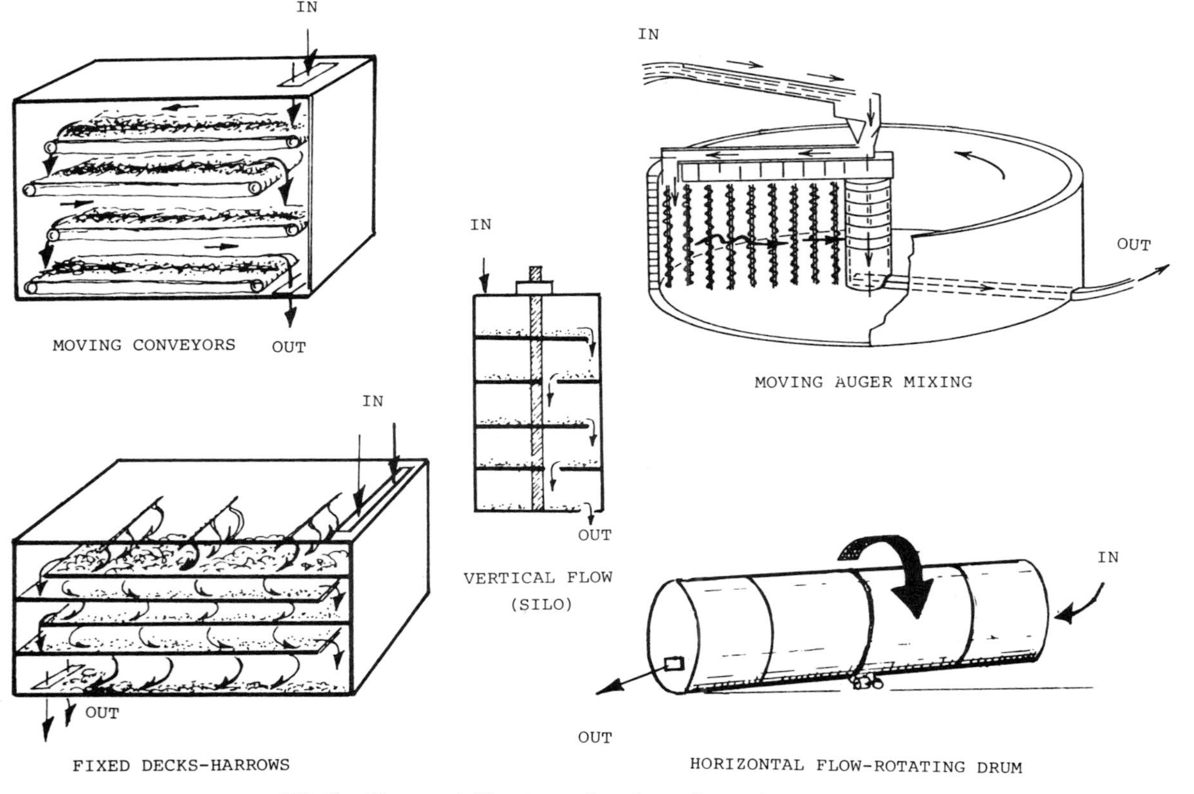

FIG. 3 Compost digesters, closed configurations.

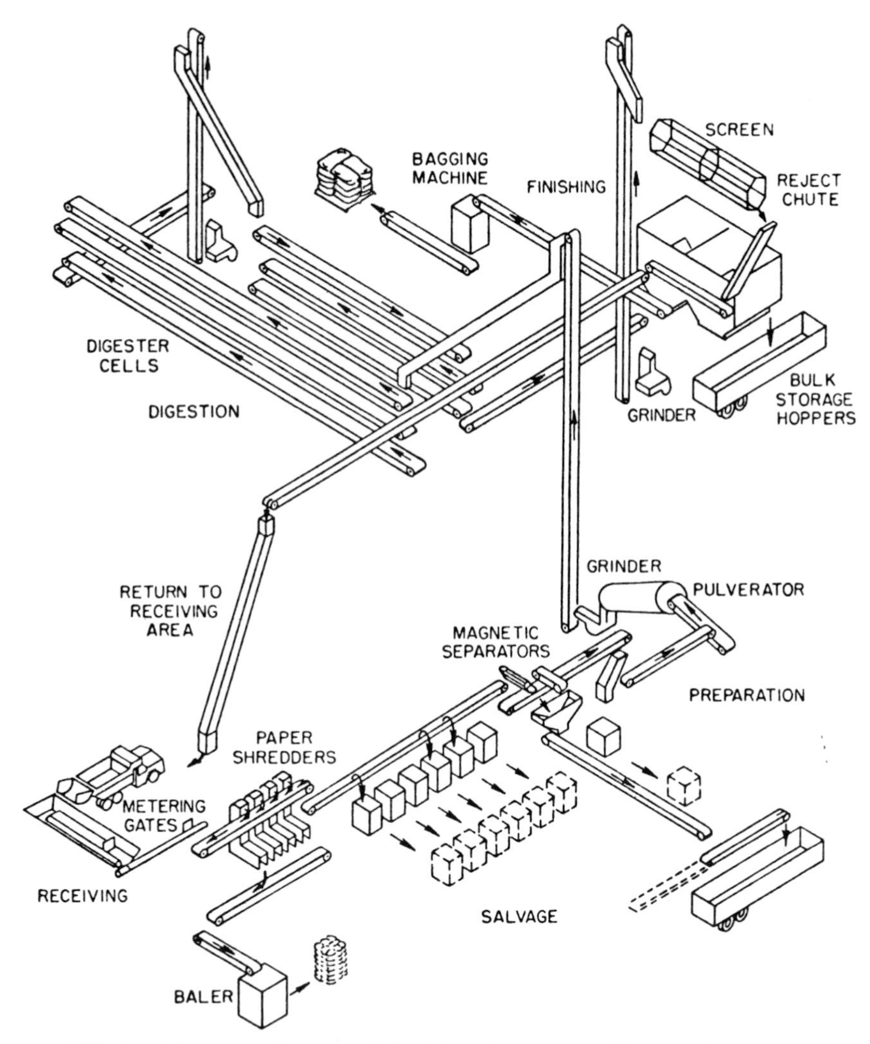

FIG. 4 Schematic diagram of the Naturizer system. From reference 33.

rakes, etc.) In the Naturizer process, as operated in St. Petersburg, Fa., by the International Disposal Corporation (see Fig. 4), the material was transferred or moved about once a day, and fans were used to provide air to the digesting mass in the reactor. To achieve both dust and moisture control, digested sewage sludge, raw sewage sludge, water or segregated wet garbage is added at the first grinding. Results from the Norman, Okla. compost plant indicate that this process can be completed in 6 days [34].

Among the few operating compost plants in the United States, perhaps the most successful one has been the research or experimental wet-pulp Fairfield-Hardy plant located in Altoona, Pa. (see Fig. 5). This plant with a daily capacity of 41 metric tons has been operated by Altoona FAM, Inc. for approximately 25 yr. Loadings range from 22.3 to 40.8 metric tons of processed solid wastes per day. Garbage, paper, and other refuse are separated from glass and cans prior to collection. Initial grinding is through use of a hammer mill. Wet pulpers are used to comminute or reduce the refuse. The pulper consists of a large bowl containing a round steel plate studded with teeth that can be made to rotate. After the bowl is partially filled with water, the steel plate is rotated at approximately 650 rpm, and the raw refuse is added. As the refuse enters the bowl it is thrown against the teeth on the rotating steel plate and shredded. Water as well as sewage sludge are added so that a slurry of about 5% solids is maintained in the pulper. The slurry is then passed through a bar screen to remove noncompostable materials such as, cans, etc. Prior to digestion the material is dewatered to about 58% moisture [12]. The water, now a rich wastewater, must also be disposed of after removal.

The digester in the Altoona operation consists of a 11.4 m circular vat to which the pulped and dewatered solid wastes are added on a semi-continuous basis. Material is added to the top of the tank to a depth of 1.8 m and mixing is achieved through the use of augers mounted from a movable bridge. The augers slowly turn through the pulp, moving the material upward and toward the center of the tank, mixing the digesting, pulped refuse. Temperatures in the digester usually run between 60–77°C. Air is introduced into the digester through the use of blowers at the bottom of the tank. Stabilization is achieved in the process after approximately 5 days detention [35]. After composting in the digester, the material is further cured for about three weeks in windrows for a period of a few weeks before it is marketed.

A number of variations to the vertical silo, shown in Fig. 3, have been used [36, 37]. As shown, the ground or shredded refuse is introduced to the silo or circular tank. Inside the digester the material may be rotated or moved through the use of revolving arms or platforms. If platforms are used, openings between the platform levels facilitate movement of the pulp through the digester. Often air is forced through the composting pulp to maintain the digestion process aerobic. After a period of detention, usually days, the composted material is withdrawn for further use.

In Europe the Dano Bio-Stabilizer has been used to a great extent [9]. The digester, shown in Fig. 3, consists of a horizontally

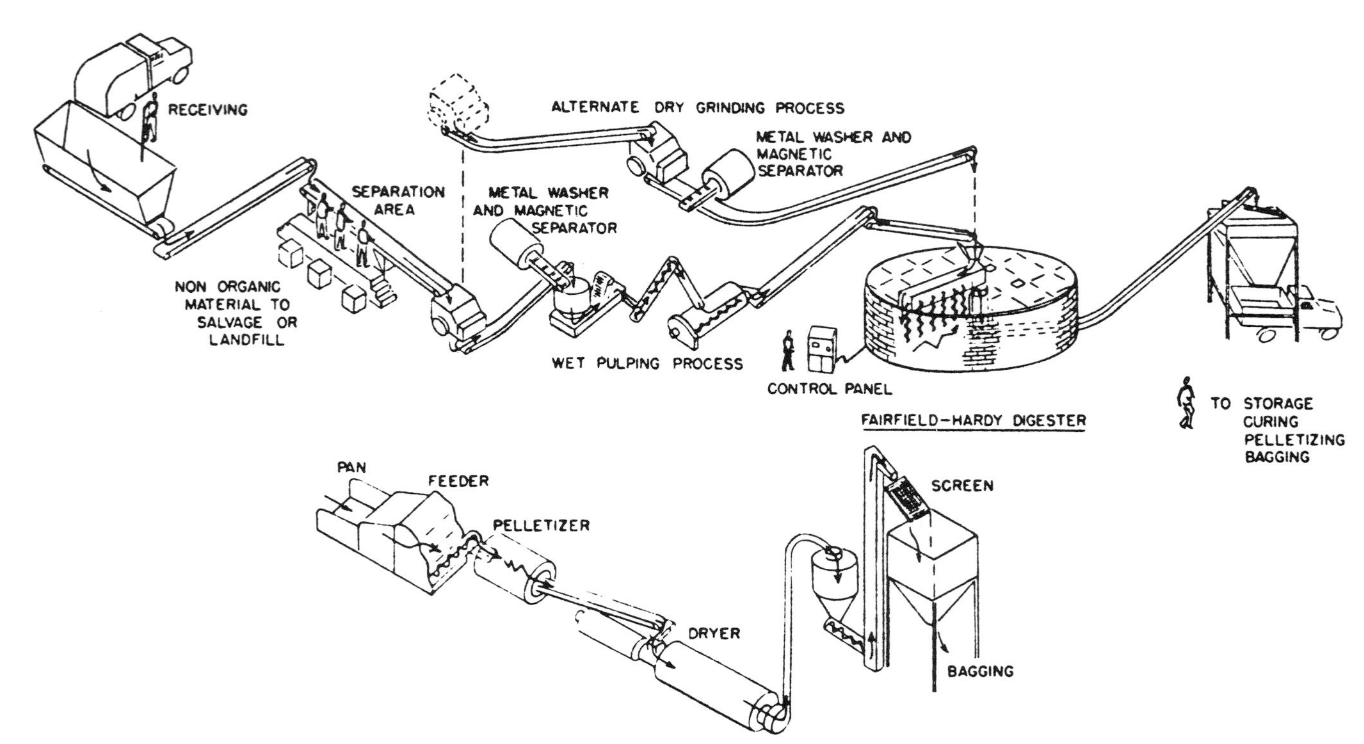

FIG. 5 Fairfield-Hardy compost system. From reference 33.

placed, rotating steel drum or cylinder set at a slope so that material introduced at the upper end can work its way through the drum when it is rotated. As in most continuous processes, moisture and aeration is employed. The drum is usually rotated at a speed of from 1.5 to 5 rpm. Final composting is achieved in open windrows. Composting can be completed in two weeks with periodic turning on the 4th, 8th, and 12th days.

One of the newer continuous-flow, mechanized plants that has been used in a major municipality is the recently closed Ecology plant, located in Brooklyn, New York [37, 38]. This was a 136 metric ton/day fixed-deck plant that employed harrows for mixing. A more complete flow diagram for this plant is shown in Fig. 2. This plant is a good illustration of a modern, totally mechanized municipal composting plant. The objectives of this plant were to produce a vehicle for fertilizer or soil conditioners.

The Ecology plant was divided into three work areas: (a) pretreatment, that is, receiving, sorting, and storage; (b) digestion or bioreactor operations; and (c) finishing, including final drying, screening, upgrading, and bagging. Pretreatment, that is, receiving, grinding, and sample preparation was carried out to correspond with refuse pickup schedules. Although the refuse was unsegregated, local regulations restricted the material placed in refuse cans mostly to garbage and paper together with bottles, cans, plastics, and rags, but excluding construction debris and larger bulky or objectionable materials (dead animals, etc.). Loading into the receiving hopper was manually assisted as well as through the use of a "cat," or at times a backhoe. A 1.6-m moving metal apron with a metal leveling device introduced the material to an inspection station where the material was inspected for any objects harmful or obstructive to the equipment (tires, large metal objects, etc.). Solid wastes reduction was through the use of two stages of shredding which effectively reduced the material to approximately 1 cm or less. An overhead magnetic head pulley was used to remove ferrous metals. The shredded material was then moved using a 79-cm conveyor belt and bucket elevators into a compartmentalized open storage tank of a capacity of 409 metric tons. The storage tank was operated so that material could be continuously removed to the digester using a traversing rotating screw located at the bottom of the storage bin. Conveyor belts and bucket-lift elevators were then used to move the material to the top of the digester tank.

The biological reactor in the Ecology plant [37, 38] consisted of eight stacked, fixed decks totally enclosed. The decks measured 49 m long and 3 m wide, with 1.2 m between decks. The digesting pulp material was moved along the fixed decks through the use of harrows, or plows, set so that one set of harrows on a chain-driven belt could be used to service paired, adjoining decks. In general, depth of digesting pulp ranged from about 16 to 39 cm. The harrows, as shown in Fig. 3, push the material forward, mixing and aerating the compost pulp as the harrows push through the material. At the end of each deck the composting pulp is pushed onto the deck below. Usually, composting to stabilization in the digester was achieved in approximately 40 h, which corresponds to deck 6. Thereafter biological activity decreases. The last two decks were used for drying. Air was continuously circulated throughout the digester. In this process, moisture as well as nitrogen and phosphorus supplements were added to digesting material. Dosages for nitrogen and phosphorus were worked out based on respirometric data [39].

D. Curing

Quite often, the digestion process is followed by a prolonged, undisturbed period of stabilization called curing. Usually this is carried out following the windrow process. During curing further microbiological stabilization takes place which generally improves the end product if it is to be used as a soil supplement or for agricultural purposes. Curing may require several weeks or months, which is determined by the end use of the composted product. If the compost is to be used for agriculture it will compete for nitrogen, depleting the nitrogen from the soil; without curing its use will be counterproductive.

Generally, in the windrow process, an additional two weeks following digestion is adequate for curing. When mechanical processes are used, the period of time required for curing may range from one week to several months. In the Ecology process (Fig. 2) no curing was employed, whereas in the Altoona plant (Fig. 5), the product is cured for three weeks [38].

E. Finishing or Upgrading

Composted municipal solid wastes, although more or less uniform in color and appearance, on close examination reveals the presence of glass, metals, plastics, wood fragments, rags, and dust. Usually the composted end product has a light to fluffy consistency, depending on the final moisture content. Although the material can be used in this form for many applications, it is frequently desirable to improve or upgrade its consistency to minimize handling problems as well as to improve quality. This can be achieved by the use of several process operations that can be used alone or in combination to impart the desired characteristics to the compost product.

Probably compaction and pelletizing are the most commonly employed processes to achieve upgrading. Compaction is usually achieved by forcing the product through large, heavy rollers that compress the product. Optimal compaction, at times, requires a given moisture content, and, at times, to enhance or increase the bulk density of the product, additives may be used. Compaction is usually followed by chopping or milling and screening to obtain proper sizing.

Municipal solid wastes, after composting, become the so-called "humus," which does not contain sufficient nutrients such as nitrogen and phosphorus to merit its use as a fertilizer. However, this humus can serve as a useful soil supplement, since it does contain some nitrogen and phosphorus as well as organic material and trace minerals. At times, composted refuse has been upgraded to fertilizer. This has been achieved through the addition of chemical fertilizers to certain levels to produce a commercial fertilizer of a specific composition using the composted refuse as carrier or vehicle. This was done at the Ecology plant in Brooklyn [38]. At other times garden or agricultural products have been prepared. In the Altoona plant [38], mostly starch is added as an adhesive and the compost mixture is granulated, dried, screened, and bagged.

F. Storage

Since compost results in a product that is tied to the agricultural planting season, the greatest demand for compost is in the early spring and fall. As a result, storage must be an integral part of design. Storage can be combined with the curing phase and the product can be stored for later upgrading or in a finished form. At times the compost is stored in piles outdoors. However, the product should be protected and kept at a more or less uniform moisture level if it is to be further upgraded or bagged.

IV. PROCESS PARAMETERS

A. Nutrients

The most important biological process parameters with respect to rate of digestion in the design and operation of continuous-flow composting operation are nutrients. It has already been noted that solid wastes are deficient in nitrogen and phosphorus and that if high rate composting is to be achieved, nutrient supplements must be added in controlled amounts. In general, soluble sources of nitrogen and phosphorus are desirable and can be readily mixed with the liquid supplements.

Although commercial fertilizers solubilize readily and are good sources of nitrogen and phosphorus, the addition of ammonium or nitrate salts as well as soluble phosphates, can also meet these requirements. Since moisture, as well as nutrients, are also required to enhance the biological process operations, as in batch operations, nitrogen and phosphorus supplements are sometimes added in the form of municipal sludges or sewage. Besides the negative aspects already noted with respect to pathogens, the use of sludges can affect the product and the process must generally be considered with respect to the physical characteristics of the end product. Activated sludge, for example, contains polysaccharide slimes than can create handling problems in the composted end product. Moreover, the continuous input of sewage or sludge creates other problems related to sludge handling as well as waste waters and other potential odor sources.

In general, the most desirable continuous-flow composting operations are obtained by the gradual addition of nitrogen and phosphorus to the digesting material. These nutrients are usually added in the early stages of the biological process. Additions, as noted, are usually based upon approximate carbon concentrations in the material. Of greater significance is soluble, available carbon (e.g., biochemical oxygen demand). In general, nitrogen is not inhibitory to the process until levels of approximately 6% are reached. Since nitrogen compounds are expensive and activity is satisfactory at low levels, nitrogen is rarely added in high concentrations. Generally, if nitrogen plus the necessary phosphorus are added gradually over a 24-h period to the incoming load, the optimum decomposition will be achieved. Thus, in a preliminary stage, it is desirable to determine the carbon, nitrogen, and phosphorus ratios that are available and calculate the amounts needed.

For satisfactory composting the available carbon to nitrogen (C/N) ratio should be from about 30–35:1. Ratios higher or lower than this range can retard the composting process. In general, ratios lower than 25:1 can lead to ammonia loss as well as increased odors, although pH is a consideration, especially at elevated alkalinity. Ratios greater than 40:1, on the other hand, are considered to be nitrogen deficient. It is to be noted that as biological decomposition proceeds, the loss of carbon dioxide gradually decreases the carbon to nitrogen ratio. It should also be noted that total carbon does not represent total available carbon. Although a C/N ratio of 30–35:1 is desirable for optimal, rapid municipal refuse composting, this is not the ratio of carbon to nitrogen that is usually present in incoming solid wastes. In the United States the C/N ratio most often found in solid wastes is about 55–60:1.

The carbon to phosphorus ratio (C/P) should be 100:1 to ensure proper growth and digestion [39]. Thus, in order to achieve rapid, optimal compost stabilization, nitrogen and phosphorus supplements are often needed to accelerate the process. Fortunately solid waste streams usually contain a sufficient amount of phosphorus for microbial growth in a composting process.

B. Moisture Content

Moisture content is the mass of water lost per unit wet mass when the material is dried at 103 °C for a specific period of time (such as 8 h or more). The theoretical optimum moisture content in composting process is 100%. The practical moisture content, however, is a function of aeration capacity of the process equipment and of the structural nature of the material being composted [23].

As a rule-of-thumb, the moisture content becomes a limiting factor when it drops below 45% because low moisture content in the pulp reduces the rate of composting, or the rate of biological decomposition.

The maximum permissible moisture content is generally considered to be the optimum moisture for a given material, although when the moisture content exceeds the optimum value, mechanical problems or localized anaerobic decomposition may happen.

The moisture of incoming solid wastes is highly variable and, in general, moisture levels of incoming solid wastes are too low to achieve high-rate composting. Moisture supplements must therefore be added. The most satisfactory moisture level for adequate composting is between 55 and 65%. The maximum permissible moisture contents for various wastes are listed in Table 2 [23].

Moisture supplements, like nutrients should be added near the beginning of the process operation. Liquid supplements to solid wastes added must be made on a calculated basis and with respect to the dry weights of the material. If additional moisture is added with the nitrogen or phosphorus supplements or is present in the sewage or sludge, this, too, must be considered in moisture calculations. In general, moisture additions must be spaced out over short periods of time. Moisture cannot be added rapidly, or in such a way as to produce areas in the

Table 2
Maximum Permissible Moisture Contents of Various Solid Wastes

Major component	Moisture content, %
Manure	55–65
Municipal solid wastes	55–65
Residential wet wastes (garbage, lawn clippings, trimmings, etc.)	50–55
Paper	55–65
Wood (sawdusts, other small chips)	75–90
Straw	75–85

pulp that are very wet or saturated since the excess moisture content may not be desirable. At times, in order to minimize heat loss in the process, hot water is added to the composting pulp. As will be noted later, the monitoring of moisture in a high-rate composting process is important.

When sewage sludge is used, it has been noted that the moisture content of the resulting mixture of sludge and solid wastes may be in excess of that required for optimal composting, in which case, the sludge must be dewatered [12]. A relationship between sludge and solid waste moisture that can be used to estimate sludge dewatering is shown in Fig. 6. The plot shown assumes that solid waste (or refuse) is received at a rate of 1.9 kg (4.2 lb) per capita per day before removal of 25% non-compostables, the solid waste and sludge are from the same population, and the sludge contains 3% solids and is generated at the rate of 54 g (0.119 lb) per capita per day.

Figure 7 [12] shows a relationship between sludge moisture and the ratio of sludge to solid waste in a population plotted for two solid waste production rates, 1.9 kg (4.2 lb), and 0.9 kg (2 lb) per capita per day. The plot assumes that water content of 60% is to be maintained in the sewage–sludge–refuse mixture, and that sewage from only 27% of the population can be handled as received at the lower rate. At the larger solid waste production rate, about 50% of the sewage sludge generated can be handled without dewatering, assuming 3% solids. Solid waste was assumed to be 35% moisture, as received (wet wt), and sludge solids are assumed to be generated at 54 g (0.119 lb) per capita per day.

Figure 8 [12], indicates that the amount of sewage sludge (assuming 3% solids) that can be used without dewatering varies in direct proportion to the moisture content of the incoming solid waste. The plot assumes that 75% of the solid waste received would be mixed with sludge and that this would result in a mixture containing 60% moisture.

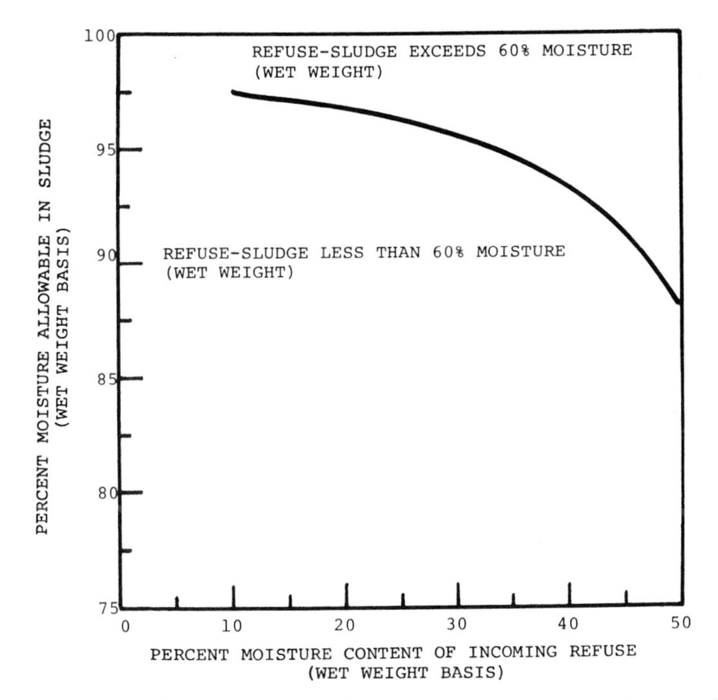

FIG. 6 Relationship between sewage sludge and refuse moisture. From reference 12.

The effect of moisture content on the oxygen consumption rate is described in Section IV.E, Aeration and Mixing.

C. Temperature

Temperature elevations that take place in a composting refuse pulp are an excellent reflection of the biochemical activity that is taking place. In the windrow method, mixing is controlled and is not usually carried out when temperature peaks are reached. In the continuous-flow method, mixing is carried out to maintain the pulp aerobic as well as to prevent excessive cooling. Generally, in order to build up temperatures, the mass should be permitted to remain undisturbed for a period of time. However, care should be taken that the mass does not become anaerobic when aerobic composting is desired. Temperature losses are also minimized through the use of insulation and warming the air if it is circulated.

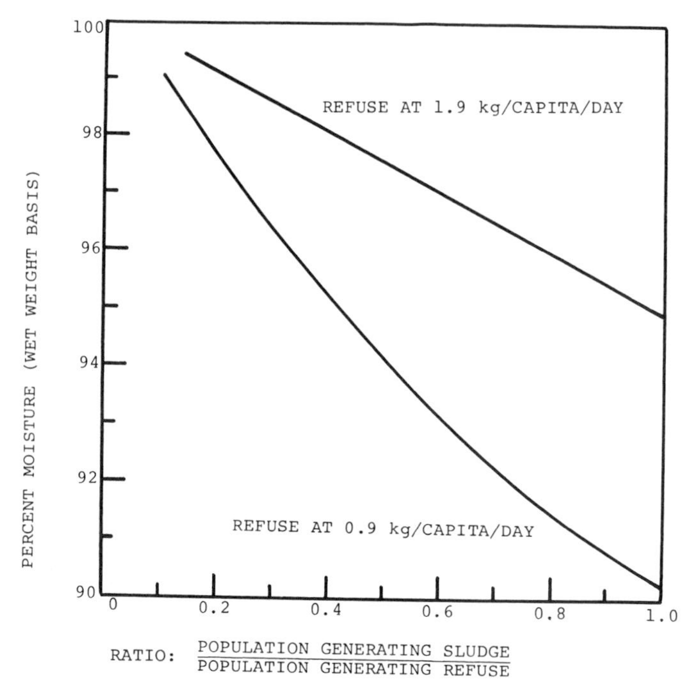

FIG. 7 Relationship between sludge moisture and population ratio of sludge to refuse at two refuse loadings. From reference 12.

Temperatures are often monitored periodically in the composting pulp as well as in the air over the pulp. Normally temperature changes occur in a consistent even pattern, but the rate of temperature rise is higher than the rate of decrease. Usually the temperature rises rapidly from ambient to a peak and then decreases. At times, a secondary less elevated temperature rise can be observed. In general, temperatures obtained in the windrow operations are usually somewhat greater than those obtained in continuous-flow operations.

Typical windrow operations usually show temperatures between 66 and 71 °C (150–160 °F). These temperatures are easily reached and maintained as shown on Fig. 9 [12]. In general, temperatures between 60 and 66 °C (140–150 °F) can be maintained for a period of about three weeks. Higher temperatures are usually observed in the center of the composting mass. Temperature–time relationships are recognized as factors in the destruction of pathogenic microorganisms.

In the open windrow process used at Johnson City, Tenn., it was found that single weekly temperature readings helped determine the progress of composting [12].

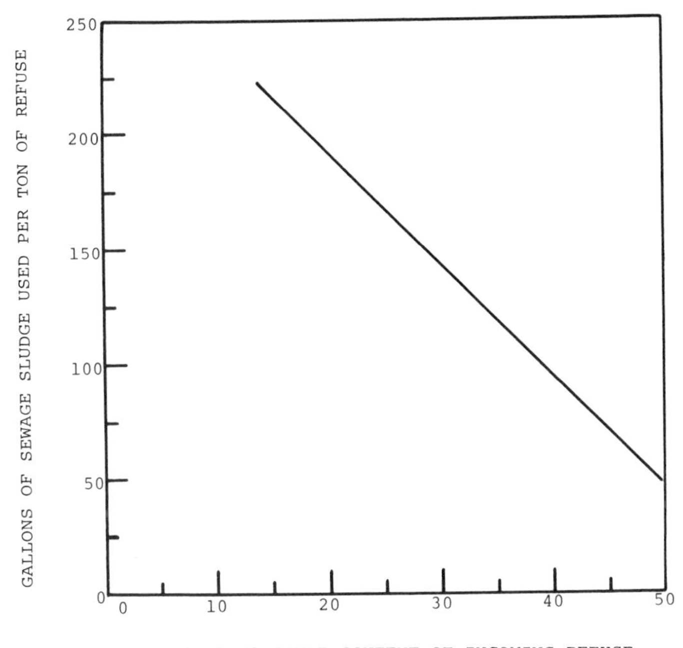

FIG. 8 Relationship of sewage sludge (3% solids) and incoming refuse moisture to produce a mixture with 60% moisture. From reference 12.

Open-tank mechanical composting methods, such as used in the Metrowaste plant at Gainesville, Fla., reached 82 °C after 6 days. Although forced aeration was used in this plant, agitation was provided only intermittently or not at all [40]. Fig. 10 shows the Gainesville compost plant.

Enclosed digesters, such as the Fairfield-Hardy process in Altoona, Pa. (Fig. 5) normally attain temperatures of from 60 to 71 °C (140–160 °F), although temperatures of 79 °C (176 °F) are reported.

Modern composting processes are designed to operate within the mesophilic (10–40 °C) and thermophilic ranges (40–70 °C). Most environmental researchers and engineers agree that for efficient mechanical composting, the operating temperature should be at least 35 °C. The range of optimum temperature for the composting process is from about 35 to about 55 °C, which covers the optimum temperatures of various types of microorganisms involved in the process. Mesophilic bacteria are more efficient than thermophilic bacteria for organic decomposition, although thermophilic decomposition occurs at a more rapid rate. Pathogenic microorganisms and weed seeds are generally

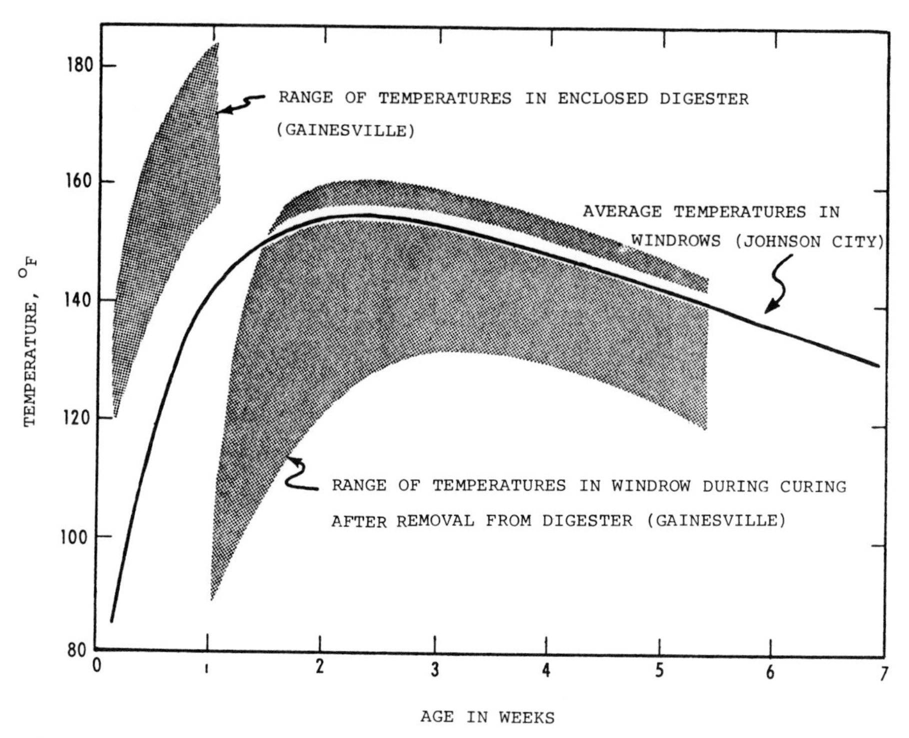

FIG. 9 Typical and comparative temperature profiles obtained from windrow and enclosed compost plants. From reference 12.

killed at thermophilic temperature range. In the overall composting process, thermophilic composting must be involved at some stage for controlling pathogens and weed seeds. Because microbial processes are inhibited, the process becomes less efficient when the temperature exceeds the thermophilic range (i.e., 70 °C).

The effect of temperature on the oxygen consumption rate is described in Section IV.E, Aeration and Mixing.

D. Hydrogen Ion (pH) Level

The optimum pH range for most bacteria is between 6 and 7.5, whereas for most fungi it can be between 5.5 and 8.0. The initial pH of a 3-day-old solid waste is usually between 5.0 and 7.0. Although pH would appear to be an important parameter, it is generally self-adjusting and not usually a major concern in municipal composting operations. Only in the initial stages pH may be variable and may even appear to require

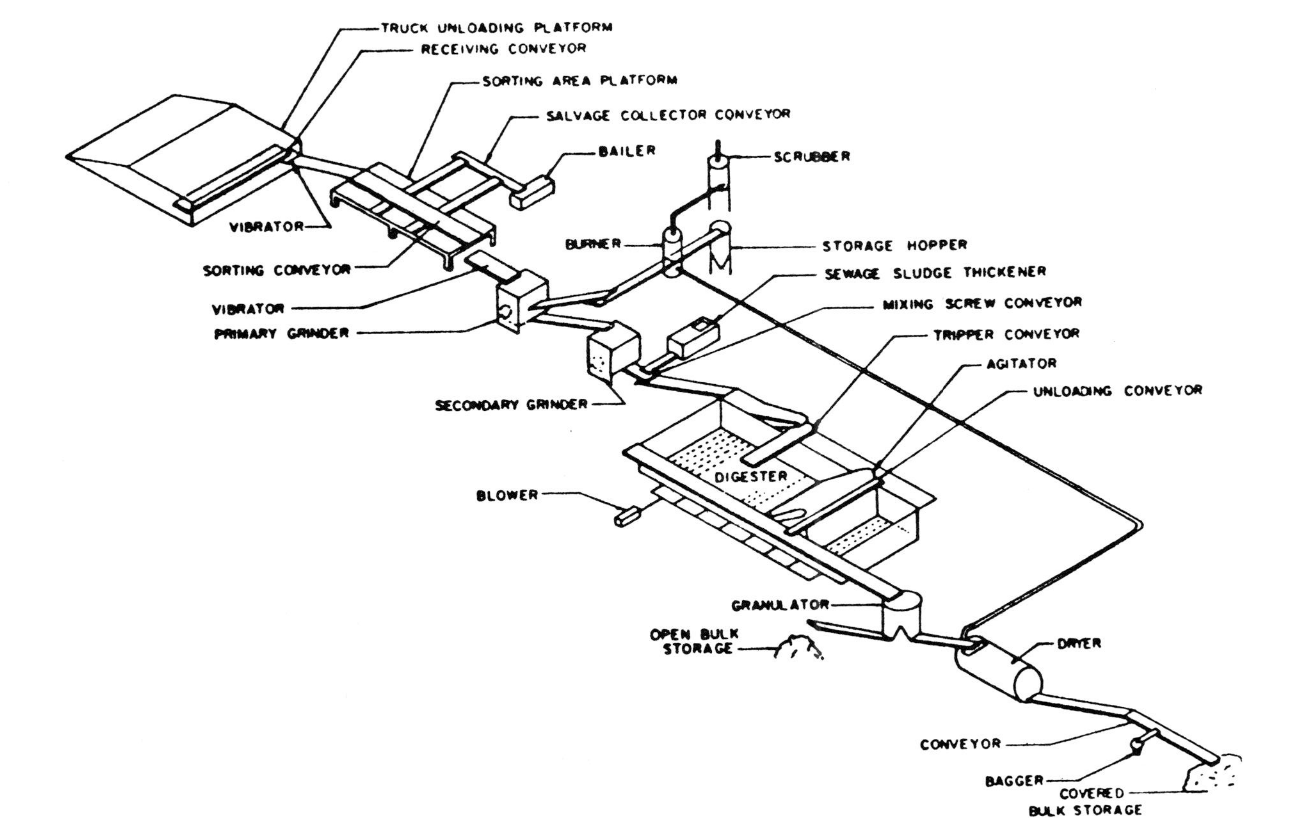

FIG. 10 Gainesville Municipal Waste Conversion Authority, Inc. compost plant, Gainesville, Florida. *From:* C. I. Harding, "Recycling and Utilization," in *Proceedings of the Surgeon General's Conference on Solid Waste Management for Metropolitan Washington, July 19–20, 1967*, USPHS Publ. No. 1729, Govt. Printing Office, Washington, D.C., 1967, pp. 105–119.

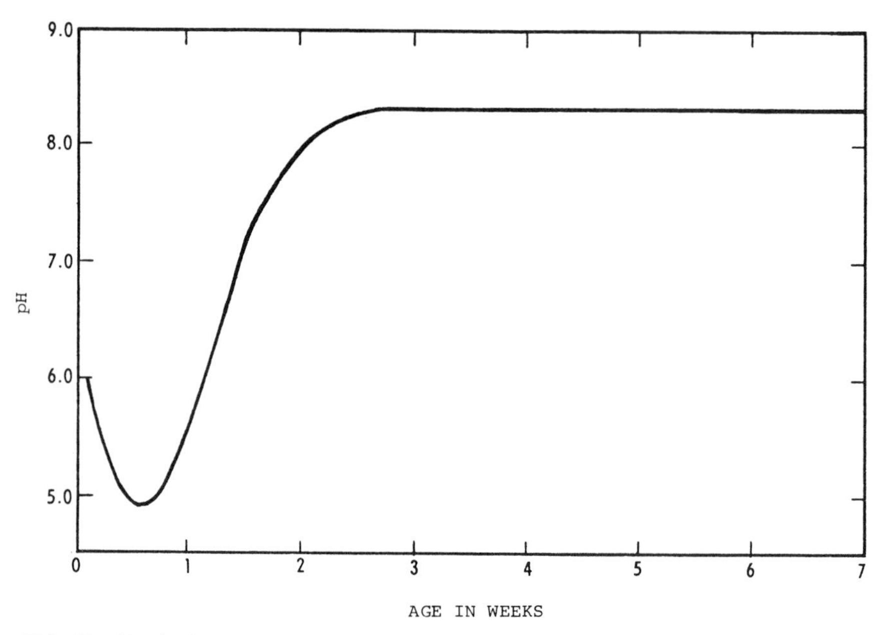

FIG. 11 Typical pH profile obtained in the windrow process. From reference 12.

adjustment. In general, as the composting process progresses, the pH becomes more stable. Furthermore, the process affects hydrogen ions, and the pH is a function of the end products that are being formed from inputs. For example, in a solid waste that contains a great deal of available carbohydrates, the pH can decrease because of the production of acid end products.

In general, however, in municipal refuse decomposition breakdown occurs in a regular sequence. Soluble carbohydrate are decomposed first and the pH drops below neutrality. As contact time is increased, these materials are consumed and proteins begin to decompose causing the pH to gradually rise. In general, however, the pH seldom drops below 5.0, or rises above 8.5. As a rule, the pH range for optimal aerobic composting lies between about 6.5 and 8.0.

Experience at the municipal compost plant in Johnson City, Tenn., indicates that the pH was usually between 5 and 7. The pH pattern obtained in a windrow method is shown as Fig. 11. As shown, the pH initially drops to 5 during the early periods of decomposition, then levels off at around 8, remaining at this value as long as the system remains aerobic.

As in the case of temperature, the pH of the compost varies with time during the composting process and is a good indicator of the extent of organic decomposition within the biomass.

Experience at the plant (Fig. 10) in Gainesville, Fla., indicates that when prolonged, anaerobic storage was employed, pH levels of approximately 4.5 were attained [40]. Reduction in pH value under anaerobic conditions is caused by acid-forming bacteria.

In a municipal solid waste that does not have an unusual substrate (e.g., high carbohydrates, proteins, acids etc.), no pH correction is normally required. However, when pH adjustments are required they may be made by adding lime, sodium bicarbonate, caustic soda, or any available dilute acid. The least expensive reagent for raising pH is lime; however, lime addition is usually accompanied by an increase in ammonia formation and consequent loss of nitrogen in the end product.

As in other biological waste treatment processes, a small amount of composting material may be used in a large volume of water for trial pH titrations. In general, after equilibration, small amounts of titrant are added to the stirred mixture and a plot of pH vs dosage is prepared. From this plot, the dosage required for pH adjustments can be calculated. To prevent localized adverse effects, the addition of acids and bases must be made with caution and dilute solutions must be added over an extended period of time instead of as a slug addition. The addition of nitrogen- or phosphorus-containing nutrients to compost can, at times, also cause changes in pH. Thus, prior to adding such materials, the pH of the mixture should be adjusted to neutrality.

E. Aeration and Mixing

Mixing or aeration in a mechanical composting operation is vital when rapid aerobic composting is desired. In mixing, the microbes are contacted with the substrate and aerobic conditions are maintained. Mixing or aeration may be achieved by a wide variety of mechanisms all of which result in movement of the pulp added to the digester. A number of continuous-flow closed configurations have been used to achieve mixing. In high-rate municipal composting, failure to aerate results in offensive, odorous conditions. Proper mixing should be sufficient to prevent anaerobic conditions and expose the digesting pulp to air, however, frequent movement of the pulp can also result in unnecessary cooling or loss of moisture.

Chrometzka [41] estimated that oxygen requirements range from 9 mm^3/g of volatile matter per hour for ripe compost, to 284 mm^3/g of

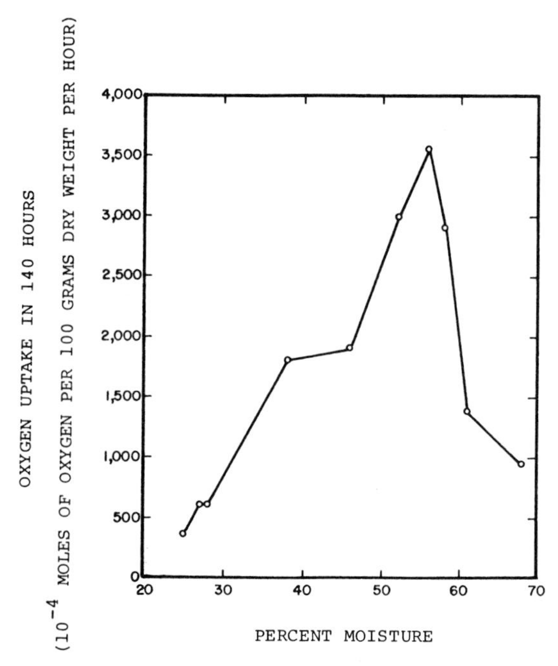

FIG. 12 Total average area under each oxygen uptake curve at various moisture contents. From reference 42.

volatile matter per hour for 4-week-old fresh compost. The 7-day-old raw solid waste required 176 mm^3/g/h. He also reported [41] that the oxygen requirement increased about 16% with increasing the moisture content of a fresh compost from 45 to 60%; therefore, the moisture content is a determinant of the oxygen requirement. Figure 12 shows the effect of moisture on the oxygen consumption rate [42]. If the waste is too dry, insufficient moisture exists for the metabolism of the microorganisms; if the waste is too moist the pore spaces are filled with water and the process becomes anaerobic due to the reduction in surface area.

Figure 13 shows the effect of temperature on the oxygen consumption rate [42]. In the figure, Line A indicates the theoretical value,

$$Y = 0.1\,(16^{0.028T}) \tag{6}$$

where Y = oxygen consumption rate, mg O_2 per hour per g volatile solids,

T = temperature, °C,

and Curve B shows the experimental value.

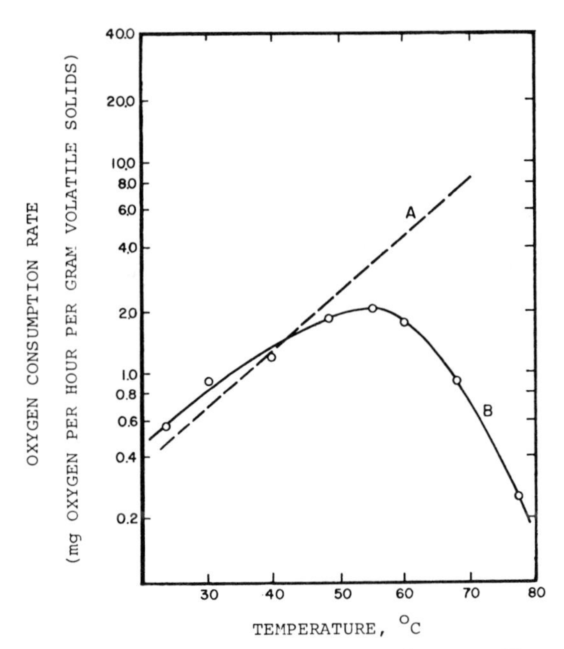

FIG. 13 Effect of temperature on oxygen consumption rate. From reference 42.

The primary reason for mixing is to equalize temperature and moisture within the digesting mass and also to prevent a channeling of air and the caking of the material; the secondary reason for mixing is to aerate. Accordingly it is far more important to use forced air for aeration than it is to have mixing more frequently than once a day in the early stages and once every several days toward the end of the aerobic composting process.

F. Genetic Traits and Seeding

When all environmental conditions are close to optimum, the rate of organic decomposition depends upon the capability of microorganisms, which, in turn, depends upon their genetic makeup. The potential determined by the genetics of the microorganisms cannot be surpassed.

A wide variety of microorganisms is present in municipal solid waste, sludges, manure, and other materials to be composted. However, there is a lag time required to increase the population of the required microorganisms (i.e., the seed) in a batch composting process between each distinct phase change. Therefore, it is often necessary to add the starter "seed" at start-up, which consists of adding to the process a portion of the pulp collected during a period of active growth, and is different from the addition of special inocula noted earlier.

In general, the addition of seed by intimate mixing at start-up reduces the time required for stabilization and minimizes the time needed for the achievement of a steady reliable process. However, seed is not always readily available. Seed can be collected and stored from a small pilot plant or from another compost plant. In general, relatively large amounts of seed must be added (from 10–33%) for a rapid start-up. Maximum benefits can be obtained if the seed is stored at low temperature and if conditions are such that the microbial population is not reduced considerably.

Laboratory scale units are often used prior to pilot-plant and full-scale design. Jeris and Regan [43], using a 1–3 kg controlled, bench-scale digester, have recently published results of 4 yr of laboratory-scale composting using municipal refuse, paper, and synthetic refuse. The environmental conditions they determined for optimum composting are shown on Table 3, taken from earlier work [44–46].

Table 3
Environmental Conditions for Optimum Composting of Mixed Solid Wastes[a]

Parameter controlled	Optimum value
Temperature, °C	59 ± 5
Moisture, %	67 ± 3
Free air space, %	32 ± 3
Seed recycle, %	50–75
Mass turnover time, days	4–5
pH	8.0 ± 0.5
Nitrogen added (as NH_3), mg/g dry mass	2.0
Phosphorus added (as P), mg/g dry mass	1.34
Air requirements, ft^3/day per lb of volatile solids	16.7–20.3
Air residue, %	10

[a] From reference 43.

V. PROCESS CONTROL

A number of process control parameters are important for composting operations to achieve and maintain optimum conditions. Except for a few publications [47, 48], little attention has been given to analytical methods for solid waste materials. Generally, most of the process control parameters are with respect to moisture additions or with respect to the microbiological stabilization process.

If the pretreated material is to be stored for any extended period of time, it should be examined periodically for evidence of anaerobic decomposition. In general, anaerobic decomposition and the rate of biological decomposition can be controlled in the initial phases by controlling the moisture content. Although not widely used, the occurrence of anaerobic conditions and a forewarning of adverse biological decomposition and products (e.g., odors) can be monitored by examining the oxidation–reduction potential (ORP) of the material. These measurements should be taken with care and generally small volumes of water should be allowed to come to equilibrium with the probe.

Biological process parameters that are important in the microbiological stabilization include nutrient supplements that must be added. These are usually controlled so that the nitrogen and phosphorus are added at the beginning of the process, and tapered down as the process moves toward completion. Moisture must be controlled closely and once optimal moisture concentration is determined it should be closely adhered to. Moisture is the most important parameter to be determined and controlled, especially during the early phases of the process, that is, during the first 25–40% of the process. For this reason rapid moisture determinations are followed by quick moisture adjustments over the appropriate phase.

Moisture can be determined by a variety of methods. The traditional method of oven drying at 103 °C, although satisfactory, is too slow in terms of operational process requirements. Very few rapid methods are discussed in the literature. Most rapid methods do not, in fact, operate at the moisture levels found in composting digesters. A rapid, inexpensive method that employs a commercially available microwave oven has been recently recommended by Cardenas [49]. Using this method, moisture can be determined in 10 min.

As noted, adequate aeration and mixing is essential. Again a delicate balance must be achieved to be worked out for the individual process. Generally, mixing or aeration must be sufficient to maintain the pulp aerobic and permit at the same time optimal biological activity to occur, while not resulting in excessive cooling of the composting mass. Usually this is determined empirically, although such factors as pulp depth and digester air temperature affect the process. Once the proper aeration or mixing rate has been worked out, however, little process control is required. Normally, mixing rates are low and a static period to allow temperature buildup to occur is permitted.

Although pH is often monitored, as indicated, it is not an important parameter for process control and is soon regulated by the decomposition of the material. Early in the processes the pH is low, but it rises as the process goes to completion.

Temperature is usually monitored in the pulp and the air over the composting pulp. Temperature data indicate the relative status or completeness of the composting process. In the process operation, the temperature rises very rapidly in the early phases and decreases as the product moves toward completion or stabilization. The temperature can be determined using a recorder by simply placing calibrated Constantine wire probes in contact with the moving pulp mass and the air. Heat is not usually added to the process, but is, instead, generated by the microorganisms themselves. In all biological reactors a certain amount of heat loss occurs. It can be replaced by maintaining an elevated temperature over that of the pulp temperature through the use of blowers. A temperature elevation of 9–12 °C (10–15 °F) in the atmosphere over the pulp to that of the pulp is satisfactory. Air is often added or recirculated to maintain the system aerobic. It must be warmed to prevent the pulp mass from cooling.

Since modern municipal composting is an aerobic process in which respiration takes place, the ratio of carbon dioxide to oxygen should be monitored. This can be carried out by the use of specific probes strategically placed in the system.

As noted earlier, it is important that the system does not enter into an anaerobic phase and for this reason the determination of the oxidation-reduction potential may be important. The proper execution of an oxidation–reduction potential reading has been the subject of experimental errors in the past. A discussion of methods can be found in the volume of Ciaccio [50].

An example of the process changes that take place in a typical high-rate mechanized compost plant is shown in Fig. 14 (data taken from the Brooklyn Ecology compost plant.) In general, the parameter that is the most sensitive and most readily reflects the progress of the composting operations includes measurement by a Warburg respirometer. In the Ecology plant short-term (3.5 h) respirometric measurements have been developed. As shown in Fig. 14, the consumption of oxygen (mm O_2 per gram of dry pulp) and the release of carbon dioxide increase very rapidly and then decrease to a value of about half that of the initial phase. A much slower response is seen in the temperature because temperature changes require time to reach equilibrium. However, respirometric measurements are difficult and require specialized equipment as well as trained technicians.

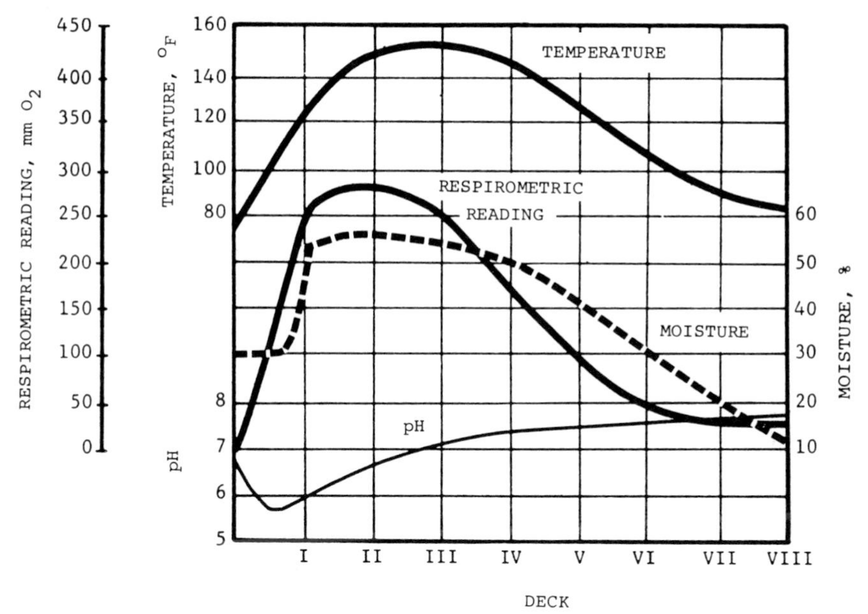

FIG. 14 Process parameter variations in ecology process.

VI. PATHOGEN SURVIVAL

A number of reviews and articles are available related to the survival of pathogens in composting materials and their final end products [9, 12, 17, 51–54]. Studies have been made of bacteria (pathogens as well as indicators), spore-forming fungi, actinomycetes, parasitic worms (including cysts, and eggs) and viruses. It has been discussed in Section IV.C that pathogenic microorganisms are rapidly destroyed at elevated temperatures. Pathogenic bacteria, for example, are rapidly killed when all portions of the composting mass are subjected to temperatures of about 60 °C or higher. In general, such microorganisms cannot survive temperatures of from 54 to 60 °C for periods longer than 30 min to 1 h.

A study by Krige [53], who investigated the presence of pathogens such as *Ascaris lumbricoides ova*, polio virus, a phage marker and coliform bacteria in South Africa, indicated that composts marketed in South Africa are relatively safe. He found all samples to be negative for *Mycobacterium tuberculosis*, *Salmonella spp* and *Shigella spp* as well as virulent polio virus added to an actively composting research plant. Krige concluded [53] that the compost produced under controlled conditions is safe from a public health view and felt that this removed many of the objections against the marketing of compost.

Wiley and Westerberg [17] added pathogens typical of those present in the intestinal tract of man and which would be present in sewage sludge. Among these were *Salmonella newport*, polio virus type I, *Ascaris lubricoides ova*, and *Candida albicans*. These investigators added these microorganisms to an active aerobic composter, to which raw sewage sludge was being added. Their results indicated [17] that after 43 h of composting no viable indicator organisms could be detected. They further noted that factors other than temperature and time might contribute to the death of pathogens (e.g., antibiotic actions or the end products from other microorganisms). They concluded that when sewage sludge was composted in the manner described in this study that a final product can be obtained which would be safe and could be investigated for use as a soil conditioner or fertilizer.

Other studies [12, 55] reported that properly managed windrow operations resulted in a product that was safe for agricultural and gardening applications. Researchers found [55] that pathogenic bacteria that may be associated with sewage sludge and municipal refuse were destroyed by the windrow process. They further noted [55] that temperatures of 49 to 54 °C were sufficient to reduce the total and fecal coliform levels to undetectable levels. However, coliform regrowth occurred when the temperature decreased during the last stages of the process. *Mycobacterium tuberculosis* and polio virus were also destroyed. *M. tuberculosis* was destroyed in 14 days at 65 °C and in all cases by the 21st day. Temperatures of 54 °C or higher were found to deactivate the polio virus.

A summary of the survival times and temperatures of some of these pathogens is shown in Table 4 [51]. Other disinfection and stabilization alternatives were reported by Wang et al. (56–59).

The USEPA review [12] indicates that there are no references in the literature that any sanitation worker has ever been infected by fungi as a result of handling solid wastes. Generally, this work, as well as those that have been noted, indicates that as long as the composting process is allowed to proceed normally with proper mixing and attainment of temperatures for stabilization, there is no danger.

VII. COST CONSIDERATIONS

Composting should generally be regarded as a resource recovery method for the production of an end product that can be marketed

Table 4
Thermal Death Points of Some Common Pathogens and Parasites[a]

Pathogens and parasites	Thermal death points and contact times
Salmonella typhosa	Death within 30 min at 60 °C; no growth beyond 46 °C
Shigella (Groups A to D)	Death within 1 h at 55 °C
Escherichia coli	Most die within 1 h at 55 °C and within 15–20 min at 60 °C
Endamoeba histolytica	Thermal death point is 68 °C
Vibrio cholerae	Very sensitive to change of environment; outside human body die in a few hours in the feces at room temperature
Trichinella spiralis	Infectivity reduced as a result of 1 h exposure at 50 °C, thermal death point 67 °C
Necator americanus	Death within 20 days at 45 °C
Ascaris lumbridcoides	Death within 20 days at 45 °C, 2 h at 50 °C, 50 min at 55 °C, 3.5 min at 60 °C
Taenia saginata	Death within 5 min at 71 °C
Mycrococcus pyogenes	Death within 10 min at 50 °C
Streptococcus pyogenes	Death within 10 min at 54 °C
Mycrobacterium tuberculosis	Death within 15–20 min at 66 °C
Corynebacterium diptheriae	Death within 45 min at 55 °C
Brucella abortus	Death within 3 min at 61 °C

[a] From reference 12.

rather than as a method of solid waste disposal. In general, in the United States, composting has not been successful as a business enterprise. Most of the compost plants built, in fact, have failed for financial reasons, usually because of an inability to dispose of the end product. The authors updated the cost data from Salvato [60] and prepared Table 5, which indicates that composting is one of the most expensive methods of municipal waste refuse disposal. However, some of the costs can be recovered through sales of the end product and/or sales of salvageable materials. Income from the recovery of paper, metals, rags, etc., varies widely.

Cost estimates reported for composting process operations vary widely. Figures are usually available for individual plants, but differences in size, equipment, operation, number of employees, shifts, land costs, disposal of product, etc., affect the final figure.

A review of costs for both windrow and enclosed-digestion plants is available from the USEPA [12], including important data on the capital costs for various capacity windrow plants [12]. If sewage or sewage sludge is to be utilized in the process, costs associated with handling

Table 5
Solid Waste Handling and Disposal Costs

Item	Cost per ton, \$	Capital cost per ton, \$
Collection	30.0–39.0	—
Sanitary landfill[a]	2.3–7.5	24.0–42.0
Incineration	12.0–24.0	24,000–37,500
Composting	15.0–30.0	24,000–36,000
Pyrolizer[b]	12.0–24.0	15,000–30,000
Transfer station	3.0–6.0	3,000–4,500
Haul by trailer and sanitary landfill	10.5–18.0	—
Compact, bale, rail to sanitary landfill	12.0–18.0	—
Rural container station[c]	21.0–30.0	—
Haul cost for 20-yd^3 compactor,[d] per mile	0.6–0.8	—
Compaction, rail haul disposal to sanitary landfill	22.5–22.8	—
Compaction	3.0–4.1	27,000–36,000
Shredding	6.0–12.0	—
Shred, compact, bale	16.2–19.5	—
Barging to sea	12.0–18.0	—

[a] Capital cost includes site, fencing, roadway, scale, hammer mill or shredder, tractors, trucks, and engineering based on annual capacity. Can be reduced by one-half if fencing, hammer mill or shredder, and miscellaneous are omitted.

[b] Experimental with solid wastes.

[c] Includes cost to build and operate.

[d] Varies with compactor or trailer size, men per crew, and actual travel time to disposal site.

such materials must also be considered. In general, costs for enclosed-digestion compost plants are similar to those for windrow operations, for example, with respect to receiving, sorting, grinding, storage, etc. However, there would be differences with respect to digester costs, and possibly, land costs, depending on whether curing and extensive storage is employed.

The capital and operating costs for an enclosed, mechanical compost plant (see Fig. 5) have been estimated by Huang and Dalton [61]. They estimate, from a preliminary analysis of other composting processes, that if no market were found for the humus (or the compost), the cost per metric ton of refuse for municipal refuse processing would increase by about 25%.

A comparison of capital costs in the USEPA review [12] on a basis of per ton refuse processed for the digestion system of a 150 ton/day windrow plant with that of an enclosed digester, indicates that capital

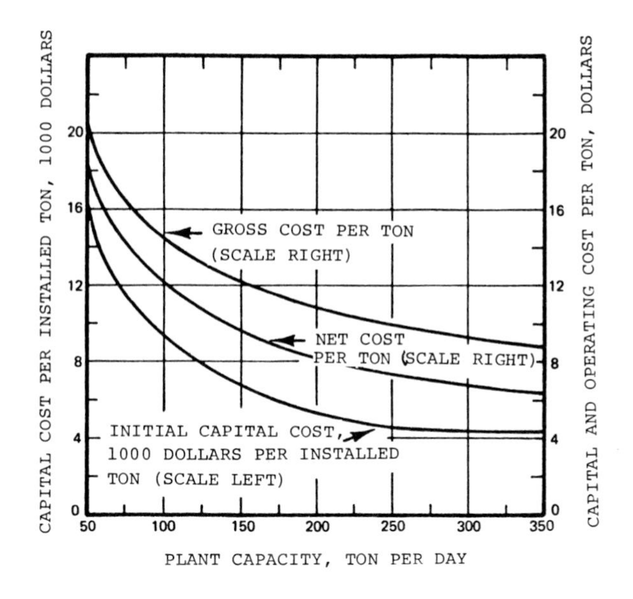

NOTES:

1. Plant capacity is normally one or two shifts per day to achieve plant capacity.
2. Gross cost trend is the owning and operating facilities without any credits.
3. Net cost trend is for owning and operating facilities considering sales of compost and salvaged materials.
4. All costs consider compost digested sludge with refuse in 1971.

FIG. 15 Composting costs. From reference 12.

costs are less for windrow compost plants. Based on the work of Kupchick [62], the 1965 estimates of costs per ton of refuse processed for European plants using the windrow method ranged from 48 to 64% of that of comparable cost figures for enclosed systems.

An added cost that must be noted is related to handling of noncompostable materials. Operating costs for high-rate compost plants is insufficient to make a reasonable comparison at this time.

Costs associated with sludge composting [63] are shown in Fig. 15. Revenues from the sale of compost products range from $1.50 to $3.50 per ton of material fed.

VIII. END PRODUCTS

In general, the quality of the end product is a function of the control over the process. In the production of compost (or humus) as a soil supplement or fertilizer vehicle, the quality of the compost is inversely proportional to the quantity of inert material such as plastic, rubber, leather, and rags. The market value of compost containing appreciable quantities of such materials is less.

The composition of the composted end product, will, of course, vary widely. However, some definite data are still available. The compost from a windrow plant, such as that in Johnson City, Tenn., according to USEPA [12], contained carbon, nitrogen, phosphorus, potassium, sodium, and calcium mostly in a combined form, whereas iron and aluminum, and possibly magnesium and copper are present primarily as uncombined metals (see Table 6). The values in Table 6 for nitrogen, phosphorus, potassium, calcium, and percent ash correspond to those found by investigators of other composts. In general, values found in the Ecology plant (Fig. 2) in Brooklyn, NY, an enclosed mechanized plant, were the same. The moisture content of refuse and sludge–refuse composts as well as the results of a hygienic evaluation made by the U.S. Government is shown in Table 7 [63].

The potential uses for the composted municipal refuse include agricultural conditioners, mulch, and fertilizers. Composted refuse has

Table 6
Elements in 42-Day-Old Compost at Johnson City[a]

	Average percent dry weight		
Element	Containing sludge, 3–5%	Without sludge	Ranges, all samples
Carbon	33.07	32.89	26.23–37.53
Nitrogen	0.94	0.91	0.85–1.07
Potassium	0.28	0.33	0.25–0.40
Sodium	0.42	0.41	0.36–0.51
Calcium	1.41	1.91	0.75–3.11
Phosphorus	0.28	0.22	0.20–0.34
Magnesium	1.56	1.92	0.83–2.52
Iron	1.07	1.10	0.55–1.68
Aluminum	1.19	1.15	0.32–2.67
Copper	<0.05	<0.03	
Manganese	<0.05	<0.05	
Nickel	<0.01	<0.01	
Zinc	<0.005	<0.005	
Boron	<0.0005	<0.0005	
Mercury	not detected	not detected	
Lead	not detected	not detected	

[a] From Ref. 12.

Table 7
Hygienic Quality of Compost[a]

Treatment method	Material	Water content, %	Maximum Temp. Achieved, °C	Hygienic evaluation	Remarks
Contour composting					
Contour spreading	sludge + solid waste	55	46	Not pathogen-free after 5 months	—
Windrow spreading	sludge	60	52	Not pathogen free after 6 months	—
Windrow spreading	solid waste	40–60	> 55	Pathogen-free after 3 weeks	—
Windrow spreading	sludge + solid waste	40–60	> 55	Pathogen-free after 3 weeks	—
Mechanical composting					
Rotating drum, Dano Process	solid waste	45–55	> 60	Pathogen-free after 6–7 days	Spore-free after 1 week of windrow composting
Rotating drum	sludge + solid waste	approx. 50	> 60	Pathogen-free after 6–7 days	Spore-free after 1 week of windrow composting
Rotating tower, Multibacto process	solid waste	40–50	> 65	Pathogen-free after 1 day	Spore-free after 1 week of windrow composting
Rotating tower	sludge + solid waste	45–55	> 65	Pathogen-free after 1 day	Spore-free after 1 week of windrow composting

[a] From reference 12.

been found to contribute to the moisture-holding capacity as well as the tilth and permeability of soil. Composted material as a soil conditioner has also been used in general landscaping as well as a top dressing, or a soil conditioner in public parks or recreation areas such as golf courses and highways and it would appear that there is a reasonably large consumer market dealing with commercial nurseries, home gardeners, mushroom growers, etc., where potting soils and top dressings are used.

Composting has also been considered more and more as a landfill because it is biologically stable. Municipal compost generally has a high resistance to compaction and thus would be an ideal landfill and require only short periods of stabilization after which the land can be reutilized or reclaimed. A potential market has been noted related to cattle feed. Because of the fibrous, cellulosic nature of municipal compost it would appear to be a natural choice for a number of products that have also been made from compost from time to time, including cardboard, insulation boards, paper boards, artificially extruded lumbers, and acoustical boards. Some municipalities have combined municipal compost with leaf and lawn clippings and other suburban-generated organic waste materials. The end product is an organic soil supplement that is available to the local population, developers, landscapers, etc. The practice has been followed to a limited extent, for example in Westchester County, NY, and as such is usually consistent with a solid waste management plan.

IX. SUMMARY

In summary, although municipal refuse composting operations in Europe have been successful, composting has not yet been established as an economical operation in the United States. The principal reason for this has been disposal of the end product. Composting for profit should not be considered as a sole method of solid waste disposal. Composting, however, would fit well with a broad-based solid waste management plan that would include other considerations such as landfill, incineration, resource recovery, and recycle. In this light, over a long, time period, composting would be a prudent, environmentally compatible method for recycling available organic materials. Process design then should be coupled with end product utilization and the availability of a steady reliable market.

X. PRACTICAL EXAMPLES [64]

A. Example 1.

Table 8 indicates physical parameters of a typical uncompacted, municipal solid waste. Determine which components of solid waste are compostable. Estimate the compost production from 2500 kg of solid waste assuming that a 40% reduction in weight of compostable material would result from the composting process.

Solution

1. Plastics, most textiles, rubber, leather, glass, metals, and dirts are essentially noncompostable. Therefore, the compostable components of solid wastes include: food wastes (15%), paper (40%), cardboard (4%), garden trimmings (12%), and wood (2%) according to Table 8.
2. The compost production from 2500 kg of uncompacted municipal solid wastes will be:

$$2500 \text{ kg } (15\% + 40\% + 4\% + 12\% + 2\%)(1\text{–}0.40) = 1095 \text{ kg}$$

B. Example 2.

The analytical data of the combustible components in a typical municipal solid waste (Table 8) are indicated in Table 9. Determine the dry composition of compostable portion of the solid waste.

Solution

1. Assume that 1000 kg of wet uncompacted solid waste is collected for analysis. Knowing the percent by weight and the moisture content of each waste component from Table 8, one can then calculate the wet and dry weight of each waste component. The composition of each waste component is computed according to the analytical data of the combustible wastes presented in Table 9. Table 10 is the computation table for this example. Final data are listed in the bottom line of Table 10.
2. Table 11 is another computation table set up to determine the dry composition of compostable portion of the solid waste. The

Table 8
Physical Parameters of Typical Uncompacted Municipal Solid Wastes[a]

Component	Percent by weight	Density		Moisture content, %	Inert residual, %	Energy content	
		lb/ft³	kg/m³			Btu/lb	kj/kg
Food wastes	15.0	18.0	288.3	70.0	5.0	2,000	4,652
Paper	40.0	5.1	81.7	6.0	6.0	7,200	16,747
Cardboard	4.0	3.1	49.7	5.0	5.0	7,000	16,282
Plastics	3.0	4.0	64.1	2.0	10.0	14,000	32,564
Textiles	2.0	4.0	64.1	10.0	2.5	7,500	17,445
Rubber	0.5	8.0	128.2	2.0	10.0	10,000	23,260
Leather	0.5	10.0	160.2	10.0	10.0	7,500	17,445
Garden trimmings	12.0	6.5	104.1	60.0	4.5	2,800	6,513
Wood	2.0	15.0	240.3	20.0	1.5	8,000	18,608
Glass	8.0	12.1	193.8	2.0	98.0	60	140
Tin cans	6.0	5.5	88.1	3.0	98.0	300	698
Nonferrous metals	1.0	10.0	160.2	2.0	96.0	—	—
Ferrous metals	2.0	20.0	320.4	3.0	98.0	300	698
Dirt, ashes, brick, etc.	4.0	30.0	480.6	8.0	70.0	3,000	6,978
Total	100.0						

[a] From Refs. 30 and 64.

Table 9
Typical Data on Ultimate Analysis of the Combustible Components in Municipal Solid Wastes[a]

Combustible component	Percent by weight (dry basis)					
	Carbon	Hydrogen	Oxygen	Nitrogen	Sulfur	Ash
Food wastes	48.00	6.40	37.60	2.60	0.40	5.00
Paper	43.50	6.00	44.00	0.30	0.20	6.00
Cardboard	44.00	5.90	44.60	0.30	0.20	5.00
Plastic	60.00	7.20	22.80	—	—	10.00
Textiles	55.00	6.60	31.20	4.60	0.15	2.45
Rubber	82.57	3.69	0.50	0.17	1.59	11.48
Leather	60.00	8.00	11.60	10.00	0.40	10.00
Garden trimmings	47.80	6.00	38.00	3.40	0.30	4.50
Wood	49.50	6.00	42.70	0.20	0.10	1.50
Dirt, ashes, brick, etc.	26.30	3.00	2.00	0.50	0.20	68.00

[a] From Refs. 30 and 64.

empirical-mole composition of the dry compostable solid waste is determined to be

$$C_{533.82} \quad H_{877.09} \quad O_{394.18} \quad N_{5.72} \quad S_{1.00}$$

C. Example 3.

Compute the theoretical amount of oxygen required to completely oxidize 1000 kg of typical municipal solid waste (Table 8) by aerobic composting process. The composition of the dry compostable solid waste has been determined in Example 2. Assume that the nitrogen is converted to ammonia (NH_3) in the first step and then the ammonia is converted to nitrate (NO_3) in the second step.

Solution

1. It is known from Example 2 that for every 1000 kg of typical wet uncompacted municipal solid waste, there are only 604.00 kg of wet compostable waste, or 465.40 kg of dry compostable waste (see Table 10). About 5.80% of dry

Table 10

Computation Table for Determination of Typical Solid Wastes' Composition

Solid wastes component	Wet weight, kg	Moisture content, %	Dry weight, kg	Composition, kg					
				Carbon	Hydrogen	Oxygen	Nitrogen	Sulfur	Ash
Food wastes	150	70	45.00	21.60	2.88	16.92	1.17	0.18	2.25
Paper	400	6	376.00	163.56	22.56	165.44	1.13	0.75	22.56
Cardboard	40	5	38.00	16.72	2.24	16.95	0.11	0.08	1.90
Plastics[a]	30	2	29.40	17.64	2.12	6.70	—	—	2.94
Textiles[a]	20	10	18.00	9.90	1.19	5.62	0.83	0.03	0.44
Rubber[a]	5	2	4.90	4.05	0.18	0.02	0.01	0.08	0.56
Leather[a]	5	10	4.50	2.70	0.36	0.52	0.45	0.02	0.45
Garden trimmings	12	60	4.80	2.29	0.29	1.82	0.16	0.01	0.22
Wood	2	20	1.60	0.79	0.10	0.68	0.00	0.00	0.03
Dirt, ashes, brick, etc.[a]	4	8	3.68	0.97	0.11	0.07	0.02	0.01	2.50
Total, Combustible	668	—	525.88	240.22	32.03	214.74	3.88	1.16	33.85
Total, Compostable	604		465.40	204.96	28.07	201.81	2.57	1.02	26.97

[a] Noncompostable.

Table 11
Determination of Empirical-Mole Composition of Dry Compostable Solid Waste

Elements and ash	Waste composition, kg	Elemental composition % by wt	Atomic weight	Ratio of atoms[a]	Ratio of atoms[b]
C	204.96	44.04	12	3.6700	533.82
H	28.07	6.03	1	6.0300	877.09
O	201.81	43.36	16	2.7100	394.18
N	2.57	0.55	14	0.0393	5.72
S	1.02	0.22	32	0.0069	1.00
Ash	26.97	5.80	—	—	—
Total	465.40	100.00			

[a] Original.
[b] Simplified.

compostable waste cannot be oxidized (see Table 11). Therefore, the amount of oxidizable dry compostable waste = 465.40 (100–5.8)/100 = 438.41 kg.

2. Write a balanced equation for the complete aerobic decomposition of solid waste by aerobic composting (Eq. 2, Sect. II.B):

$$4\,C_aH_bO_cN_dS_eP_f + (4a + b - 2c + 5d + 6e + 5f)O_2 \longrightarrow 4a\,CO_2 + (2b - 2d - 4e - 6f)H_2O + 4d\,NO_3^- + 4e\,SO_4^{2-} + 4f\,PO_4^{3-} + (4d + 8e + 12f)H^+ \quad (2)$$

in which $a = 533.82$, $b = 877.09$, $c = 394.18$, $d = 5.72$, $e = 1.00$, and $f = 0.00$. The empirical-mole composition of oxidizable dry compostable solid waste is:

$$C_{533.82}\ H_{877.09}\ O_{394.18}\ N_{5.72}\ S_{1.00}$$

3. The theoretical amount of oxygen required = $(4a + b - 2c + 5d + 6e + 5f)$ kg-moles O_2/4 kg-moles waste = $(4 \times 533.82 + 877.09 - 2 \times 394.18 + 5 \times 5.72 + 6 \times 1.00 + 5 \times 0.00)$ kg-moles O_2/4 kg-moles waste
= 2258.61 kg-mole O_2/4 kg-moles $C_{533.82}H_{877.09}O_{394.18}N_{5.72} \times S_{1.00}$
= 2258.61 × 32 kg O_2/4 (533.82 × 12 + 877.09 × 1 + 394.18 × 16 + 5.72 × 14 + 1.00 × 32) kg waste
= 72275.52 kg O_2/54807.56 kg waste
= (1.32 kg O_2/kg waste) 438.41 kg waste
= 578.14 kg O_2

D. Example 4.

Determine the amount of air required to compost 1000 kg of dry compostable solid wastes. Assume that the initial composition of the raw material to be aerobically composted is given by $[C_{534}H_{877}O_{394}N_6S_1]_5$, that the final composition is estimated to be $[C_{534}H_{877}O_{394}N_6S_1]_2$, and that 400 kg of material remains after the aerobic composting process. Air composition (in volume fractions) is assumed to be: carbon dioxide, 0.0003; nitrogen, 0.7802; oxygen, 0.2069; water, 0.0126. Assuming ideal gases, the volume fractions may be taken as mole fractions and are equal to the percentages of volume divided by 100. The air composition, as given, is for rare gases included with nitrogen and with moisture content corresponding to 70% relative humidity at 60 °F (15.6 °C). Air of this composition has a weight of 28.7 kg/kg-mole of total gas.

Solution

1. The amount of oxygen required for the aerobic stabilization of municipal solid wastes can be estimated by using Equation 1 (Sect. II.B) in which $a = 5(534) = 2670$, $b = 5(877) = 4385$, $c = 5(394) = 1970$, $d = 5(6) = 30$, $e = 5(1) = 5$, $f = 0$, $u = 2(534) = 1068$, $v = 2(877) = 1754$, $w = 2(394) = 788$, $x = 2(6) = 12$, $y = 2(1) = 2$, and $z = 0$.
2. Determine the moles of material present initially and at the end of the aerobic composting process.
   kg-moles present initially:
   $1000 \text{ kg}/[2670(12) + 4385(1) + 1970(16) + 30(14) + 5(32)] \text{ kg} = 0.0145932$
   kg-moles present at end:
   $400 \text{ kg}/[1068(12) + 1754(1) + 788(16) + 12(14) + 2(32)] \text{ kg} = 0.0145932$
3. Determine the kg-moles of material leaving the composting process per mole of material entering the process
   $n = \text{kg-moles of } C_uH_vO_wN_xS_y/\text{kg-moles of } C_aH_bO_cN_dS_e$
   $= 0.0145932/0.0145932$
   $= 1.0$
4. Determine the values of r, s, and t.
   $r = 0.5[b - nv - 3(d - nx)]$
   $= 0.5[4385 - 1(1754) - 3(30) + 3(1)(12)] = 1288.5$
   $s = a - nu = 2670 - 1(1068) = 1602$
   $t = 2[e + f - ny - nz] = 2[5 + 0 - 1(2) - 1(0)] = 4$

5. Determine the amount of oxygen required.
   Oxygen requirement per 1000 kg of solid waste
   $= 0.5[nw + 2s + r - c + t]$kg-moles of O_2/kg-mole of waste
   $= 0.5[1(788) + 2(1602) + 1288.5 - 1970 + 4]$kg-moles of O_2/kg-mole of waste
   = (3314.5 kg-moles of O_2/kg-mole of waste)(0.0145932 kg-mole of waste)
   = 48.37 kg-moles of O_2
6. Determine the air requirements.
   kg-moles of air required per 1000 kg of solid wastes
   = 48.37 kg-moles of O_2
   = 48.37/0.2069 = 233.78 kg-moles of air
   kg of air required per kg of dry compostable solid waste
   = (233.78 kg-mole of air)(28.7 kg of air/kg-mole of air) ÷ (1000 kg of dry compostable solid waste)
   = 6.71 kg of air per kg of dry compostable solid waste

E. Example 5.

Determine the required amount (kg/day; m^3/day) of wet sewage sludge with a solids content of 3% that must be added to a typical municipal solid waste (indicated in Table 8) in order to achieve the desired combined moisture content of 60% for aerobic composting of solid waste and sludge. Assume the municipal solid waste generation rate is equal to 2.95 kg per capita per day (6.5 lb per capita per day), and the population is 10,000. The specific gravity of the wet sewage sludge is approximately equal to 1.0. The glass, tin cans, nonferrous metals, and ferrous metals are recovered from the municipal solid wastes for reuse before aerobic composting.

Other noncompostable materials, such as plastics, textiles, rubber, leather, dirt, and ashes, however, remain in the composting process.

Solution

1. Set up a computation table (see Table 12). Both the percent by weight data and the moisture content data are obtained from Table 8, in which the glass, tin cans, and metals (total 17%) are recovered for reuse before aerobic composting. The wet weight data are calculated based on an arbitrarily assumed sample weight of 100 kg. The total dry weight of solid waste components is calculated to be 61.66 kg. Finally the moisture content of the solid waste sample is calculated:

Table 12
Determination of Moisture Content for Typical Solid Wastes to be Composted[a]

Component	Percent by weights	Wet weight, kg	Moisture content, %	Dry weight, kg
Food wastes	15.00	15.00	70	4.50
Paper	40.00	40.00	6	37.60
Cardboard	4.00	4.00	5	3.80
Plastics	3.00	3.00	2	2.94
Textiles	2.00	2.00	10	1.80
Rubber	0.50	0.50	2	0.49
Leather	0.50	0.50	10	0.45
Garden trimmings	12.00	12.00	60	4.80
Wood	2.00	2.00	20	1.60
Dirt, ashes, brick, etc.	4.00	4.00	8	3.68
Total	83.00	83.00		61.66

[a] Based on an as-delivered sample weight of 100 kg.

$$\text{Moisture content } (\%) = \left(\frac{83.00 - 61.66}{83.00}\right) 100$$
$$= 25.71$$

2. Calculate the daily amount of solid waste delivered to a compost plant.
   Total amount of raw solid waste generated
   = 10,000 person (2.95 kg per capita per day)
   = 29500 kg/day
   Amount of solid waste delivered to the compost plant
   = (29500 kg/day) (0.83)
   = 24485 kg/day
   (Note: 25.71% moisture)
3. Calculate the dry weight of solids, SW_{dry}, and the weight of water, SW_{water}, in the waste material.
   $SW_{dry} = (1 - 0.2571)\ (24485 \text{ kg/day})$
   $= 18190 \text{ kg/day}$
   $SW_{water} = 0.2571\ (24485 \text{ kg/day})$
   $= 6295.09 \text{ kg/day}$
4. Solve for weight of wet sludge solids SL_{wet}, that must be added to achieve a final moisture content of 60%. It is known that:

$$SW_{wet} = SW_{dry} + SW_{water} = 24485 \text{ kg/day} \quad (7)$$

$$SL_{wet} = SL_{dry} + SL_{water} \quad (8)$$

$$SL_{water} = (1.00 - 0.03)\ SL_{wet} \tag{9}$$

$$0.60 = (SL_{water} + SW_{water})/(SL_{wet} + SW_{wet}) \tag{10}$$

From Eqs. 7 and 8, the following relationship is obtained:

$$SL_{water} = 32.33\ SL_{dry} \tag{11}$$

From Eqs. 7, 8, 10 and 11, Eq. 12 is established:

$$0.60 = (32.33\ SL_{dry} + 6295.)/(32.33\ SL_{dry} + SL_{dry} + 24485) \tag{12}$$

Thus,
SL_{dry} = 680.94 kg/day
SL_{water} = 32.33 SL_{dry} = 22014.82 kg/day
SL_{wet} = SL_{dry} + SL_{water} = 22695.76 kg/day
(Note: sludge at 3% solids)

5. Check calculations.
Dry solids
SW_{dry} = 18190 kg/day = 38.6%
SL_{dry} = 680.94 kg/day = 1.4%
Water
SW_{water} = 6295.09 kg/day
SL_{water} = 22014.82 kg/day

Subtotal = 28309.91 kg/day = 60.0%

Grand total 47180.76 kg/day = 100.0%

6. Final answers
Wet sewage sludge (at 3% solids) needed
= 22695.76 kg/day
= 22.69576 m^3/day.

REFERENCES

1. D. Mosher and R. K. Anderson, *Composting Sewage Sludge by High-Rate Suction Aeration Techniques*, 50 pp. USEPA Publ. SW-614D, Washington, D.C., 1977.
2. D. J. Ehreth, A. B. Hais, and J. M. Walker, in *Composting of Municipal Residues and Sludges*, p. 6, Information Transfer, Inc. Rockville, Maryland, 1977.
3. W. L. Gaby, *Evaluation of Health Hazards Associated with Solid Waste/Sewage Sludge Mixtures*, USEPA Publ. EPA-670-2-75-023, Washington, D.C., 1975.

4. E. Spohn, *Compost Sci.* **18** (2), 14 (1977).
5. A. J. Howard, *R. Soc. Arts* **84**, 25 (1935).
6. F. K. Jackson and Y. D. Wad, *Indian Med. Gaz.* **69** (1969).
7. J. L. Ravoni, J. E. Heer, and D. J. Hagerty, *Handbook of Solid Waste Disposal*, pp. 7–56, Van Nostrand Reinhold Co., New York, 1975.
8. G. Beccari, U.S. Pat. 1,329,105 (1922).
9. H. B. Gotaas, *Composting*, World Health Organization, Geneva, Switzerland, 1956.
10. E. Eweson, *Univ. Kan., Bull. Eng. Architect.* **29** (1953).
11. W. A. Westrate, G., *Public Cleans.* **41**, 491 (1951).
12. A. W. Breidenbach, *Composting of Municipal Solid Waste in the United States*, USEPA, Publ. SW-47r, Washington, D.C., 1971.
13. C. G. Goleuke, *Composting*, Rodale Press, Emmaus, Pa., 1972.
14. C. G. Goleuke, and H. B. Gotaas, *Am. J. Public Health* **44**, 339 (1954).
15. J. S. Wiley, *J. Water Pollut. Control Fed.* **34**, 80 (1962).
16. P. R. Krige, *J. Inst. Sewage Purif.* Part 3, 312 (1964).
17. J. S. Wiley and S. C. Westerberg, *Appl. Microbiol.* **18**, 994 (1969).
18. G. Farkasdi, *Compost Sci.* **6**, 11 (1965).
19. W. Obrist, *Compost Sci.* **6**, 27 (1966).
20. C. B. Davey, *Soil Sci. Soc. Am. Proc.* **17**, 59 (1953).
21. S. A. Wilde, *Forest Prod. J.* **8**, 323 (1958).
22. J. C. Wiley, in *Waste Treatment*, (P. C. G. Isaac, Ed.), p. 349, Pergamon Press, Oxford, England, 1960.
23. D. G. Wilson, *Handbook of Solid Waste Management*, pp. 197–225, Van Nostrand Reinhold Co., New York, 1977.
24. P. H. McGauhey and C. G. Goleuke, *Reclamation of Municipal Refuse by Composting*, Tech. Bull. No. 9, University of California, Berkeley, 1953.
25. G. N. Fesenstein, J. Lacy, F. A. Skinner, P. A. Jenkins, and J. Pepys, *J. Gen. Microbiol.* **41**, 389 (1965).
26. B. E. Kane and J. T. Mullins, *Mycologia* **65**, 1087 (1973).
27. E. D. Schroeder, *Water and Wastewater Treatment*, pp. 179–216, McGraw-Hill, New York, 1977.
28. L. K. Wang, D. Vielkind, and M. H. Wang, *Ecol. Model.* **5**, 115 (1978).
29. G. Niese, *Arch. Mikrobiol.* **34**, 285 (1959).
30. G. Tchbanoglous, H. Theisen, and R. Eliassen, *Solid Wastes*, 621 pp. McGraw-Hill, New York, 1977.
31. L. K. Wang, J. C. Rivero de Aguilar, and M. H. Wang, *J. Environ. Manag.* **8**, 13 (Jan. 1979).
32. J. Olds, *Compost Sci.* **9**, 18 (1968).
33. H. L. Drobney, H. E. Hull, and R. F. Testin, *Recovery and Utilization of Municipal Waste*, USEPA Publ. SW-10c, 1971.
34. P. H. McGauhey, *Compost Sci.* **1**, 508 (1960).
35. K. L. Schultze, *Compost Sci.* **5**, 5 (1965).
36. J. R. Snell, *Consulting Eng.* **4**, 49 (1954).
37. Anonymous, *Environ. Sci. Technol.* **5**, 1088 (1971).
38. R. R. Cardenas, Jr., *Proc. of 1973 Solid Waste Conf.*, p. 285, Worcester Polytechnic Institute, Worcester, Mass., 1973.
39. R. R. Cardenas, Jr. and S. Varro, Disposal of Urban Solid Wastes through Composting, in *Processing Agricultural and Municipal Wastes* (G. Inglett, Ed.), p. 183, AVI Publ. Co., Westport, Conn., 1973.

40. Gainesville Municipal Waste Conversion Authority, Inc., *Gainesville compost plant; and interim report*. Cincinnati, U.S. Department of Health, Education, and Welfare, 1969.
41. P. Chrometzka, *Determination of the Oxygen Requirement of Maturing Composts*. International Research Group on Refuse Disposal, Inf. Bull. No. 33, August, 1968.
42. B. G. Liptak, *Environmental Engineers Handbook*, Vol. 3, 1130 pp., Chilton Book Co., Radnor, Pa., 1974.
43. J. S. Jeris and R. Regan, Optimum Conditions for Composting, in *Solid Wastes*, p. 245 (C. L. Montell Ed.), Wiley, New York, 1975.
44. J. S. Jeris and R. W. Regan, *Compost Sci.* **14**, 10 (1973).
45. J. S. Jeris and R. W. Regan, *Compost Sci.* **14**, 8 (1973).
46. J. S. Jeris and R. W. Regan, *Compost Sci.* **14**, 16 (1973).
47. U.S. Environmental Protection Agency, *Physical, Chemical and Microbiological Methods of Solid Waste Testing*, Publ. EPA 6700-73-01 National Environmental Center, Cincinnati, Ohio, 1973.
48. American Public Works Association, *Municipal Refuse Disposal*, 3rd ed., Public Administration Service, Chicago, Ill., 1970.
49. R. R. Cardenas, Jr., *Compost Sci.* **18** (1), 14 (1977).
50. L. L. Ciaccio, R. R. Cardenas, Jr., and J. S. Jeris, Automated Methods in Water Analysis, in *Chemical and Microbiological Analysis in Water and Water Pollution* (L. L. Ciaccio, Ed.), Vol. IV, Chap. 27, Marcel Dekker, New York, 1972.
51. K. Jalal, *Compost Sci.* **10**, 20 (1969).
52. T. G. Hanks, *Solid Waste/Disease Relationships—A Literature Survey*, Rept. SW-1c, U.S. Health, Education and Welfare, Public Health Service, Solid Waste Program, Washington, D.C., 1967.
53. P. R. Krige, *J. Inst. Sewage Purif.* Part 3, 215 (1964).
54. J. Wiley, *J. Water Pollut. Control Fed.* **34**, 80 (1962).
55. M. T. Morgan and F. W. MacDonald, *J. Environ. Health* **32**, 101 (1969).
56. L. K. Wang, M. H. Wang, and G. G. Peery, *Water Sewage Works*, **125**, 30 (July, 1978).
57. L. K. Wang, M. H. Wang, and G. G. Peery, *Water Sewage Works*, **125**, 58 (August, 1978).
58. M. H. Wang, L. K. Wang, G. G. Peery, and R. C. M. Cheung, *Water Sewage Works*, **125**, 99 (September, 1978).
59. M. H. Wang, L. K. Wang, G. G. Peery, and R. C. M. Cheung, *Water Sewage Works*, **125**, 33 (October, 1978).
60. J. A. Salvato, *Environmental Engineering and Sanitation*, Wiley, New York, 1972.
61. C. J. Huang and C. Dalton, *Energy Recovery From Solid Waste*, Vol. 2, Technical Report, NASA Contr. Rept. CR-2526, Washington, D.C., 1975.
62. G. J. Kupchick, *Bull. World Health Org.* **34**, 798 (1966).
63. U.S. Environmental Protection Agency, *Process Design Manual for Sludge Treatment and Disposal*, USEPA 625-74-006, Washington, D.C., Oct., 1974.
64. L. K. Wang and M. H. Wang, *Characterization and Composting of Municipal Solid Wastes*, Technical Report No. SIT-ME-0179A, Stevens Institute of Technology, Hoboken, N.J., January, 1979.

6

Materials and Energy Recovery

P. Aarne Vesilind

Department of Civil Engineering, Duke University, Durham, North Carolina

Norman C. Pereira

Monsanto Company, St. Louis, Missouri

I. INTRODUCTION

Solid waste management has through the years consisted of collection and disposal. Technological advances have for the most part been in collection equipment design, incinerator efficiency, and landfill operation. In short, the objective has been to "pick it up and put it down," at the lowest possible cost.

Only within the last few years, as nonreplenishable materials are becoming scarce, and as one city after another uses up landfill space, has effort been directed toward an alternative system, namely the recovery of useful and marketable materials from solid waste.

In this chapter, the objectives and rationale for resource recovery are discussed first, followed by a section on resource recovery operations dealing with the various operation and processing modules that can be designed into a total resource recovery facility. Next, some actual case studies are presented discussing installations that are currently operational and have provided valuable operating and cost data. The section deals with the economics of resource recovery. The va factors that have a bearing on resource recovery economics are discussed, along with what incentives could be provided in order to make resource recovery a market-oriented industry.

Throughout this chapter, the emphasis is on the utilization of urban or municipal refuse as the principal raw material for resource recovery.

II. RESOURCE RECOVERY OBJECTIVES

The flow of materials in our society may be illustrated by the schematic diagram shown in Fig. 1. The centers of interest are the two blocks marked "industrial sector" and "consumer sector." The industrial sector represents all activity that involves the consumption of materials in the production or manufacturing of finished products and goods. The consumer sector, in turn, represents the consumption by the general public of the goods and products output from the industrial sector. It will be noted from Fig. 1 that the industrial sector has three distinct inputs: (a) raw materials, which are gleaned from the face of the earth and injected into refining and subsequent manufacturing or fabricating operations, (b) industrial scrap from manufacturing and fabricating operations that is subsequently recycled back into the industrial sector, and (c) materials recovered from the waste output of the consumer sector. The principal output from the industrial sector, i.e., manufactured goods, also becomes the principal input to the consumer sector. The consumer, in turn, has three options regarding the use of the manufactured goods: (a) these materials may be disposed of after use, i.e., the waste can be burned, landfilled, composted, etc.; (b) the material may be treated and collected after use, and in the process sufficient quantities recovered to be recycled within the industrial sector; or, (c) the material may be reused for the same or different purpose within the consumer sector (the last option is the second input to the consumer sector in Fig. 1).

Using glass as an example of a material and product, we can illustrate how the schematic in Fig. 1 can be applied. Sand and other raw materials (X_{RM}) are used to manufacture glass, most of which is then sold to the consumer as finished products (X_M). Some of this glass is defective or for other reasons unusable for consumer use and is recycled (X_{RIS}) as industrial scrap (known as "cullet" in the glass industry). The consumer uses the glass, and after having served its usefulness (either as bottles or other products) disposes of the glass as waste (X_{CD}) in the usual manner. Some of this glass is recycled, that is,

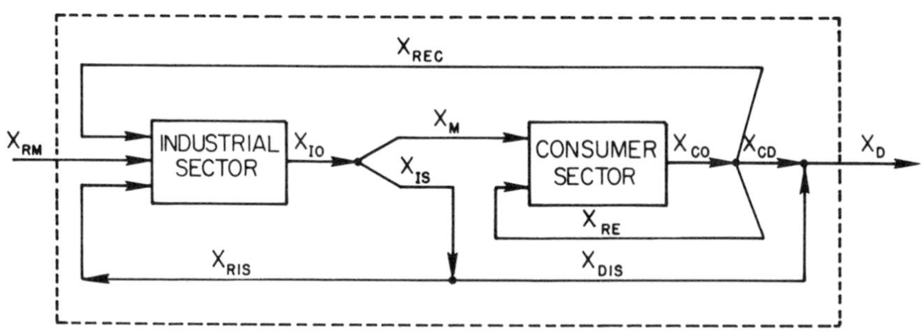

FIG. 1 Schematic representation of materials flow: X_{RM} = raw materials input; X_{IO} = total materials output from industrial sector; X_{IS} = industrial scrap; X_{RIS} = recovered industrial scrap; X_{DIS} = disposed industrial waste; X_M = manufactured goods; X_{CO} = total waste output from consumer sector; X_{RE} = reused post-consumer waste; X_{REC} = recovered post-consumer waste; X_{CD} = disposed post-consumer waste; X_D = total disposed waste.

it is returned (X_{RE}) to be reused (refillable bottles, for example), and some of the glass might be used for other purposes such as home canning or for storage. Some glass might be recovered (X_{REC}) and reinjected into the manufacturing process. Such recovery takes place by a conscious effort of the community or other organizations that collect and recycle glass products through the system, sometimes at a financial loss.

Other materials could be substituted and discussed similarly using the schematic diagram in Fig. 1. This diagram is, in effect, a simplified illustration of materials flow in our society.

Of interest to us are the quantity of raw materials that enter the system, denoted by X_{RM} on the schematic, and the quantity of waste materials to be disposed of, denoted by X_D.

It may be argued that both the X_{RM} and X_D quantities should be as large as possible, and certainly there are many benefits to be gained by increasing these quantities. For example, a large quantity of raw materials injected into the manufacturing process represents a high rate of employment in the raw materials industry. This can have a residual effect of creating cheaper raw materials and thus reducing the cost of manufacturing.

A large X_D component is also beneficial in the sense that the waste disposal industry (which includes such diverse elements as the local junkman and the large corporations that manufacture heavy equipment for landfills) has a key interest in the quantity of material that people dispose of. Thus a large X_D component would mean more jobs in this industry.

However, large X_{RM} and X_D components can have detrimental effects that easily outweigh these benefits. First of all, the effect of having a large raw material input can mean a detrimental effect on our environment in the sense that a great deal of nonreplenishable raw materials are extracted, often using methods that disregard environmental considerations. A classic example is strip mining. A large X_D component can similarly have a significant environmentally detrimental effect in that large land areas (landfills) are needed for the disposal of waste, or that disposal by incineration may cause serious air pollution problems in local situations.

A large X_{RM} component can eventually lead to a depletion of natural resources. The mineral resource outlook for the United States is certainly not optimistic (Table 1). Many minerals are already scarce and those that are in abundance today can hardly sustain our population for more than two or three generations. More significantly, within this period we shall become highly dependent on foreign nations for much of our essential raw materials (Table 2).

National security is based on the nation's ability to achieve self-reliance in raw materials. We have already seen the problems that can be created by relying on other countries for such necessities as crude oil. Without doubt, cartels will be developed by nations that have large deposits of other nonreplenishable materials, and it can be expected that the cost of such products as aluminum, tin, and rubber will increase

Table 1
Projected U.S. Resources[a]

	Annual consumption (1970)	Reserves	Estimated supply years
Tin	74,000 t	46,000 t[b]	1
Copper	2,060,000 t	88,500,000 t	30
Iron	137,000,000 t	2,000,000,000 t	15
Petroleum	4,000,000,000 bbl	45,000,000,000 bbl[c]	8
		530,000,000,000 bbl[d]	80
Gas	22×10^9 ft^3	275×10^9 ft^3	8
Cryolite	21,000 t	~0	0
Alumina	15,000,000 t	50,000,000 t	3

[a] From Teller [1].
[b] 1000 tons = 907 metric tons.
[c] Known.
[d] Projected.

Table 2
U.S. Dependence on Imported Metals[a]

	Percentage imported			
Metal	1950	1970	1985[b]	2000[b]
Aluminum	64	85	96	98
Chromium	n.a.[c]	100	100	100
Copper	31	0	34	56
Iron	8	30	55	67
Lead	39	31	62	67
Manganese	88	95	100	100
Nickel	94	90	88	89
Potassium	14	42	47	61
Tin	77	n.a.[c]	100	100
Tungsten	n.a.[c]	50	87	97
Zinc	38	59	72	84

[a] From Teller [1].
[b] Projected. From reference 9, © 1974, by the American Association for the Advancement of Science.
[c] n.a. = not available.

substantially. The dependence on other nations for these materials can thus result in the loss of national integrity.

In addition, if other nations were to attain the standard of living that developed nations have at present, the raw materials supply of the world would probably not be adequate to meet the demand. Furthermore, the thrust for greater production would mean extracting concentrated raw materials from the earth and converting them into products that, when disposed of over a wide land area, would spoil the environment and at the same time make recovery and reuse of these products impractical.

The prime consideration against increased X_{RM} and X_D† is not merely the depletion of large quantities of raw materials, but also the associated problem of disposing of large quantities of waste. This creates the esthetic problem of an environmental insult, and simultaneously denies resources to future generations.

It should therefore not be difficult to argue that in our schematic of Fig. 1, the components X_{RM} and X_D should be reduced to the smallest

† To see how X_{RM} and X_D are interrelated, consider the system enclosed by the dashed lines in Fig. 1. A materials balance on such a system shows that $X_D = X_{RM}$. Hence a reduction in X_{RM} implies a reduction in X_D (if there is to be no accumulation within the system) and vice versa.

quantities practical, and that we should try to redesign our industrial decision-making, and our economic and value systems, in order to achieve this end.

In the following subsections, the various options available for achieving low X_{RM} (raw material consumption) and X_D (waste disposal) are analyzed by considering macroscale material balances based on the materials flow in Fig. 1.

A. Options for Reducing Raw Material Consumption

Consider a materials balance in the industrial sector of the schematic in Fig. 1. Equating all inputs to this sector to its outputs gives

$$\begin{aligned} X_{RM} + X_{REC} + X_{RIS} &= X_{IO} \\ &= X_M + X_{IS} \\ &= X_M + X_{RIS} + X_{DIS} \end{aligned}$$

solving for X_{RM} gives

$$X_{RM} = X_M + X_{DIS} - X_{REC} \tag{1}$$

An analysis of Eq. (1) indicates how raw material consumption can be reduced. One possibility for achieving a low X_{RM} component is to decrease the amount of manufactured goods, X_M. This results in a lower standard of living and necessitates a redesign of products in such a way as to use less materials and energy. Although a "low energy and materials living style" may appeal to a few, it is clearly unacceptable to the majority. The quantity of manufactured goods would only be reduced by an economic depression, changes in consumer preference, or by government edict, such as restrictive legislation on packaging.

The second possibility, according to Eq. (1), is to reduce the amount of disposed industrial scrap, X_{DIS}, that is essentially a waste component. This calls for more efficient consumption of raw materials whereby less X_{RM} is consumed in order to produce a specific amount of X_M.

It should be noted that the recycled industrial scrap component, X_{RIS}, is rather conspicuously absent from Eq. (1). Recycled industrial scrap that consists of home scrap (waste material reused within an industrial plant), and prompt industrial scrap (clean industrial waste material used immediately by another industry) has obvious economic advantages to industry, but increasing this component would not aid in minimizing raw material consumption. This is not to say that the recycling of industrial scrap is unimportant; in fact, it is the ultimate goal of any industrial operation to produce as little scrap as possible.

The third possibility for reducing X_{RM}, according to Eq. (1), is to increase the recovery of materials, X_{REC}. If this is accomplished, the total amount of goods manufactured by industry need not change, but it would be possible to reduce the raw materials input to this system and at the same time the amount of materials destined for disposal. This strategy thus seems to be not only feasible but also economically, politically, and practically, the most attractive.

B. Options for Reducing Disposed Waste Quantities

Looking at the other end of the schematic in Fig. 1, the total disposed waste component, X_D, a materials balance shows that

$$X_D = X_{CD} + X_{DIS} \tag{2}$$

Equation (2) indicates that the total disposed waste is composed of two components, i.e., postconsumer waste and industrial waste. Examining Fig. 1 it is obvious that one of these components, X_{DIS}, can be minimized by increasing X_{RIS} in the industrial sector. The other component, X_{CD}, can be further analyzed by considering a materials balance at the output from the consumer sector, i.e.,

$$X_{CO} = X_{CD} + X_{REC} + X_{RE} \tag{3}$$

thus

$$X_{CD} = X_{CO} - X_{REC} - X_{RE} \tag{4}$$

Hence for a fixed output from the consumer sector (X_{CO} constant), the postconsumer solid waste (and thus the total disposed waste X_D) can be reduced by increasing X_{REC} and X_{RE}. It is clear that the only two methods of reducing the quantity of materials to be disposed of is to increase the recovery component and/or the reuse component. We have already commented on the feasibility of increasing the recovery component, and the same basic consideration holds for reuse.

C. Feasible Alternatives for Reducing Raw Material Consumption and Solid Waste Generation

From the arguments presented above, it seems reasonable to conclude that the principal alternatives available to achieve low raw material consumption (X_{RM}) and low waste generation (X_D) are:

1. Waste reduction (reduction of X_{CD})
2. Increased reuse (increasing X_{RE})
3. Increased recovery (increasing X_{REC})

1. *Waste Reduction*

Waste reduction means preventing waste at its source by either physically redesigning products or otherwise changing consumption and waste generation patterns.

Waste reduction can be achieved by (a) increasing the lifetime of a product, and (b) reducing the amount of material used per product without sacrificing its utility.

Some examples of potential savings from increased product lifetime are as follows: An 80% shift to refillable beer and soft drink containers that would make at least 12 trips to the bottling plant and back, is not unreasonable under the Oregon-type bottle legislation. Better tires for automobiles that would last 60,000 miles or more instead of 20,000 miles, certainly seem to be in the near future. In almost all cases, products such as appliances, furniture and household electrical goods can be redesigned so as to increase their life by 10–15% at minimal cost. If this increase were achieved for beverage containers, tires, appliances, furniture, etc., a reduction of 20 million tons of postconsumer waste, or a 10% reduction in the 1985 projected waste generation figures is possible [2].

The automobile is another example. The average life of a passenger car in the United States is about 10 yr (much less than in other countries). The weight of a car is about 4000 lb (1800 kg) and increasing the life by only 2 yr to an expected life of 12 yr by 1990 would achieve a saving of about 6 million tons (5.5 million metric tons) of steel, 150,000 tons (136,000 metric tons) of aluminum, and 150,000 tons (136,000 metric tons) of zinc.

The reduction in material use per product can most readily be achieved by redesigning some of the packaging that is currently used in marketing operations. For example, a drawn and ironed steel can results in a saving of about 25–30% in materials over the common seamed tin can. Redesign of the automobile to save only 5% of the steel would, over one year, save about 350,000 tons (318,000 metric tons) of steel.

It should be emphasized that these proposed reductions and changes in product materials and design would involve an alteration of the existing production and marketing structure of numerous companies, as well as a significant change in consumer habits. Therefore any enforcement of rules or regulations must take place with the full knowledge of the economic repercussions.

It is not too unlikely that some governmental intervention will be forthcoming in the future, especially if this appears to be the only practical solution to the natural-resource depletion and waste-generation problems; any legislation for the purpose of reducing waste is, in the opinion of many, an unattractive option. Thomas Jefferson's dictum that "that government is the best that governs the least" still seems to hold even in the case of solid waste and resource management.

On a long-term basis, rising energy and material costs will force waste reduction, even without overt government action. In the meantime, the solid waste currently being generated has to be handled by means other than the landfill and incinerator.

2. *Reuse*

The product reuse option is applicable to the broad and increasing category of consumer goods designed to be used once and then discarded. Reusable products could be substituted in many instances. At the present time, many of our products are reused in the home. For example, paper bags obtained in the supermarket are often used for bagging refuse in order to carry it from the house to the trash can. Newspapers are rolled up in order to make fireplace logs, and coffee cans are used to hold nails. All of these are examples of reuse. Unfortunately, none of these provide a great economic impact on the total quantities of materials consumed or disposed of.

Packaging beverages (beer and soft drinks) in refillable containers would constitute a major form of reuse. The role that beverage containers play in four components of waste (ferrous metals, aluminum, glass, plastics) is shown quantitatively in Table 3. By 1985, beverage containers will make up 31% (1.9 million tons) of the ferrous metal cans in all municipal waste in the United States; aluminum beverage cans will constitute 86% of all aluminum cans in waste (0.80 million tons); glass one-way bottles 52% of total glass containers at 7.8 million tons per year; and plastics 9% of blow-molded containers (0.1 million tons). On a composite basis, beverage containers would represent 45.4% of all containers made from these four materials, or 10.6 million tons annually in the USA.

Beverage containers of glass, plastic, steel, and aluminum constitute at present 6% of total waste generation; this figure will rise to 6.5% by 1980, and decrease to 6% by 1985 [3, 4].

The advisability of an Oregon-type bottle law is still hotly debated. The benefits that can be derived from such a mandatory deposit system include not only reductions in litter and solid waste quantities (obvious direct local benefits), but also national benefits in terms of savings in materials and energy. In addition, beverages in refillable containers are

Table 3
Summary[a] of Packaging and Beverage Container[b] Waste Generation for Ferrous, Aluminum, Glass, and Plastics, 1972–1985
in 1000 tons[c]

Material and container type	Year 1972	1975	1980	1985	Beverage containers as % of total
Ferrous metal wastes	11,100	11,700	13,100	15,200	
Cans	5,560	5,625	5,760	6,075	
Beverage cans	1,536	1,615	1,800	1,900	31.3
Aluminum wastes	900	1,200	1,500	1,800	
Cans	403	587	813	930	
Beverage cans and ends	340	500	700	800	86.0
Glass wastes	13,200	14,500	16,400	16,600	
Bottles	12,200	13,400	15,100	15,100	
Beverage bottles	6,010	6,865	7,798	7,800	51.7
Returnable bottles only	879	842	794	794	
Plastics wastes	4,500	5,700	8,400	11,000	
Blow-molded containers	458	610	920	1,270	
Beverage containers	11	22	36	110	8.9
Totals					
Materials	29,700	33,100	39,400	44,600	
Containers	18,621	20,222	22,593	23,375	
Beverage containers	7,897	9,002	10,334	10,610	45.4
Beverage containers, % of total containers	42.4	44.5	45.7	45.4	
Total solid waste generated	130,000	140,000	160,000	180,000	
Containers, %	14.3	14.4	14.1	13.0	
Beverage containers, %	6.1	6.4	6.5	5.9	

[a] From Franklin [4].
[b] Beer and soft drink.
[c] 1 million tons = 907,200 metric tons.

generally cheaper. The disadvantages from the imposition of a container deposit are a potential decline in sales, consequent unemployment, and reduced tax revenues.

A number of excellent studies have been concluded in the last few years to delineate the economic impact of such programs being instituted on a nationwide scale [5, 6]. A comprehensive report on the Oregon bottle law concluded that [6]:

1. Beverage container litter has been decreased by an estimated 66%.

2. Beer and soft drink sales have neither fallen below the level of the year prior to enactment nor risen as in previous years.
3. Job losses have occurred in the container manufacturing and canning industries.
4. Jobs have been created in the brewing, soft drink, and retail sectors of the economy.

3. *Recovery*

The concepts of recovery (or resource recovery, the main topic of this chapter) are manifested in various related fields of activity. Principally, these concepts comprise (a) recovering materials from refuse in a relatively pure form suitable for use as raw material for products similar to those discarded; (b) recovering materials that because of their previous application or contamination cannot be totally reclaimed but may be utilized in lower grade applications; (c) changing, in form and substance, large portions of the waste into new products; and (d) burning waste directly to produce energy or converting waste into a storable fuel such as oil or gas [7].

The main target for recovery operations is the vast quantity of mixed municipal and urban solid waste. An examination of Tables 2

Table 4
Expected Ranges in Mixed Municipal Refuse Composition[a]

Component	Composition (% of dry weight)[b] Range	Nominal	
Metallics	7–10	9.0	Mechanical recovery
Ferrous	6–8	7.5	Mechanical recovery
Nonferrous	1–2	1.5	Mechanical recovery
Glass	6–12	9.0	Mechanical recovery
Paper	37–60	55.0	Mechanical recovery
Newsprint	7–15	12.0	Mechanical recovery / Conversion recovery
Cardboard	4–18	11.0	Mechanical recovery / Conversion recovery
Other	26–37	32.0	Mechanical recovery / Conversion recovery
Food	12–18	14.0	Conversion recovery
Yard	4–10	5.0	Conversion recovery
Wood	1–4	4.0	Conversion recovery
Plastic	1–3	1.0	Conversion recovery
Miscellaneous	< 5	3.0	Conversion recovery

[a] From Drobny *et al.* [8].
[b] Moisture content: 20–40%.

and 4 in Chapter 1, regarding quantities and compositions of discarded municipal solid wastes, certainly points to the large potential for resource recovery. This recovery potential can be divided into two categories: (a) materials available for essentially mechanical separation and reuse in a relatively pure form consisting of inorganic materials; this is referred to as the "mechanical recovery" fraction; and (b) organic materials which can be recovered only through conversion; this is labeled "conversion recovery" fraction (see Table 4). Conversion usually leads to some form of a derived product such as compost, fiber, or wallboard building material, or more importantly, a source of energy. It should be noted that in Table 4 paper appears under "mechanical recovery" as well as "conversion recovery." The reason is that when paper is mixed with refuse it becomes contaminated with dirt, grease, and other materials that make it unacceptable for mechanical recovery and more amenable to conversion recovery.

Thus, depending on the composition of the refuse stream, recovery is essentially a two-phase process. The first phase is termed a "front end system" and refers to mechanical materials recovery (metals, glass, paper), and the second phase is labeled "back end system" and deals with the recovery of the remainder or organic portion, i.e., the bulk of the waste material (about 80%), either by conversion to a fuel or as raw material for a product (see Fig. 2).

a. *Materials Recovery.* Many of the "mechanical recovery" components of municipal solid waste can be recovered and recycled for subsequent use, among the most important being paper, steel, aluminum, and glass.

The amount of paper in the municipal solid waste stream is staggering. About 60 million tons (55 million metric tons) of paper enter the solid waste stream annually, of which only about 15% is recovered. It is estimated that about 45 million tons (41 million metric tons) per year could be recovered economically from the solid waste stream without the use of new technology. This recovery would include about 8 million tons newspaper, 12 million tons corrugated cardboard, 10 million tons printing and writing paper, and 15 million tons wrapping and other types of paper.

The wastepaper market is highly dependent on the scarcity of virgin fiber and prices have fluctuated from a high of \$60/ton to as low as \$5/ton. Increased paper recycling depends to a large extent upon commitments by the paper industry to use wastepaper on a day-to-day basis rather than only when virgin fiber is unavailable.

Paper can be recovered mechanically from mixed wastes by wet or

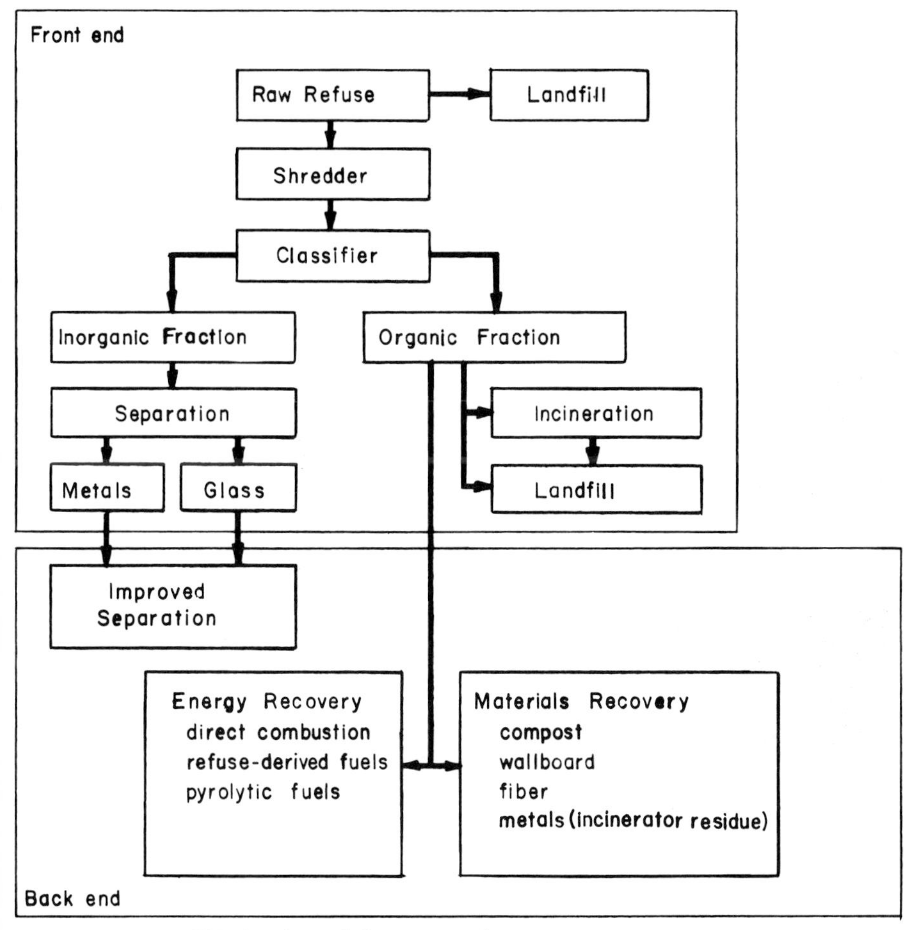

FIG. 2 A modular approach to resource recovery.

dry separation. Wet separation has been demonstrated at Franklin, Ohio (see case study below). Operational systems for dry separation of paper fibers exist in Europe, and attempts are being made to introduce such technology in the United States. The principal means of paper recovery, however, is by source separation and separate collection. Source separation is defined as the setting aside of recyclable waste materials at their point of generation (e.g., home, office, commercial places of business). This separation is followed by transportation from the point of generation to a secondary materials dealer or directly to a manufacturer. Source separation of paper is feasible primarily for newspapers from homes, corrugated containers from commercial and industrial establishments, and printing and writing papers from offices [10].

Ferrous metal recovery represents a significant source of revenue because (a) there is a significant quantity of ferrous metal, about 7–8%, in the municipal solid waste stream, with a typical value of about $12–20 per ton; (b) ferrous metals are relatively easy to extract; and (c) ferrous metals are reused in steelmills, and used in the precipitation of copper from low-grade copper ore. The most important is steel recovery.

The principal sources of raw materials for steel recovery are steel cans and junk autos. In 1973 about 4 million tons (3.5 million metric tons) of steel cans were generated in the United States, and only about 2% were recovered [2]. Annual discard of vehicles approached 10.5 million units in 1975, and based upon current trends is expected to reach 11.6 million by 1980, and 13.2 million by 1985. Autos account for 85% of the total, with trucks and buses accounting for the rest. At present, about one-fourth of the recycled steel scrap is from discarded automobiles. After replacement parts and other valuable components have been removed, the stripped hulks are processed into steel-making scrap by baling, shredding and subsequent magnetic separation, or further cleaning by incineration. A typical junk automobile weighing 3600 lb (1600 kg) could yield approximately 2500 lb (1100 kg) of steel and 500 lb (230 kg) of cast iron, the rest being copper, zinc, aluminum, and about 400 lb (180 kg) of nonmetallics.

Cans may be recovered from composite municipal solid waste or they may be collected separately (source separated as in the case of paper). After recovery, steel cans must go through a detinning operation before processing in the steel furnace. Detinning of cans is a process in which a caustic solution (sodium hydroxide) containing an oxidizing agent (sodium nitrate or nitrite) is used to remove both the tin used for the seams and the underlying iron–tin alloy from the steel. After washing, the steel scrap is virtually free of tin and is compressed into large bales weighing up to 600 lb (270 kg) or more. The use of aluminum ends on steel cans presents a problem to detinners since the aluminum reacts violently with the detinning solution and causes a loss of caustic solution. With aluminum present, a two-stage detinning operation is carried out to keep caustic and oxidizing agent consumption to a minimum.

Scrap steel, which sells for about $50–100/ton can be used by the steel industry as part of the metallic charge to the basic oxygen furnace (BOF), or as the primary charge to electric furnaces. In the BOF, the scrap is used not only as a source of metallics but also as a coolant for controlling the temperature. Consequently, scrap usage cannot be arbitrarily changed without adjusting other variables to maintain the thermal balance required for the fast, smooth, trouble-free operation characteristic of the process. One method to increase the percentage of scrap usage in the BOF is to preheat the scrap charge. It would be realistic to increase scrap usage to 40% of the total metallic charge (current usage about 30%) by preheating. This would allow scrap to replace 10% of the iron now used in the BOF.

Since the heat source for the electric furnace is electricity, thermal balance requirements do not limit the amount of scrap consumed by the electric furnace. In 1971, 99% of the metallic charge to the electric furnace consisted of scrap. The scrap consisted of in-house waste and high-grade industrial waste resulting from metal discarded at various stages in manufacturing as well as postconsumer ferrous wastes [11, 12].

Aluminum constitutes less than 1% of the solid waste stream; it has, however, a value of over $200 per ton. About 1 million tons (0.9 million metric tons) of aluminum is discarded annually (50% cans, 30% foil). In addition a 3600 lb junk automobile yields about 50 lb scrap aluminum. Only about 4% of aluminum is currently being recovered, although aluminum can recycling operations have claimed 15–20% recovery in some urban areas.

Secondary aluminum is not as pure as primary aluminum because the removal of metallic impurities, except for magnesium, by the usual melting and refining procedures is difficult and uneconomic. Hence, the quality and type of aluminum scrap largely determines the alloy produced. Therefore the use of secondary aluminum is usually limited to the manufacture of castings.

Aluminum end products are classified into: (a) wrought products, such as sheet, plate, rolled and continuous cast rod and bars, and wire extrusions and forging, and (b) castings, including sand, permanent mold, and die castings. In general, wrought products require a lower degree of impurities than cast products because alloying agents such as copper and silicon reduce the ductility of aluminum. However, if the scrap from one type of wrought product can be segregated, it can be remelted into ingots that can be used to produce more wrought product.

Aluminum scrap is used from: (a) cans remelted into wrought ingots from which more cans can be produced, (b) scrap from municipal solid wastes remelted into a low magnesium content casting alloy, and (c) scrap from junk autos remelted into a low magnesium content casting alloy [11, 13].

The fourth component of some economic value is glass, an ideal material in the sense that it is clean, and can be reprocessed many times without loss in structural strength or attributes. Of the 13 million tons (12 million metric tons) of glass that are discarded annually only about 3% is currently recovered. Glass represents 6–10% of the municipal solid waste stream, and its value is $13–22 per ton. Color-sorted glass may sell for more, depending on market location; however, the economics of color sorting still remain questionable.

In glassmaking, cullet (scrap glass) is in some ways preferable to virgin raw materials because its use reduces fuel consumption and refractory wear. The industry generally limits the use of cullet in the glass formula to approximately 20% by weight, although 80–100% cullet formulations have been used.

A mechanical recovery system that achieves materials recovery of only the inorganic portion of solid waste (about 20% of the incoming waste), and disposes of the organic fraction by conventional means such as landfill or incineration, is clearly undesirable. The conversion recovery or organic fraction of municipal solid waste is a sizable portion (approximately 80%) and some form of recovery has to be exercised on this fraction in order that the total recovery system pay for itself. Materials recovery from the organic fraction usually results in compost or paper fiber as products.

In the past 20 years various composting plants have been established in the United States, but most of these have been closed because of a lack of a viable market for the compost. At present, only one plant at Altoona, Pa., has been known to be operating on a regular basis. In spite of the current pessimistic outlook there still may be a future for composting. Changing environmental philosophies, government regulations, economics, and expanding uses may make composting more broadly applicable in the future. Several systems currently being proposed are heavily oriented toward composting for special markets [14, 15, 16].

Paper fiber is recovered through operations such as the Franklin, Ohio, wet-separation system. In this plant, approximately 20 tons (18 metric tons) of paper fiber can be recovered for every 100 tons (91 metric tons) of solid waste processed; the plant capacity is 150 tons (136 metric tons) of waste per day. Other systems for fiberboard and wallboard recovery have also been proposed [17]. At the present time (1979), fiber recovery is not viable due to the lack of adequate fiber markets.

The Bureau of Mines has extensively studied the recovery of materials from municipal incinerator residues. Incinerator residues are a wet complex of metals, glass, slag, charred and unburned paper, and ash containing various mineral oxides. Research is being focused on developing various separation techniques for recovering and separating the residues into fractions that, if necessary, could be further treated to yield products suitable for recycling. Continuous screening, grinding, magnetic separation, shredding, gravity separation, and other separation techniques are being combined into a processing train in order to recover iron concentrates, aluminum copper-zinc composites, clear and colored glass fractions, and carbonaceous ash tailings [18, 19].

Tables 5 and 6 estimate materials recovery from postconsumer municipal waste for the period 1971–1975. The quantities shown exclude materials recovered from obsolete scrap sources such as demolition debris and junk automobiles. It is evident that overall recovery is not great—about 8 million tons per year or 6% of gross municipal discards. Also, waste paper recycling dominates the recovery statistics, comprising about 88% of total recovered tonnage. The amount of aluminum recovered has also increased rapidly as a result of aluminum can recycling programs initiated by the aluminum and brewery industries.

b. *Energy Recovery.* As pointed out above, a resource recovery system should be extended to recover value not only from the inorganic fraction but from the organic fraction as well. The most promising approach to effectively using the potential economic values in the organic fraction appears to be in burning it for its energy content or its conversion to either a gaseous, liquid, or solid fuel [21].

The potential for energy recovery from solid waste is significant. For example, in 1975, 128 million tons (116 million metric tons) of municipal solid waste was disposed of, 80% of which was combustible, yielding a Btu value equal to about 520,000 bbl oil per day, assuming municipal solid waste has a heating value of about 4300 Btu/lb (10 MJ/kg)

Table 5
Trend in Material Recovery from Post-Consumer Municipal Waste, 1971–1975[a]
(in millions of tons)

Year	Gross discards	Material recycled: Quantity	Material recycled: % of discards	Net waste disposed of
1971	132.9	8.1	6.1	124.8
1972	139.3	8.8	6.2	130.5
1973	144.2	9.6	6.7	134.6
1974	144.1	9.4	6.5	134.8
1975	136.1	8.0	5.9	128.2

[a] Source: reference [20].

Table 6
Trends in Material Recovery from Post-Consumer Municipal Waste, 1971–1975, by Type of Material[a]

Material recycled	1971	1972	1973	1974	1975
Paper and paperboard	7495	8075	8730	8430	6830
% of gross paper and board discards	15.9	16.0	16.5	16.3	15.5
Aluminum	20	30	35	52	87
% of gross aluminum discards	2.4	3.2	3.4	5.0	8.7
Ferrous metals[b]	140	200	300	400	500
% of gross ferrous discards	1.3	1.4	2.4	3.4	4.4
Glass	221	273	306	327	368
% of gross glass discards	1.8	2.1	2.3	2.5	2.7
Rubber (including tires and other)	257	245	219	194	189
% of gross rubber discards	8.9	7.9	6.8	6.1	6.9
Total materials	8133	8825	9590	9400	7975
% of gross nonfood product waste	9.5	9.6	10.1	10.0	9.3
% of total post-consumer waste	6.1	6.2	6.7	6.5	5.9

[a] In thousands of tons. Source: reference [20].

[b] These estimates for ferrous metal recycling are highly inferential and preliminary. There are no regularly collected statistics on this category. EPA estimates are based in part on work by the Resource Technology Corporation for the American Iron and Steel Institute regarding magnetic separation facilities.

on an as-received basis and that crude oil has a net heating value of approximately 110,000 Btu/gall. Recognizing that much of the waste generated in rural areas is not economically recoverable, the combustible materials from standard metropolitan statistical areas (SMSAs) will still result in an equivalent of over 400,000 bbl oil per day, one-third the initial output of the trans-Alaskan pipeline. This is equivalent to about 10% of all the coal used by the utilities in the United States and 5% of all the fuel used. While it is convenient to translate BTUs into barrels of oil, this is not realistic. The production of the BTUs from refuse requires a certain amount of effort and expenditure of energy. Secondly, the refuse cannot be used in all of the many ways that oil can, and thus it cannot be a one-to-one substitution [18].

The energy content per unit weight of the solid waste combustibles fraction depends on the method of separation, material species content, percentage of combustibles, and amount of moisture. For example, dry paper has an HHV (higher heating value) of approximately 8000–8700 Btu/lb (18.61–20.24 MJ/kg), rubber and leather 7500–12,800 Btu/lb

(17.45–29.77 MJ/kg), and plastics 16,000–18,000 Btu/lb (37.22–41.87 MJ/kg). Additional data on typical composition of refuse and the associated calorific content is provided in the Appendix to Chapter 3.

The technology for achieving energy recovery is already operational or being developed (see Table 7), and hence many options are open to towns and municipalities for recovering energy from refuse. The prime factor to bear in mind is that the refuse must be converted into energy in a form that is acceptable to either a utility or an industrial energy consumer. Recognizing this, the most attractive options available for energy recovery are:

1. Direct combustion of the solid waste, with or without shredding, for the production of steam or electricity.

The direct combustion of raw municipal solid waste for energy recovery is by no means a new concept. There are a number of refractory-wall incinerators with waste heat recovery boilers, dating from the 1950s, operating on this basis. This technology has since the late 1960s been superseded by the waterwall incinerator. In this type of incinerator, the furnace walls consist of vertically arranged metal tubes joined side-by-side with metal braces. Radiant energy from burning solid waste is absorbed by water passing through the tubes (see Chapter 3, and the Saugus, Massachusetts case study in this chapter).

2. Separation of the combustible (organic) fraction by shredding and classification, followed by combined firing of the shredded waste with pulverized coal in electric utility boilers.

When solid waste is shredded and subjected to air classification, most of the inorganic materials remain with the heavier fraction, and the organic, or combustible components are found with the light fraction. From the viewpoint of the fuels market, such "front-end" processing yields a combustible material known as refuse-derived fuel (RDF) for which fairly narrow specifications can be prepared, defining physical properties, heat of combustion, and percentages of water and ash.

The composition of a typical RDF is given in Table 8. Note the largely organic makeup although a very small amount of glass and ceramics are also present. The heating value of this fuel is about 6000 Btu/lb (14 MJ/kg) which, by means of an optional drying step, can be upgraded to exceed 8000 Btu/lb (18.6 MJ/kg). A comparison of coal and a typical RDF is given in Table 9.

The burning behavior of the air-classified light fraction has been proven for suspension burning with auxiliary fuel. Rather minor changes or additions in existing solid-fuel combustion equipment will permit ready introduction of this waste fraction into the burning zone, where typically 10–20% of the total heat generated is supplied by the

Table 7
Summary of Resource Recovery Facilities Implementation, 1976[a]

Location[b]	Type[c]	Capacity, TPD	Products/markets	Startup date
Altoona, Pennsylvania	Compost	200	Humus	1963
Ames, Iowa	RDF	400	RDF, Fe, Al	1975
Blytheville, Arkansas	MCU	50	Steam/process	1975
Braintree, Massachusetts	WWC	240	Steam/process	1971
Chicago, Illinois (Northwest)	WWC	1600	Steam/industry	1972
N-E. Bridgewater, Massachusetts	RDF	160	RDF/utility	1974
D-Franklin, Ohio	Wet pulp	150	Fiber, Fe, glass, Al	1971
Harrisburg, Pennsylvania	WWC	720	Steam/sludge drying	1972
Merrick, New York	RWI	600	Electricity	1952
Miami, Florida	RWI	900	Steam	1956
Nashville, Tennessee	WWC	720	Steam heating & cooling	1974
Norfolk, Virginia	WWC	360	Steam/navy base	1967
Palos Verdes, California	Landfill methane recovery	1 MMft^3/day	Gas utility & Fe	1975
D-St. Louis, Missouri	RDF	300	RDF coal-fired utility	1972
Saugus, Massachusetts	WWC	1200	Steam/process	1976
N-S. Charleston, West Virginia	Pyrolysis	200	Gas, Fe	1974
D-Baltimore, Maryland	Pyrolysis	1000	Steam/utility; Fe, glass	1975
G-Baltimore County, Maryland	RDF	550	RDF, Fe, Al, glass	1976
Chicago, Illinois (Crawford)	RDF	1000	RDF/utility	1975
Hempstead, New York	Wet process RDF	2000	Electricity/utility; Fe, Al, glass	1978
Milwaukee, Wisconsin	RDF	1000	RDF, corrugated, Fe	1975
D-Mountain View, California	Landfill methane recovery	1 MMft^3/day	Gas utility	1979
N-New Orleans, Louisiana	Materials	650	Nonferrous, Fe, glass, paper/secondary materials industries	1976
Portsmouth, Virginia (Shipyard)	WWC	160	Steam/heating loop	1976
D-San Diego County, California	Pyrolysis	200	Liquid fuel utility; Fe, Al, glass	1977

[a] From reference [20].
[b] D—EPA demonstration grant; G—EPA implementation grant; N—non-EPA demonstration facility.
[c] RDF—Refuse-derived fuel; WWC—Waterwall combustion; RWI—Refactory wall incinerator with waste heat boiler; MCU—Modular combustion unit.

Table 8
Typical Refuse-Derived Fuel Composition by Waste Component[a]

Component	Wt. %
Paper	55.0
Food wastes	16.0
Yard wastes	13.7
Glass, ceramics	2.7
Wood	2.6
Textiles	2.5
Leather and rubber	1.8
Miscellaneous	3.9

[a] From Maaghoul [22].

Table 9
Comparison of Coal and Refuse-Derived Fuel[a]

Content, wt. %	Coal fired at Union Electric's Meramec Station	Typical subbituminous coal	Refuse-derived fuel
Moisture	10.0	21.0	18.0
Ash	9.0	6.0	14.0
Chlorine	0.05	—	0.07
Sulfur	3.4	0.5	0.10
As fired, Btu/lb	11,315	9570	5,784
(MJ/kg)	(26.3)	(22.2)	(13.4)

[a] From Maaghoul [22].

RDF and the remainder by pulverized coal. This constitutes the so-called supplementary fuel concept and has been carried out at the Meramec Station of Union Electric in St. Louis (see case study below).

The air-classified light fraction may also be compressed into cubes or briquettes. Used as a solid fuel, this variety of RDF is particularly adapted for stoker and spreader-stoker furnaces where fuels are burned on grates rather than in suspension.

A major drawback of all solid RDFs is the rather high ash content. Most of this ash finds its way into the boiler stacks and limits the use of RDF to facilities equipped with adequate pollution control equipment. Also, RDF cannot be fired in systems designed only for gas or oil. RDF can best be used in coal-fired power plants with installed air pollution equipment.

3. Pyrolysis of the refuse for the production of steam for heating or cooling systems.

4. Pyrolysis of the refuse for the production of storable fuels such as oil and low Btu gas.

Pyrolysis requires raising the refuse contents to a temperature at which the volatile matter will distill or boil off, leaving carbon and inert matter behind. The carbon and volatiles do not burn in the process owing to an intentional deficiency of air in the reactor. Volatile matter may be burned off as waste in a secondary chamber to which air is added, with subsequent heat recovery, or the offgases may be cooled and condensed to selectively recover oils and tars. Alternatively, the gases may be cleaned and used as a gaseous fuel. The subject of pyrolysis is further presented in the next section of this chapter, and in Chapter 3.

As can be seen from Table 7, the conversion of solid wastes into energy is no longer in the theoretical stages of development, but has moved into the beginning phases of commercial application. Based on energy recovery systems existing or planned at the present time, it is projected that by the early 1980s almost 30 cities and counties around the country should be operating the equivalent of thirty-six 1000 ton/day plants, recovering an estimated 85 trillion Btu values per year, equivalent to about 50,000 bbl oil per day.

III. RESOURCE RECOVERY OPERATIONS

The technique of materials or energy recovery is in essence the judicious selection of a series of operations, that, when placed in a proper sequence will yield a useful product from the solid waste. Many of the operations discussed below have been successfully applied in several refuse processing projects [23–29], and enough technical and economic operating data are available to make their inclusion in processing facilities feasible.

The selection and integration of diverse operations into a coherent whole, capable of processing solid waste economically and reliably, and in a socially and ecologically acceptable manner, requires a great deal of synthesis, analysis, and engineering judgement.

A. Shredding

The term "shredding" has replaced "pulverizing" and "milling" as the overall description for reducing the size of refuse pieces. It may eventually be replaced by the more accurate term "size reduction" or "mechanical volume reduction" as used in Chapter 2.

It has been suggested that shredding is the first really new solid-waste management idea to be implemented in 2000 years, or since the Romans started using central dumps for disposal of community wastes [30]. This may be exaggerated, but it is nevertheless true that size reduction has certainly had a significant impact on the solid waste field. Size reduction is central to almost all present recovery schemes, and can be accomplished in wet pulpers (such as the Franklin, Ohio, installation discussed below) or by dry brute force shredders such as hammermills, flail mills, or grinders.

Shredders have been used for size and volume reduction prior to landfilling [31] or energy and materials recovery [32–34]. Shredding produces homogeneous particles from which glass, paper, and metals can be extracted for reuse; the remaining organic refuse can be more effectively incinerated for energy conversion. Shredders are also widely used in the reclamation of steel from junked automobiles [35].

The product from a shredder can be described by a number of semi-empirical relationships. One of the best seems to be the Gaudin and Meloy model [36],

$$B = 1 - \left(1 - \frac{x}{x_1}\right)^r$$

where B is the fraction finer than a size x, x_1 is the size of the largest piece passing the shredder, and r is a constant (≈ 7 for refuse) [37].

Comparison of this equation and data from actual shredded refuse is shown in Fig. 3 [38].

Several general statements can be made of shredder performance based on empirical data. These relationships are shown in Fig. 4. For example, product particle size decreases with increasing shredder speed. On the other hand, with increasing moisture content of the refuse, particle size increases. This is to be expected since a sloppy refuse does not shear but "squeezes" through the grates.

The energy requirements of shredders are substantial. Not only is energy required for the hammer, but energy is wasted in heat and refuse movement. It has been estimated that the typical shredder uses only 3–10% of its energy input for useful work [39].

Shredders are evaluated in terms of "specific energy," or the kWh of energy per ton of refuse processed. The effect of feed rate and particle size of the feed is shown schematically in Fig. 4. Note that there seems to be an optimum feed rate for a shredder in terms of the efficiency, and a finer initial feed will require less energy to produce a given product.

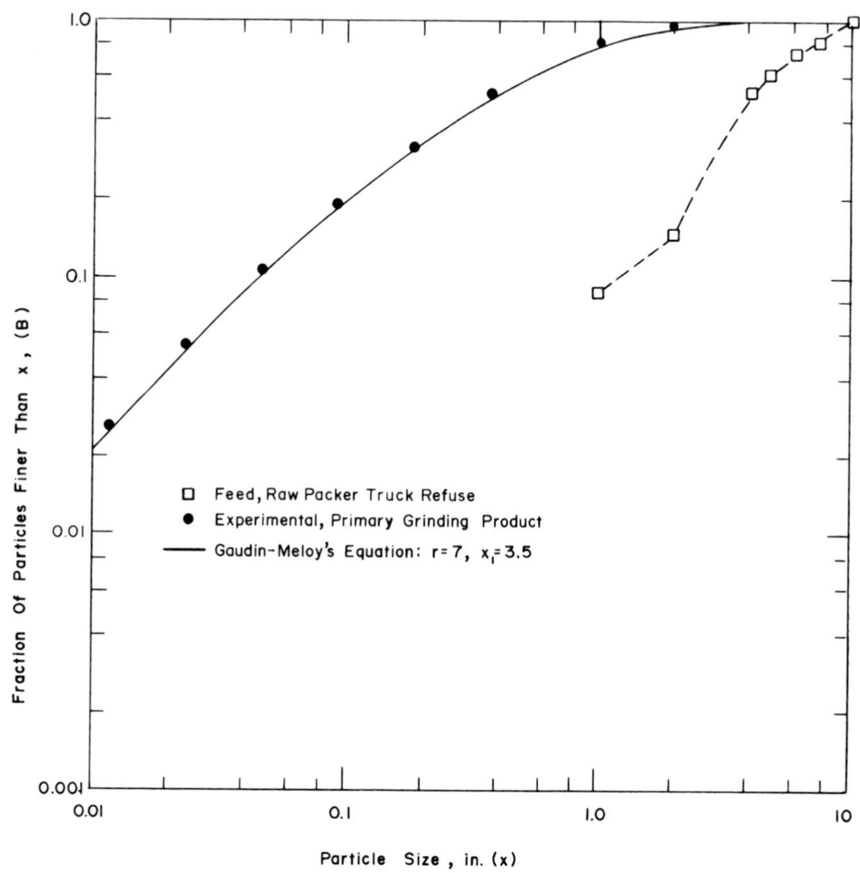

FIG. 3 Typical particle size distributions of shredded refuse.

The wear of hammers is probably the most serious drawback in the application of shredders to refuse processing. Many installations run for two shifts per day and use a third shift for rebuilding or refacing the hammers.

Another problem with dry shredding is the possibility of explosions. In order to avoid this, the shredder compartment can be flooded with a fire retardant gas whenever the air pressure spikes [40]. Alternatively, explosion doors are provided that can be blown out and easily replaced. The best explosion prevention is the careful screening of the refuse as it enters the shredder feed conveyor. Shredding is the subject of Chapter 2 in this book.

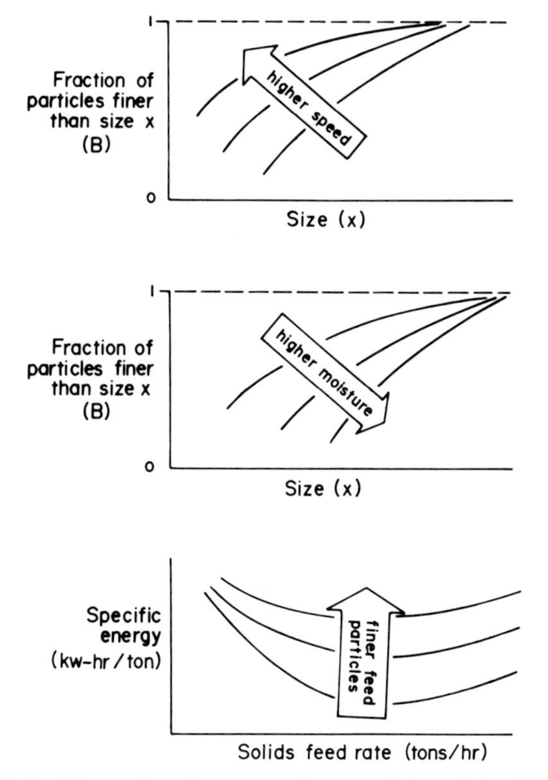

FIG. 4 General performance characteristics of shredders.

B. Separation

The homogenization and size reduction of solid waste by shredding is followed by the separation step wherein different components of the shredded solid waste are isolated by mechanical or other means. It is an integral part of any solid waste system where mineral recovery, fuel preparation, composting, or other resource recovery techniques are to be employed. A variety of methods by which separation of the various components of shredded refuse may be accomplished is discussed in the following paragraphs.

1. *Air Classification*

Air classification is one of the most promising separation methods. Generically, an air classifier subjects a stream of solid waste to a countercurrent or crosscurrent stream of air with the result that the heavier, mostly inert, components (metals, glass, dense plastics, rubber, stones, and organics) are unaffected by the air flow and fall through, and the lighter combustible components (paper, film plastic, fabric, and some wood) are entrained by the air current and exit with the upflowing air.

Many types of air classifiers have been used, some of which are shown in Fig. 5. A stream of air always carries the light fraction into a trap (such as a cyclone) and allows the heavy fraction to drop down.

Mixed municipal refuse is usually divided into the heavy fraction (20–30 wt %) and the light fraction (70–80 wt %). This ratio can be changed by varying the operating conditions of the classifier or the characteristics of the feed. The decision as to whether a particle will go up or down depends on the aerodynamic properties of the particles at the given operating conditions.

Since the feed to air classifiers is heterogeneous, the size, shape, and density are important factors. Density is influenced significantly by moisture content, a major concern in the day-to-day operation of a classifier; for example, the density of paper increases considerably in wet refuse.

The size of the particles entering the air classifier depends on the operation of the shredder. Small grate openings produce a much finer product. As noted earlier, however, the size distribution can vary considerably.

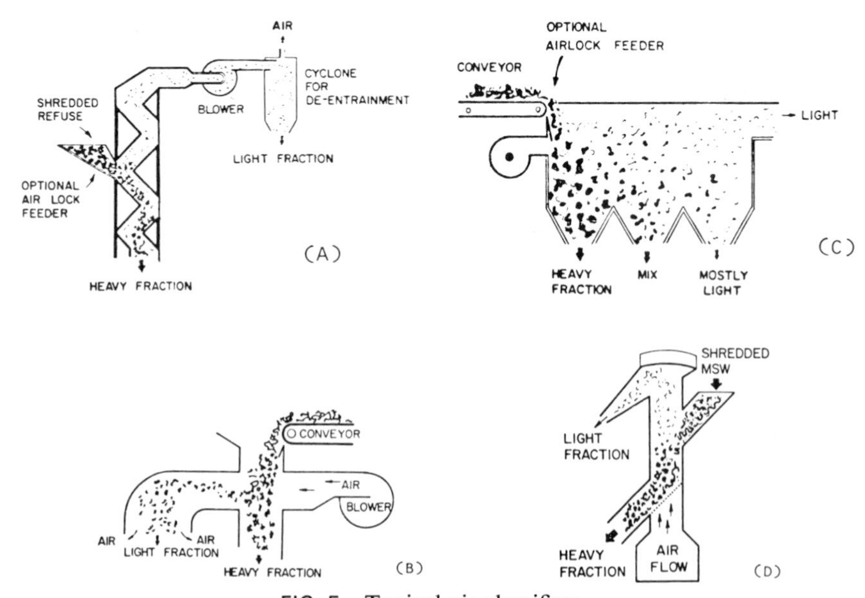

FIG. 5 Typical air classifiers.

Several methods have been suggested for calculating a single value which would be characteristic of the particle size distribution. Widely used is the calculation of an "average diameter,"

$$D = \sum X_i D_i$$

where X_i is the weight fraction of particles of diameter D_i. Another approach is to calculate a "characteristic size" which is defined by the Rosin-Rammler model [37] as:

$$B = 1 - e^{-(x/x_0)^n}$$

where B is the fraction of particle sizes smaller than x, n is a constant, and x_0 is the "characteristic size." If the size distribution data are plotted on a log-log scale as $\ln[1/(1-B)]$ vs. x, the characteristic size is read off the plot as $\ln[1/(1 - B)] = 1$, and defined as the size corresponding to 63.2% of the particles (by weight) that pass this size sieve. This approach for size characterization has considerable merit since both a single size value (x_0) and a distribution function n are defined. Present indications are that for mixed municipal refuse, n is 1 [37].

The measurement of particle size, x, is difficult since irregularly shaped particles have no single "diameter." In some cases the largest dimension of the irregularly shaped particles is used to define size. In air classification, a shape factor, either the "sphericity" or "effective diameter," is used to determine the aerodynamic behavior of particles. These are defined as

$$\text{sphericity} = \frac{\text{surface area of the particle}}{\text{surface area of sphere with equal volume}}$$

$$\text{effective diameter} = \frac{\text{diameter of the particle}}{\text{diameter of a sphere with equal volume}}$$

Although the air flow in a classifier is hardly laminar, the well-known Stokes relationship has been used to attempt to describe behavior.

$$v = \frac{D^2 g(\rho_s - \rho)}{18\mu}$$

where v is the air velocity equal to the settling velocity of the particle, D is the particle diameter (or "effective diameter" to take shape into account), ρ_s is the density of the particle, ρ is the air density, and μ is the air viscosity.

The application of this equation to air classification has not been very successful, however, and several empirical formulas have been suggested. The most direct is the Dallavelle model [41]:

$$v = \left(\frac{10.7\gamma}{\gamma + 1}\right) D^{0.57}$$

where γ is the specific gravity, D is particle diameter in millimeters, and v is air velocity in meters per second.

Even more empirical is a relation that was developed by plotting the percent of the light fraction of refuse against the air velocity in the classifier. Data for mixed municipal refuse in three different cities and for four different shredders show a straight line relation, such that

$$R = v/20$$

where R is the percent of the light fraction of refuse and v is the air velocity in feet per minute [42].

Several types of classifiers are currently being developed or demonstrated. The zig-zag air classifier (Fig. 5A) consists of a vertical zig-zag column with an upward air flow. Air forces the lighter components upward, whereas the heavier components drop to the bottom into a collection bin. The special feature of such a classifier is the turbulence resulting from the zig-zag path that pulls the refuse materials apart so that each piece can be acted upon independently.

The second basic air classifier is the horizontal air flow system. Here solid waste is dropped vertically through a horizontal air flow (Fig. 5B). Lighter items are blown laterally past a divider, whereas the heavier components fall vertically through the air stream.

Other air classification configurations are illustrated in Figs. 5C and 5D. Various modifications of these basic types are possible. For example, three vertical air classifiers in a row can draw lightest materials from the first column into the second column which has a different air velocity. The next heavy items fall through, and again the lightest materials go into the third column.

2. *Wet Classification*

Wet classification is the method of separating solids in a liquid–solid mixture into fractions according to size or density.

An important method of wet classification is the rising current separator which functions on the same principle as rising air classifiers except that the medium is water rather than air. Originally designed for use in coal preparation plants, these separators show great potential for ferrous metal and glass recovery from municipal solid waste [43]. The rising current separator creates an effective fluid specific gravity of greater than 1.0 by a precisely controlled rising current of water. Materials having a sink velocity less than the controlling upward velocity float to the top and are skimmed off, while those with greater velocity sink. A given particle sink/float determination is not entirely dependent on its density but is also influenced by its size and shape (i.e., drag characteristics), and the effective density and viscosity of the liquid. The separator is designed so that the rising current velocity can be readily adjusted in operation to precisely control separating performance.

Air and wet classification are gross separation schemes, that is, heavier fractions (inorganics) from light (organics). Downstream from these operations other separations are usually required in order to achieve finer separations which yield materials such as glass, aluminum, and ferrous and nonferrous metals. Some of these techniques are discussed below.

3. *Magnetic Separation*

Theoretically, any solid placed in a magnetic field is affected in some way. Either the solid is repelled (diamagnetic solid) or it is attracted by the magnetic field (paramagnetic solid). The paramagnetic solids are strongly or weakly magnetic. The method of separating solids by means of a magnetic field is called "magnetic separation."

Magnets have been used for many years to separate ores and to remove ferrous scrap from mixed shredded metal. Magnetic separators applied to solid waste processing are usually belt magnets or drum magnets. The belt magnet can be placed either in the same direction as the materials flow (Fig. 6A) or at a 90° angle (Fig. 6B). The drum type magnet (Fig. 6C) seems most widely used in scrap steel separation (for example, shredded auto bodies).

Two general types of belt magnets are in use. One type consists of a magnetic pulley over which a conveyor belt carries the solid waste stream, as shown in Fig. 6A. As the waste stream comes within the pulley's magnetic field, the magnetic material is attracted and held to the belt until it reaches the underside, passes out of the magnetic field, and is separately discharged. The nonmagnetic material is discharged over the pulley in a normal trajectory. A divider arrangement is usually installed as indicated in the figure. An adjustable divider permits optimal positioning for specific conditions of magnetic and nonmagnetic material discharge. Pulley sizes usually vary from 8 through 36 in. (203–914 mm) whereas belt widths vary from 8 through 60 in. (203–524 mm). Belt

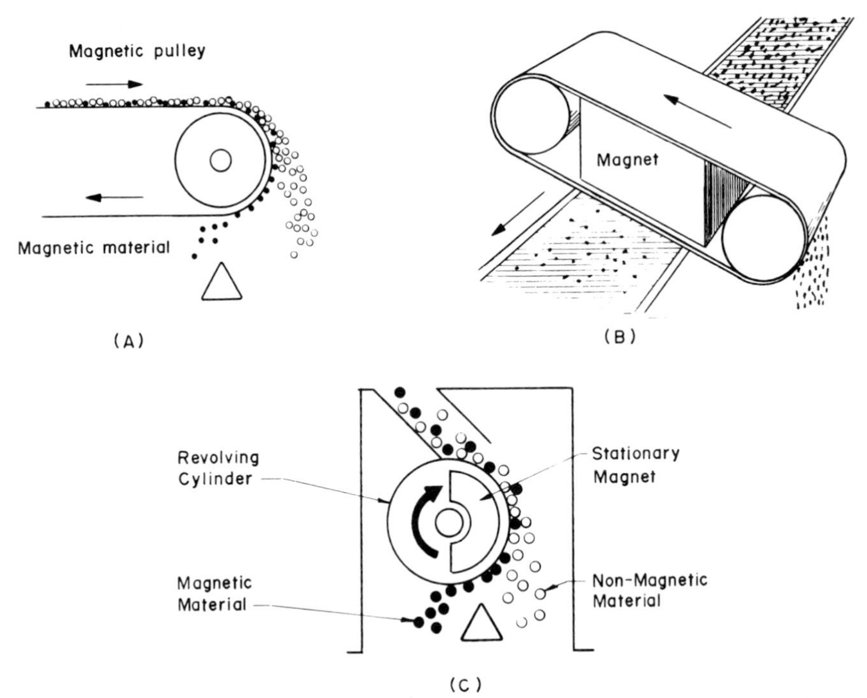

FIG. 6 Magnetic separation: (A) pulley-type belt magnet, (B) suspended-type belt magnet, (C) drum magnet.

speeds range from about 100 to about 350 ft/min (30–105 m/min) with material handling capacities ranging from about 400 ft^3/h for the smaller pulley and belt sizes, to about 55,000 ft^3/h for the larger sizes (11–1555 m^3/h).

The second type of belt magnet consists of a fixed magnet within a pulley-driven belt loop, positioned above a point where the wastes pass on a continuous conveyor belt. The magnet picks the magnetic materials off the feed belt and discharges them to the side, as shown in Fig. 6B, whereas the nonmagnetic materials continue along the feed belt.

The drum magnet consists of a stationary magnet inside a rotating cylinder or drum. As shown in Fig. 6C, the feed material reaches the drum and is acted upon by the magnetic field, whereby the magnetic material holds onto the drum. As the rotating drum carries the material through the stationary magnetic field, the nonmagnetic material falls freely from the drum, while the magnetic materials are firmly held until carried beyond a divider and out of the magnetic field. Typical drum sizes range from 12 to 36 in. (305–915 mm) in diameter with widths of

12 to 60 in. (305–1524 mm). The drum speeds vary from 45 rpm for the smaller drums to about 20 rpm for the larger drums. Material handling capacities range from about 1000 ft^3/h to as high as 25,500 ft^3/h (28–720 m^3/h).

For best results, the magnetic separation of ferrous materials from municipal waste should be preceded by some form of shredding and air classification. In this way, the ferrous metal particles will be rendered discrete in the feedstream to the magnetic separator, and their physical separation will be effected more efficiently.

Magnets used in magnetic separators are either electric or permanent. Electromagnets use insulated copper or aluminum wire windings around a soft iron core energized with direct current. Improved insulations have increased the temperature limits of electrocoils to 430 °F (220 °C). Permanent magnets do not require external energization. Special permanent magnetic alloys can produce a magnetic field at a constant level indefinitely after initial charging unless they are exposed to demagnetizing influences.

Magnets are not very expensive (relatively speaking) and are cost effective if the material has already been shredded. A belt magnet with a nominal capacity of about 50 tons/h costs (at 1978 prices) about $30,000. About 5% ferrous materials (2.5 tons) can be expected. At the 1978 market price of $30/ton of dirty ferrous scrap the magnet is paid for in about 30 days of double-shift operation.

Various demonstration projects are currently underway to magnetically separate various nonferrous metals such as chromium, aluminum, zinc, and copper. Theoretically, these metals will react to magnetic fields of different intensities, and this concept is used to effect separations of these metals [44].

4. *Screening*

Screening is a method of separating a mixture of solids according to size only. The solids are dropped on a screening surface; the material that passes through the screen openings is referred to as the "fines" or product, whereas the material that is retained on the screen is called the "tails" or reject. A single screen can make but a single separation into two fractions. Material passed through a series of screens of different sizes is separated into sized fractions, i.e., fractions in which both the maximum and minimum particle sizes are known.

Screen size is specified by "mesh" which is the number of openings per linear inch, counting from the center of any wire to a point exactly one inch distant, or the opening between the wires specified in inches or millimeters. Aperture or screen-size opening is the minimum clear space between the edges of the opening in the screening surface and is usually given in inches or millimeters. The open area of square mesh wire cloth is calculated as

$$P = \frac{O^2}{(O + D)^2} \times 100 = (OM)^2 \times 100$$

where P is the percentage of open area, M is the mesh, O is the size of opening, and D is the diameter of wire.

For example, a 20-mesh screen has a 0.0328 in. (0.833 mm) opening or aperture, and the wire diameter is 0.0172 in. (0.437 mm). By definition, a 20-mesh screen has 20 openings per inch and 20 wires separating these openings such that $20 \times 0.0328 + 20 \times 0.0172$ equals 1 in. The percent open area is

$$P = \frac{(0.0328)^2}{(0.0328 + 0.0172)^2} \times 100 = 43\%$$

or,

$$P = (OM)^2 \times 100 = (0.0328 \times 20)^2 \times 100 = 43\%$$

Similarly for a 100-mesh screen, $O = 0.0058$ in. (0.1473 mm), $D = 0.0042$ in. (0.1067 mm), and thus $P = 33.6\%$, whereas for a 3-mesh screen, $O = 0.263$ in. (6.68 mm), $D = 0.07$ in. (1.78 mm), and thus $P = 62.4\%$. Thus, the smaller the mesh size, the larger the percent open area of the screen.

Many types of screens are commercially available of which the vibrating and trommel screens appear to be most widely used in processing refuse.

Vibrating screens are especially known for their large capacities and high efficiencies. The vibrations may be generated either mechanically or electrically. Mechanical vibrations are usually transmitted by an eccentric or unbalanced shaft although other means are also available. Electrical vibrations are supplied by electromagnets; these screens have intense vibrations of 1500–7200 vibrations per minute of low amplitude. Ordinarily no more than three decks are used in vibrating screens (see Fig. 7A). A 12×24-in. screen draws about $\frac{1}{3}$ hp; a 48×120-in. screen draws 4 hp.

A trommel or revolving screen consists of a barrel frame surrounded by wire cloth or a perforated plate, open at both ends and inclined at a slight angle (see Fig. 7B). The material to be screened is delivered at the upper end; the fines or undersized material passes through the perforations whereas the oversized material is discharged at the lower end.

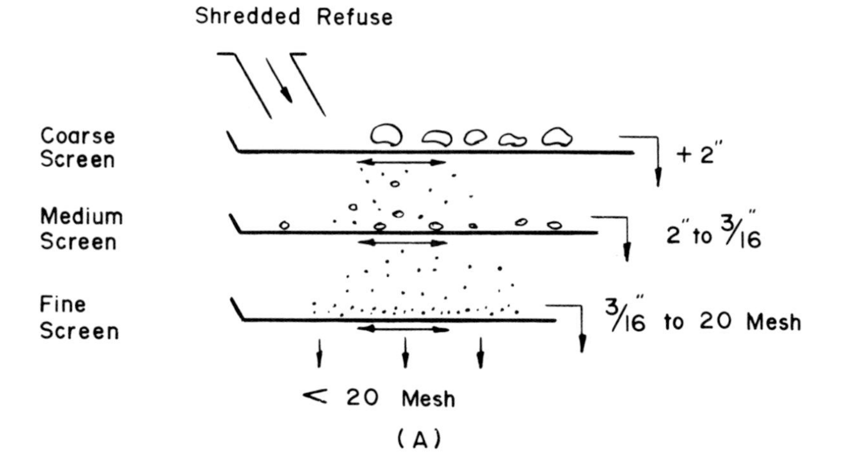

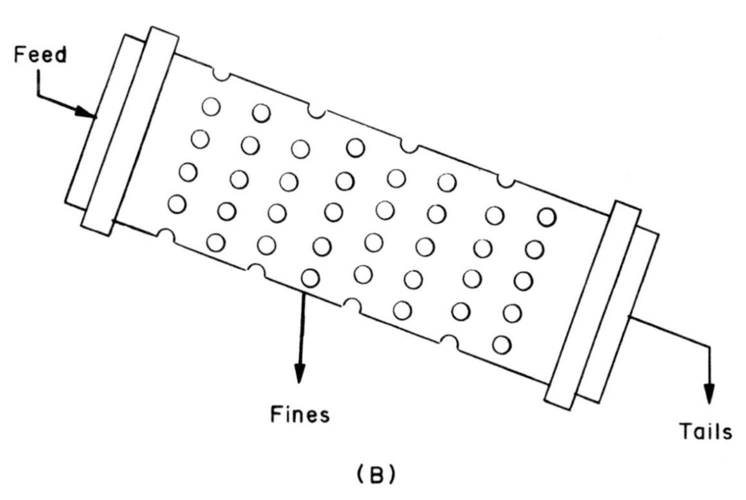

FIG. 7 Screening: (A) 3-deck vibrating screen, (B) trommel screen.

These screens revolve at about 15–20 rpm, and are relatively low capacity, low efficiency operations.

The effectiveness of a screen or screen efficiency is a measure of how closely it separates two differently sized materials. A common measure of screen efficiency is the ratio of the amount of undersize material passing through the screen to the amount of undersize in the feed. A similar definition in terms of oversize may also be formulated.

The capacity of a screen is measured by the mass of material fed per unit time per unit area of the screen. It is usually expressed in tons/(h · ft^2).

Capacity and efficiency are opposing factors, i.e., increasing one results in decreasing the other. In practice some balance between the two is aimed for.

In general, the efficiency of a screening operation at a given capacity depends on the screening operation itself, whereas the capacity is controlled by varying the rate of feed to the screen.

Screen mesh size also has an effect on capacity. It is a well-known rule of thumb that the capacity of a screen, in mass per unit time, divided by the mesh size should be constant for any specified conditions of operation.

As particle size is reduced, screening becomes more difficult and the capacity and efficiency are consequently low.

5. *Froth Flotation*

Froth flotation is used to recover glass from a mixture of glass, ceramics, stone, brick, and metals [43, 45]. Usually the glass should be finely crushed (10–28 mesh) since coarser feed cannot be suitably mixed and suspended by a flotation unit. Also, the feed should be pulped to a solids content between 25% and 35% by weight. The waste mixture is put in a tank of water where a special chemical, called a "promoter" or "collector," renders the glass particles air attractive and water repellant. With vigorous agitation and aeration in the presence of a frothing agent, the air-attractive particles become attached to air bubbles and rise to the surface where they collect as a froth and are skimmed off.

A disadvantage of flotation is the introduction of chemicals (collectors, frothers) to the system. The consumption of these chemicals may vary from 0.1 lb/ton to as high as 5–10 lb/ton. The recovery and regeneration of these chemicals or their disposal can be a costly operation.

Superior flotation results depend on the fine consistency of the waste pulp. Thus fine-size reduction is critical prior to flotation. In addition, for proper selectivity a definite contact time is required between the chemical reagent (collector) and the waste pulp.

The tonnage handled by flotation equipment varies with the pulp density of feed and time required for the flotation operation, i.e. time required for the air bubble to become attached to a particle and float to the surface. The number of cells required for a specific job can be calculated from the following expressions [46]:

$$n = \frac{T \times tpd \times d}{1440\ V}$$

where n is number of cells, V is the volume of a single cell, in ft^3, T is the flotation time, in min, *tpd* is the dry tons waste treated per 24-h day, and d is the volume of pulp (waste and water) containing 1 ton dry solids, in ft^3.

6. *Jigging*

A jig is a mechanical device used for separating materials of different specific gravities by the pulsation of the medium in which the materials are suspended [43]. The liquid pulsates or jigs up and down causing the heavier material to sink to the bottom, while the lighter material rises to the top. The jig essentially consists of a submerged screen that supports a bed of waste particles. The bed is partially suspended in the water at regular intervals by dropping the screen (movable-screen jig) for a short distance or by forcing a current of water up through the screen (fixed-screen jig). The pulsations occur at high frequency (100–200 strokes/min) and are of short amplitude (0.25–1 in.). After each pulsation the bed settles back and eventually stratifies with a lighter fraction at the top and the heaviest fraction at the bottom. This method has been used for separating fine glass particles from metals and other wastes; it consumes large quantities of water.

7. *Optical Sorting*

Glass particles that are too large for froth flotation can usually be subjected to optical sorting whereby clear glass is separated from colored (green-amber) glass and ceramics. Such a separation is necessary if the quantity of the clear glass product is to be maintained.

A schematic of the optical sorter is shown in Fig. 8. The particles are individually dropped into a screening box equipped with a light source and a photocell. The photocell compares the reflectivity of the particle with a colored tab, and if the reflection from the particle matches that from the colored tab the particle falls into a bin. If it does not match, however, an air jet is activated which pushes the particle off its trajectory and into a different bin. In addition, the colored mixture (green-amber) can be run through a color sorter with a different reference background to further separate the two colors.

8. *Heavy-Medium Separation*

After ferrous metals have been extracted from the waste stream, other metals may be removed by heavy-medium separation [43, 45]. The sink-float, heavy-liquid, or heavy-medium process uses a liquid

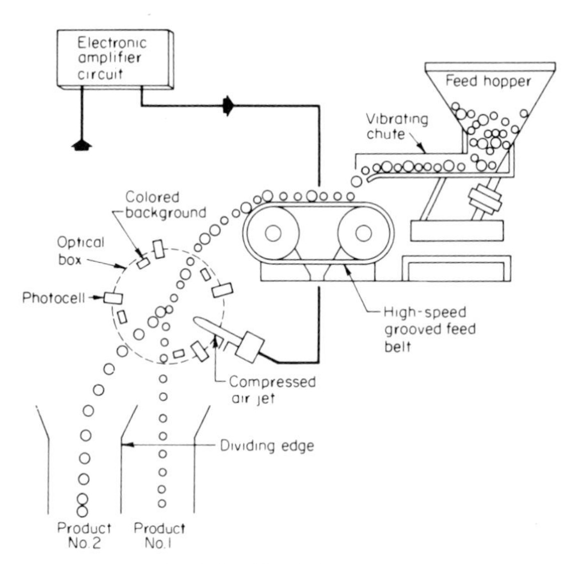

FIG. 8 Optical sorter.

sorting medium, the density of which is between that of the light and that of the heavy material. A separation is possible by merely adding the solids to the liquid, stirring the slurry, and removing a "sink" and a "float" fraction.

Since metals are heavier than water, it is necessary to add fine particles of ferrosilicon (sp. gr. = 6.5), magnetite (sp. gr. = 5.17), or some heavy organic liquids in order to give the liquid medium a specific gravity of at least 1.3–3.5.

The essential components of a heavy–medium separation process are feed preparation, heavy–medium separation, removing overflow and underflow, and recovering and cleaning the medium for reuse. The maximum size particle that can be treated is usually limited by the materials handling equipment. Pieces of ore 10–12 in. in size have been separated in the minerals industry. The minimum size is about 20 mesh, since fine particles tend to make medium recovery difficult, and slimes adversely affect the viscosity of the liquids.

A significant advance in heavy-medium separation is the application of centrifugal action to the heavy medium via a hydrocyclone. The material to be treated is pulped with the densifying medium and fed tangentially through the regular feed inlet of the cyclone. Separation occurs in the cone-shaped part of the cyclone by the action of centrifugal and centripetal forces. The heavier material leaves through the apex opening, whereas the lighter fraction leaves at the overflow top orifice.

A disadvantage of heavy-medium separation is that the medium becomes more dense and surface tension increases with repeated use. Therefore, a material that would normally sink may, indeed, float if the surface tension is large enough. This problem may be solved by churning the medium or by running the waste stream through a shredder to produce a particle size that optimizes heavy-medium flotation.

9. *Electrostatic Separation*

When a charged particle enters an electrostatic field, the particle is repelled by one of the electrodes and attracted by the other, depending on the sign of the charge on the particle. This principle has been used to separate aluminum from the heavy fraction and paper from plastics. When damp, paper conducts electricity and jumps from a rotating drum to an electrode and can be collected. Plastics are brushed off from the drum and collected separately. The disadvantage of this method is the low capacity of the unit and distribution across the drum.

The major mechanisms by which a surface charge is imparted to the particles to be separated, are conductive induction and ionic bombardment.

In conductive induction, a mixture of conductive and dielectric particles are fed onto a grounded rotating surface. While starting their rotation on the grounded surface the particles are carried past an active electrode whereby they receive a surface charge. The conductive particles quickly assume the potential of the rotor opposite to the electrode and therefore become attracted to the electrode; the dielectrics are polarized and therefore attracted to the rotor, repelled by the electrode, and continue on the rotor surface to be brushed off into a collection bin (Fig. 9A).

Ionic bombardment is the strongest form of electrification. Here the charge is supplied by a high voltage beaming or ionic electrode. The charge on the conductive particles is distributed immediately, and these particles are free to leave the rotor's surface as they approach a static electrode. The dielectrics do not lose their charge, are held to the rotor surface, and continue on to be brushed away into a collection bin. Thus, conductive and dielectric particles are collected separately (Fig. 9B).

The number of methods and devices used for sorting and separating solid waste components is staggering and grows every day. The operations presented above are technically the most promising, and their economic feasibility will be dictated by the value of the resources that they separate and recover.

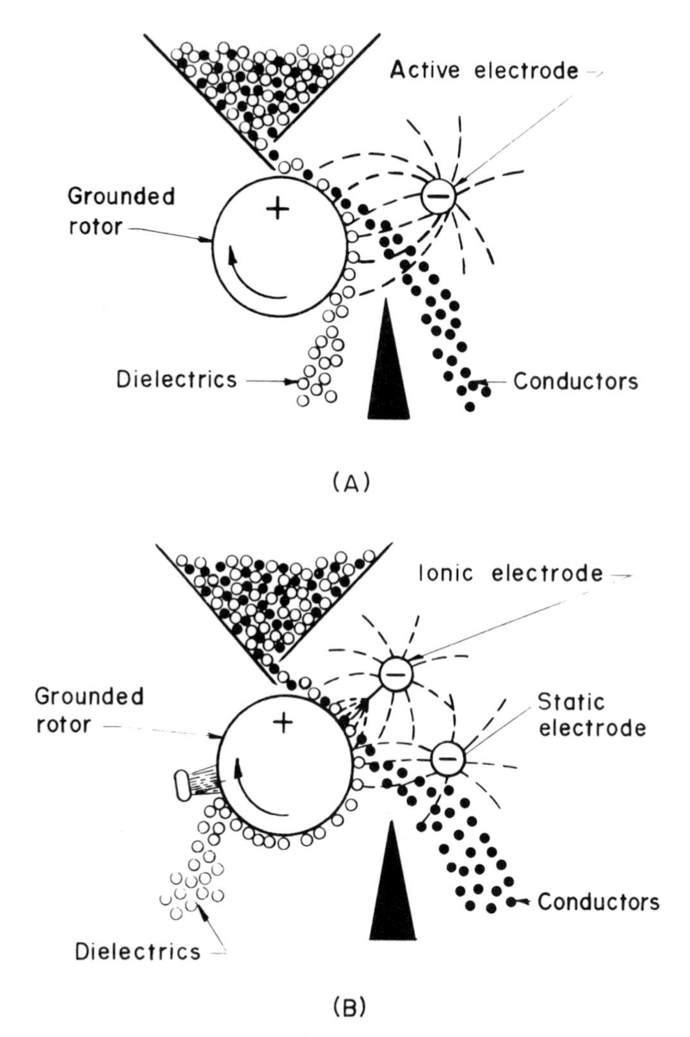

FIG. 9 Electrostatic separation: (A) conductive induction, (B) ionic bombardment.

As an illustration of how some of the above separation techniques could be utilized to synthesize a materials recovery system refer to Fig. 10, which shows a flow sheet for a proposed front end recovery system [9, 47]. Such a system would recover five fractions from municipal solid waste: bundled paper, ferrous metals, glass, aluminum, and a mixture of other nonferrous metals. It would leave as residue the organic fraction (for either disposal or backend recovery) and a small inert

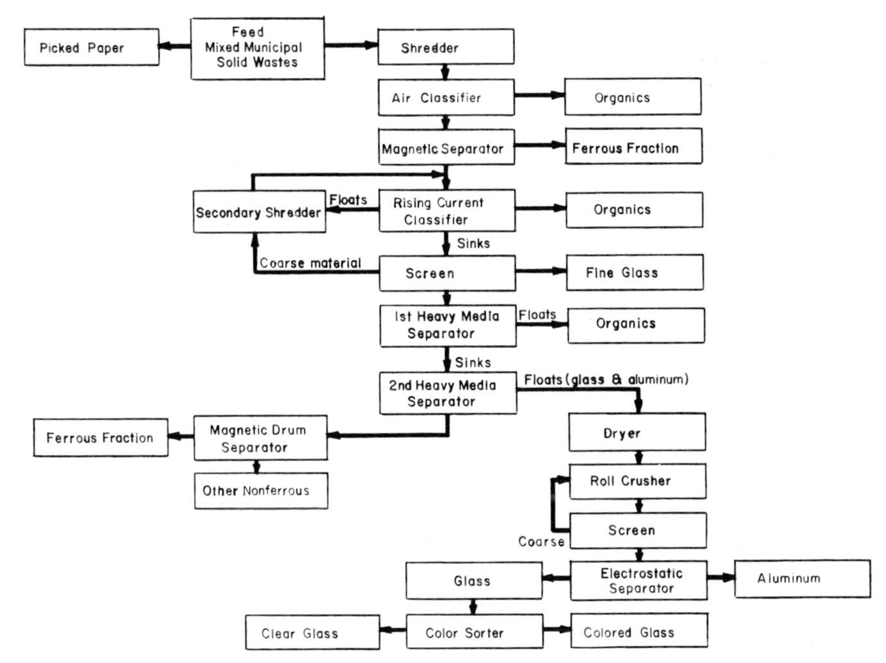

FIG. 10 Processing scheme for separating materials from mixed refuse.

fraction consisting of rubber, heavy plastics, bone, grit and sludges from the processing which may be disposed by landfilling.

Later in this chapter case studies are presented of demonstrated resource recovery systems.

C. Incineration

Combustion in the presence of excess air is an increasingly important unit operation. Strictly speaking, incineration is not a single operation since it involves many distinct steps such as loading, air flow, combustion, electrostatic precipitation, etc. It is a complicated process and is covered in detail in Chapter 3.

Modern incinerators are of the continuously fed, mechanically stoked type, and are so arranged that feed and residue discharge openings are continuously sealed while process materials flow through the openings. This improves combustion and air pollution performance.

The well-established concept of suspension burning is frequently applied in modern incineration. It may be achieved by suspending the burning fuel (in this case solid waste) in the gas stream in the combustion enclosure, or by suspending it in the gas stream and another medium, such as the fluidized bed. Suspension burning results in cost savings owing to a lack of grates and mechanical stokers, more complete combustion with less excess air, and more rapid response to automatic control systems. In addition, fluidized beds have large heat capacities and retain much of the sensible heat over a period of time. This is beneficial during intermittent opertion as start-ups are quite rapid.

Another new type of incineration is the high-temperature or slagging process. These systems operate at about 3000 °F (1650 °C), as compared to 1800 °F (980 °C) for conventional incinerators, a temperature sufficiently high to melt (or slag) metals and ceramics. Residue discharged from the reactor is a viscous heavy fluid which can be air or water quenched.

Heat-recovery equipment can be incorporated into conventional and slagging incinerator systems. Conventional systems may incorporate this feature through a downstream-waste heat boiler associated with a refractory furnace, or by building the furnace with water-cooled walls as a boiler. The boiler also includes a convection section or boiler bank, and either type unit would cool the flue gas from the furnace temperature (1800 °F) to 500 or 600 °F. The slagging process is not compatible with the water-cooled furnace construction, but such systems have been successfully equipped with downstream waste-heat boilers.

Waterwall linings offer several advantages over the use of refractory linings. Refractory furnaces normally need air about 200% or more above the stoichiometric combustion air requirement, whereas waterwall lined furnaces require only about 50%. As a consequence of the lower air requirements, smaller air handling and pollution control equipment is sufficient. Reduction in combustion air results in higher furnace temperature and therefore improved incineration efficiency.

D. Pyrolysis

The process of pyrolysis, used in steel production for many years, consists of thermal decomposition at high temperatures in the absence of oxygen. In steel production, the manufacture of coke by a pyrolytic process is a necessary step; another example is charcoal, the end product of wood pyrolysis.

Pyrolysis did not gain a foothold in solid waste processing until stringent air pollution standards were imposed on incinerators. Because of the need of large amounts of excess air (sometimes as much as 400% of stoichiometric requirements) in order to achieve efficient combustion and lower temperatures, air pollution control devices on incinerators

Component	Composition (% of dry weight)[b] Range	Nominal	
Metallics	7–10	9.0	Mechanical recovery (Metallics through Paper: Other)
Ferrous	6–8	7.5	
Nonferrous	1–2	1.5	
Glass	6–12	9.0	
Paper	37–60	55.0	
Newsprint	7–15	12.0	Conversion recovery (Newsprint through Miscellaneous)
Cardboard	4–18	11.0	
Other	26–37	32.0	
Food	12–18	14.0	
Yard	4–10	5.0	
Wood	1–4	4.0	
Plastic	1–3	1.0	
Miscellaneous	<5	3.0	

turn out to be expensive items, often doubling the cost of incineration. Pyrolysis, on the other hand, does not produce excessive gaseous end products, and thus requires only modest control equipment.

Unlike combustion in excess air, which is highly exothermic and produces primarily heat and carbon dioxide, pyrolysis of organic material is analogous to a distillation process and is endothermic. Generally speaking, pyrolytic destruction requires raising the fuel to a high temperature (1000–3000 °F), where, because of the lack of oxygen, a chemical breakdown of the waste organic material occurs. The volatile matter distills off, leaving behind carbon and inert material. The carbon and volatiles do not burn because of a lack of air. The products from pyrolysis are (a) a gas mixture consisting primarily of methane, hydrogen, carbon monoxide, and carbon dioxide; (b) a "tar" or "oil" that is liquid at room temperature and consists of organic compounds such as acetic acid, methanol, and acetone; and (c) a "char" consisting of almost pure carbon plus any inerts (glass, metals, stones) that enter the process. All three products can be converted to energy and can be used as a fuel. Typical gas heating values of 100–600 Btu/ft^3 (3.7–22.4 MJ/m^3) can be achieved; liquid heating values are in the 10,000–11,000 Btu/lb (23.3–25.6 MJ/kg) range; solid fuel values are 6000–9000 Btu/lb (14–21 MJ/kg).

1. *Chemistry*

The process of pyrolysis can be illustrated by the thermal decomposition of cellulose,

$$(C_6H_{10}O_5)_n \xrightarrow{\Delta} H_2 + CO + CH_4 + CO_2 + C_xH_y$$

where the product C_xH_y represents the hydrocarbons formed. The type of hydrocarbons formed depends on the reaction time, temperature, pressure, and the presence of catalysts.

The term pyrolysis as applied to solid waste processing today has become a catchall that includes gasification and starved-air combustion in addition to pyrolysis. Consequently the processes that typically occur in the pyrolysis of solid wastes could include some or all of the following:

1. Drying; the incoming solids are dried by driving moisture out.
2. Pyrolysis; this is a three-step process determined by the rate of heating:
   (a) Initial decomposition of solids in which loosely bonded molecules such as H_2O, CO_2, and CO, are driven off.

(b) More extensive decomposition of the solids whereby organic liquids such as tars, oils and aromatics result.
(c) Final decomposition of the solids and liquids where simpler products such as H_2, CH_4, C_2H_6, etc. are driven off leaving only char.
3. Char gasification by CO_2, H_2O, or H_2.
4. Combustion of char for pyrolysis and gasification heat source.

The most important reactions affecting gas-phase products during the pyrolysis stage are described by the following:

$$CO + H_2O \rightleftharpoons CO_2 + H_2 \quad \text{water–gas shift reaction}$$
$$CH_4 + H_2O \rightleftharpoons CO + 3H_2 \quad \text{steam–hydrocarbon reaction}$$

The reactions that affect liquid-phase and solid-phase products are far more complex.

Char gasification is mainly described by the following reactions:

$$C + 2H_2 \rightarrow CH_4 \quad \text{hydrogen-carbon reaction}$$
$$\left.\begin{array}{l} C + H_2O \rightarrow CO + H_2 \\ C + 2H_2O \rightarrow CO_2 + 2H_2 \end{array}\right\} \quad \text{Steam-carbon reactions}$$
$$C + CO_2 \rightarrow 2CO \quad \text{carbon-carbon dioxide reaction}$$

Of the above six reactions only the hydrogen–carbon reaction is exothermic and the water–gas shift reaction is slightly exothermic. The remainder are highly endothermic. One way to provide the heat necessary for these endothermic reactions is through char combustion as described by the following reactions:

$$C + \tfrac{1}{2}O_2 \rightarrow CO \quad \text{partial combustion}$$
$$C + O_2 \rightarrow CO_2 \quad \text{complete combustion}$$

A comprehensive review of pyrolysis chemistry may be found in Ref. [48].

2. *Reactor Types*

Although several types of reactor schemes have been proposed for the pyrolysis of solid waste, the most successful designs include the shaft, rotary kiln, and fluidized bed.

Shaft reactors are conceptually the simplest and are of two types; the vertical flow and horizontal flow type. There are several variations of the vertical flow reactor but the most prominent are the moving-bed and entrained-bed reactors. In the moving-bed reactor the feed material (solid waste) enters at the top and descends to the bottom under its own weight. The gases produced during pyrolysis are removed either at the top or at the bottom depending on whether the reactor is operated countercurrently or cocurrently.

The entrained-bed reactor is a cocurrent flow reactor in which the solid feed is carried upward by a gas stream at high velocity. This type of reactor is also referred to as the pneumatic conveying type. Such reactors are characterized by short residence times and therefore most reactions are surface phenomena or the solids must be very small so that heat and mass transfer are essentially instantaneous resulting in what is also known as "flash pyrolysis."

The horizontal flow reactor utilizes a feed conveyor system which mechanically transports the solid waste through the reactor and the solid waste is in turn continuously pyrolyzed from the conveyor system. The vertical flow reactors must be provided with feed mechanisms, residue discharge, and gas take-off mechanisms. In the horizontal flow reactor these mechanisms are part of the reactor.

The rotary kiln is a rotating cylinder slightly inclined to the horizontal. Typical length to diameter ratios vary from 4 to 10 and rotation speed is 0.25 to 2 rpm. Feed material enters the higher end and progresses to the lower end because of rotation and inclination of the reactor. The gas flow may be either countercurrent or cocurrent to the solids flow. Good mixing of gaseous and solid reactants is achieved, and gaseous products once formed do not have to escape through thick layers of solids or char as in the moving-bed reactor, and consequently fewer complex gas-solid reactions occur.

The fluidized-bed reactor consists of a bed of solid particles such as sand, suspended by an upward flowing gas stream. The expanded suspended mass represents a boiling liquid. The solid waste feed is injected into the hot bed of agitated solids and is pyrolyzed rapidly due to the rapid heat transfer rate of such reactors. The main advantage of this reactor is its temperature homogeneity.

For any pyrolysis reactor system, it is necessary to heat the reactants up to the temperature at which thermal decomposition occurs; the manner in which heat is supplied, further distinguishes between pyrolysis systems.

Two distinct heating methods are used. In the direct method, heat is supplied by the partial combustion of the product gases, char, or supplemental fuel directly in the pyrolysis reactor. In this case oxygen or air must be supplied, and the product gas will contain substantial amounts of carbon dioxide and steam with a resulting reduced heating value. This is not true pyrolysis since oxygen is introduced into the system. Directly heated reactors may also include steam injection for char gasification. The use of air in such reactors results in the formation of nitrogen oxides that are difficult to remove from the emissions.

In the indirect method of heating, the heating zone is separated from the pyrolysis zone. The pyrolysis takes place in a single reactor. A portion of the pyrolysis product is then sent to a different reactor where it is combusted with air to produce heat. The heat is then used to heat the pyrolysis reactor by conduction through or radiation from the reactor wall, or by recirculating a heat carrier such as sand or pebbles between the combustion and pyrolysis vessels. Wall heat transfer is generally inefficient due to large resistances from refractory linings, slag coatings and corrosion problems. The use of a separate medium though desirable from a heat transfer viewpoint, can present severe solids transfer and separation problems.

In the moving-bed reactor, direct heating is achieved by introducing oxygen or air at the bottom to burn the char product reaching the bottom of the reactor. The hot combustion gases then flow upward causing gasification, pyrolysis and drying in ascending zones within the reactor. Indirect heating may be achieved by burning pyrolysis gas and conducting the heat to the reactor through the reactor walls. In such a case only pyrolysis and drying are experienced within the reactor.

The entrained-bed reactor is usually heated indirectly by heating the char leaving the reactor top in a separate heater and recycling this hot char through the reactor along with fresh feed.

Rotary kilns are heated similarly to moving-bed reactors. In addition they are amenable to indirect heating by a circulating heat medium such as pebbles or ceramic balls. Kilns generally provide for increased heat transfer.

Fluidized-beds may be heated directly by introducing air or oxygen into the reactor. The oxygen reacts more readily with the pyrolysis gas than the char, reducing gas yield. Indirect heating may be achieved by burning part of the pyrolysis gas or char in a separate unit and by transferring this heat through a recirculating medium such as heated sand which may be easily added to and removed from the reactor.

Table 10 summarizes the various reactor types and their characteristics.

3. *Product Yields and Distribution*

The distribution of product yield between gaseous, liquid and solid phases as well as the composition of each phase can be altered by manipulating certain key process variables such as temperature, heating rate, reactor residence time, etc. Very little quantitative information

Table 10

Pyrolysis Reactor Characteristics and Heating Methods[a]

Reactor	Solid–fluid contacting alternatives			Temperature gradients	Heat transfer rate		
						Indirect heating	
	Countercurrent	Cocurrent	Crosscurrent		Direct heating	Wall transfer	Recirculating heat carrier
Moving bed	×	×	×	High	Moderate	Low	Moderate
Entrained bed		×		Moderate	—	Low	Moderate
Horizontal bed			×	Moderate	—	Low	Moderate
Rotary kiln	×	×		Moderate	Moderate	Low	High
Fluidized bed	Thoroughly mixed contents			Negligible	High	Low	High

[a] Adapted from references [49] and [50]. From reference 9, © 1974, by the American Association for the Advancement of Science.

exists on the relation between these variables for solid waste pyrolysis. Many laboratory, pilot, and demonstration pyrolysis units have been successfully constructed and operated [51] and some qualitative generalizations can be made.

The decomposition of organic matter during pyrolysis begins at about 350 °F (180 °C), although processes may operate at temperatures as high as 3000 °F (1650 °C). At very high temperatures (3000 °F) the most important products are gas and inert slag. The gas consists of low-molecular-weight hydrocarbons in addition to hydrogen, carbon monoxide, and carbon dioxide. The slag consists of a fused mass of solid residue and its flow properties will be determined by feed characteristics and reactor temperature. Only directly fired, moving-bed reactors with countercurrent flow are operated in a slagging mode.

At high temperatures (1400–1500 °F) and relatively long residence times noncondensable gas yields are maximized and the resulting solid phase is a heterogeneous char. Concurrent flow of solids and gases in a rotary kiln allows these operating conditions. Another method for producing high gas yield is fast heating to high temperatures. This can be best obtained in a fluidized-bed reactor that is noted for high heat transfer rates.

At moderate temperatures (900–1000 °F) the gas phase becomes richer in higher-molecular-weight hydrocarbons. The yield of condensible organics is maximized when the moderate temperature levels are achieved via rapid heating and the resulting gases and vapors are immediately quenched. Both the entrained-bed and fluidized-bed reactors are suitable for such operating modes.

Slow heating to moderate temperatures results in maximum production of char. Based upon heat transfer characteristics, the horizontal flow reactor appears best suited for char production because of its relatively slow heating.

It should be borne in mind that it is a complex task to predict the operating conditions that achieve specific product yields for heterogeneous wastes having variable chemical composition and structure. The above paragraphs should be viewed simply as generalizations, with actual experimentation being the only accurate way to correlate product yields and distributions with feedstock and operating conditions.

Chapter 3 in this book contains some pertinent data on the effect of heating rate and temperature on the composition and phase distribution of pyrolysis yields from municipal refuse.

Pyrolysis systems have been designed to accept both untreated and pretreated refuse. The latter usually involves shredding and air classification (and incidental metal removal). Most pyrolysis systems are well suited for processing the organic fraction remaining after front-end materials recovery.

Drying is a preliminary processing operation often undertaken to treat high-moisture-content feedstocks. The high water content of many agricultural and forestry residue feedstocks may lead to great volumes of wastewater (in excess of 75 gal./ton of raw feedstock) condensed from the product fuel gas, if these feedstocks are not adequately dried prior to processing [52].

At least two systems have been developed for handling untreated waste which, of course have a significant cost advantage. However, large particles are introduced to the reactor and long residence times are required (hence larger reactors). Furthermore, the glass and metal can cause slagging and corrosion problems in the reactor.

Several pyrolysis process variations are currently commercially available, and one of the larger systems is the Monsanto Landgard System built by the City of Baltimore. This unit is designed to process 1000 tons/day of municipal solid waste. This process is presented as a case study in a later section.

The Occidental Flash Pyrolysis process pyrolyzes finely ground organics from refuse to produce a liquid fuel. The process was developed by the Occidental Research Corporation (formerly Garrett Research and Development Company) and is the subject of an EPA demonstration at San Diego County, Cal. It uses an entrained-bed pyrolysis reactor with the pyrolysis occurring at about 900 °F (480 °C). A cyclone removes char from the vapors at the exit of the pyrolysis reactor. The char-free vapor is quickly quenched to 175 °F (80 °C) by an oil spray and the resulting pyrolytic fuel oil is decanted from the quench oil. Part of the char is recycled to the reactor as a heat-carrier, after external heating to 1400 °F (760° C).

The off-gases from the reactor contain by weight 20% char, 40% oil, 30% gas, and 10% water. The process is designed to handle 200 tons/day of refuse and produce approximately 40 gallons (150 L) of liquid fuel (equivalent in heating value to about 30 gallons, 114 liters, of No. 6 fuel oil) from each ton of solid waste processed.

The Union Carbide Purox system is designed to produce a marketable fuel gas by using a slagging pyrolysis reactor, i.e., a vertical moving-bed reactor, approximately 10 ft (3 m) in diameter and 30 ft (9 m) high. Shredded and magnetically separated refuse is fed at the top and passes down in counterflow to hot gases, then pyrolyzes in the 500 °F to 1500 °F (260–815 °C) temperature range. The char and ash move downward into the hearth area, where all the char is burned to carbon oxides with pure oxygen (hence the name Purox). Approximately 0.2

tons of oxygen are consumed for each ton of raw refuse processed at the front end. The hearth temperatures are around 3000 °F (1650 °C), which result in slagging ash.

The product gas from the Purox process has a heating value of around 300 Btu/scf (11.2 MJ/m^3). The 200 ton/day unit produces about 1.20 billion Btu of gas daily. The almost nitrogen-free gas is cleaned as it is discharged from the gasifier and when dry has a composition of 33% H_2, 47% CO, 14% CO_2, 4% CH_4, 1% other hydrocarbons, the rest (about 1%) being nitrogen and argon. This gas may be burned to produce steam, used as a chemical feedstock, or transported and used to supply process heat.

The Andco-Torrax process system commercialized by Carborundum Environmental Services resembles the Purox system, except that pure oxygen is replaced by a simple air feed. The hot off-gases contain a high percentage of nitrogen, are not useful as a chemical feedstock, but are valuable as a fuel gas. These gases leave the rector at a temperature of 700 °F (370 °C), are burned in an afterburner and sent to a heat recovery boiler. No cleaning of the gas leaving the reactor is attempted, instead the cleanup occurs after combustion, heat recovery, and steam generation.

The product gas from the Andco-Torrax process has a heating value of approximately 150 Btu/ft^3 (5.6 MJ/m^3). This process has been successfully marketed in Europe and a 220 ton/day (200 metric ton/day) plant has been operating in Luxembourg to furnish approximately 45,000 lb/h (20,400 kg/h) steam at 500 psia (3450 kPa) and 725 °F (385 °C) for a turbine electric plant. Two more similar plants are being operated in France and Germany.

An important advantage of pyrolysis is its ability to process plastic and rubber wastes that are the bane of incinerators and are not readily processed by other techniques. Pyrolysis of plastics, developed by Union Carbide, converts polyolefins, polyvinyl chloride, polystyrene, and their copolymers into waxy solids, viscous liquids, or gases, depending on residence time and operating temperatures.

Pyrolysis of tires (at 500–900 °C) pioneered by Firestone Tire and Rubber Co. and the Bureau of Mines Coal Research Center, yields approximately 45% char that can be used as filtering medium, asphalt filler, or a smokeless fuel; the balance is a liquid and gas mixture of hydrocarbons which may be used as fuel (see Chap. 1).

E. Composting

Composting is a process in which microorganisms decompose components such as paper, food scraps, and other organic materials. This process normally occurs under aerobic conditions at a temperature of about 140 °F (60 °C). After a period of time (five days to three weeks, depending on the process) the mass cools and turns into compost that has both soil conditioning and plant nutrient values. Chapter 5 is devoted entirely to the subject of Composting.

F. Chemical–Biochemical Conversion

Chemical–biochemical processes that have been suggested for converting municipal solid wastes into usable products include anaerobic digestion, fermentation, hydrolysis, hydrogenation and esterification, wet oxidation and biophotolysis [53]. Many of these processes are in the laboratory or small pilot plant stages and their economic and technical feasibility at commercial levels is not yet established. Nevertheless, they are valid candidates for resource recovery operations; some of the more promising processes will be discussed briefly.

1. *Anaerobic Digestion*

This is a process in which microorganisms feed on organic waste matter under anaerobic conditions, producing methane and carbon dioxide along with a humus-like slurry [54].

The anaerobic digestion of organic wastes is normally considered to occur in two stages. In the first stage, complex organics that are usually in the solid state are broken down to simple organics that are usually in the liquid state. This occurs when extracellular enzymes convert complex carbohydrates, proteins and fats into simple organic acids, aldehydes and alcohols. The following two-step equation illustrates the first stage biochemical reactions:

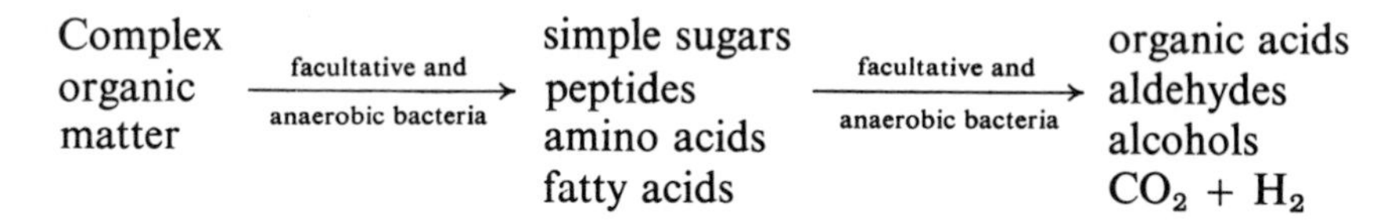

In the second stage, the end products of the first stage are metabolized to methane and carbon dioxide (along with a biologically inert organic residue) by an anaerobic group of bacteria, commonly referred to as methanogenic (methane-forming) bacteria. During this stage, the organic acids and other first stage end products, plus all the soluble nitrogenous compounds present, are converted to amines, ammonia, acid carbonates, along with gaseous products of carbon dioxide and methane. Also odorous mercaptans, skatole, indole, and hydrogen sulfide are formed as byproducts. The amines, ammonia, and carbonates are finally broken down to the final gaseous products of carbon dioxide and methane. The methane content may be 55–75% by volume, depending on process conditions. The following equations describe the second stage biochemical reactions:

$$\begin{matrix} \text{Organic acids} \\ \text{aldehydes} \\ \text{alcohols} \end{matrix} + \begin{matrix} \text{soluble} \\ \text{nitrogenous} \\ \text{compounds} \end{matrix} \xrightarrow[\text{bacteria}]{\text{methane forming}} \begin{matrix} \text{amines} \\ \text{ammonia} \\ \text{acid carbonates} \\ CO_2,\ CH_4 \end{matrix} + \begin{matrix} \text{mercaptans} \\ \text{indole} \\ \text{skatole} \\ H_2S \end{matrix}$$

and

$$\begin{matrix} \text{Amines} \\ \text{ammonia} \\ \text{acid carbonates} \\ CO_2,\ CH_4 \end{matrix} \xrightarrow[\text{bacteria}]{\text{methane forming}} CH_4 + CO_2 + \text{traces of other gases}$$

The major environmental and process parameters that affect the operation of an anaerobic digestion system are temperature, pH, volatile acids, retention time, toxic substances, and nutrients. These parameters exhibit an interdependent interaction with one another and consequently, changes in one parameter can be expected to be accompanied by changes in one or more of the other parameters. This attests to the complexity of the process and the operational difficulties that may be expected in the event of a process upset.

The commonly used temperature range is the so-called mesophilic range, 85–95 °F (30–35 °C). More intensive anaerobic biological activity (and consequently reduced digestion time) takes place in the thermophilic range, 130–140 °F (55–60 °C).

The most desirable pH range is between 6.6 and 7.2. A low pH indicates a rise in acid concentration, which in turn indicates that the methanogenic bacteria are not utilizing the volatile acids as fast as the acetogenic (acid-forming) bacteria are producing them. When this happens, lime is commonly added to raise the pH and restore normal operation.

High levels of volatile acids are attributed to an imbalance between the acetogenic and methanogenic bacteria. The acetogenic bacteria have a much higher growth rate than the methanogenic bacteria and consequently process conditions must be maintained in such a manner that no imbalance is created between these two groups of bacteria. For example, an increase in volatile solids loading rate, washout of methanogenic bacteria owing to reduced hydraulic retention time, presence of substances toxic to methanogenic bacteria, or undue temperature swings, are all process conditions that can lead to increased volatile acids.

The doubling time for methanogenic bacteria has been reported as being two to four days in the mesophilic range. At higher temperatures this value is less. A stable operation requires a hydraulic retention time of approximately twice the doubling time of the microorganisms. Operation at shorter retention times will result in washing out the methanogenic bacteria faster than they can regenerate.

The microorganisms present in the digestion system are susceptible to certain toxic substances. Materials that inhibit anaerobic digestion due to toxicity are (a) ammonia gas which at low pH is converted to the toxic ammonium cation, (b) heavy metals such as chromium, copper, zinc, and nickel at sufficiently high concentration, (c) certain organics such as pesticides, solvents, etc.

Organic wastes that are made up mainly of carbohydrates and fats and lack nutrients will require the addition of nutrient supplements for proper anaerobic digestion. Such nutrients mainly provide nitrogen, phosphorus and traces of certain inorganic salts.

Anaerobic digestion is well suited for treating sewage sludge, animal manures, and other solid wastes of high moisture content such as crop residues [55]. In fact, the Los Angeles County Sanitation Districts' Joint Water Pollution Control plant at Carson, California, utilizes sewage sludge to produce approximately 6 million ft^3/day (170,000 m^3/day) of digester gas, or 143 million Btu/h (2510 MJ/min) based on the lower heating value of the methane in the gas. The methane content is approximately 64% of the digester gas volume on a dry basis. Part of the digester gas is utilized for in-house power generation and steam production, while the balance has been sold, since 1962, to a local oil refining company to fire small boilers and for crude heaters [56].

The fact that the organic fraction of municipal refuse is composed of essentially the same constituents as are found in sewage sludge has led some researchers to investigate the anaerobic digestion of organic refuse on a laboratory scale [57, 58], and the results have been encouraging.

Typically solid waste is shredded and classified and the lighter fraction is blended with chemical nutrients, or sewage is added, to provide the necessary nitrogenous compounds. Water is added during blending to give a mixture of 10–20% solids, and the pH is adjusted to about 7. The slurry is then heated in a mixed digester and depending on the temperature and retention time in the digester, various gas yields are experienced. Gas production increases with an increase in temperature and retention period from a low of 1.4 ft^3/lb dry solids at 4 days retention time and 95 °F to 4.96 ft^3/lb dry solids at 30 days retention time and 140 °F. The offgas contains approximately 60% methane and the rest is carbon dioxide. The solids are reduced by about 50% in volume.

The various components of a digestion system such as digestion tanks, gas cleaning units and gas handling equipment are commercially available. However, since anaerobic digestion of municipal refuse has not been demonstrated on a large scale there is a definite risk factor that the system will not perform as current research suggests.

A number of problems need to be solved before the process can be shown to be a commercially viable source of fuel from refuse. The slow nature of the digestion process requires long retention times (10–20 days) and makes it very difficult to control. The nonhomogeneous nature of municipal refuse makes the process subject to various toxins; also, buffering pH can require large quantities of lime. Furthermore, some unique process-related problems arise in mixing a slurry containing 10–15% solids, and also in the solid/liquid separation of the fibrous residue from the digester. In addition, some useful function must be found for the approximately 50% residue that results from the process.

The Department of Energy has funded a demonstration plant at Pompano Beach, Florida which will use an anaerobic digestion process to convert 100 tons/day of a pulverized organic solid waste from a 65 tons/h shredding facility [59]. The plant was built in mid-1978 at a cost of approximately $3.7 million and will be run as a 2–4 year experiment to determine the commercial feasibility of the conversion process.

One aspect of anaerobic digestion that has received keen interest over the past few years, from the resource recovery point of view, is the subject of methane extraction at solid waste landfills.

It is common knowledge that within the confines of a solid waste landfill where organic wastes are deposited, there occurs a normal decomposition process which produces a mixture of gas consisting primarily of methane and carbon dioxide. This decomposition process is very similar to the biochemical processes occurring in an anaerobic digester, with one significant difference. Whereas the digester can be operated with a certain degree of control over the bacterial activity, the landfill is a haphazard operation. Consequently, a well-operated digester may yield 4–6 ft^3 of gas/lb refuse (0.25–0.4 m^3/kg), whereas gas yields within a landfill would, at best, be a hundred times lower.

In spite of this significant yield disparity it is certain that enough gas will be produced in even the least active landfill sites to necessitate preventing the gas from migrating laterally to adjacent property where it might accumulate and be accidentally ignited. Thus the need for landfill gas evacuation has been well recognized.

The first significant efforts at recovering the evacuated landfill gas were initiated by the Sanitation District of Los Angeles County at its Palos Verdes sanitary landfill in 1971. Extensive tests run on a 100-ft deep well located in the landfill indicated that gas with an approximate composition of 50% methane and 48% carbon dioxide, and a heating value of around 500 Btu/ft^3 (18.6 MJ/m^3), could be produced. An internal combustion engine, powered with gas from the landfill, was used to drive a centrifugal blower which produced 300 ft^3/min (8.5 m^3/min) of gas from the well [60]. Subsequently, a facility to process about 2 million ft^3/day (56,600 m^3/day) of landfill gas was built in the summer of 1975. This facility also consists of a molecular sieve processing plant which separates carbon dioxide from the raw gas and delivers pipeline-quality gas. Since January of 1977, this extraction facility has operated steadily at a throughput of 1.2 million ft^3/day (34,000 m^3/day) of raw gas. The final product gas is injected into Southern California Gas Company's distribution grid.

A similar demonstration project is currently underway at the Mountain View landfill in the San Francisco Bay Area. The $850,000 project is partly funded by the EPA ($270,000) with the rest provided by Pacific Gas and Electric Company.

The Mountain View site was of interest to EPA because, unlike the Palos Verdes landfill, which is 100–140 ft deep, it is a shallow landfill with an average depth of 40 ft. Thus, it is more representative of landfills throughout the U.S.

The gas collection system in this project is comprised of 18 wells, 35 ft deep, over a 30-acre area. Gas production is estimated to be 1 million ft^3/day (28,300 m^3/day) with an estimated life, at this production rate, of 10 years. Tests run on the raw gas have indicated a typical dry composition of 52% methane, 38% carbon dioxide, 9% nitrogen, and 0.5% oxygen. The nitrogen and oxygen result from a small but unavoidable intrusion of air into the landfill during the gas extraction process. The higher heating value (HHV) of the raw gas is between 425 and 500 Btu/ft^3 (15.8–18.6 MJ/m^3) depending on the degree of air intrusion.

The raw gas will be purified by a molecular sieve process, whereby both carbon dioxide and moisture will be removed. After processing, the purified gas production will be about 600,000 ft^3/day (17,000 m^3/day) and the HHV will rise to between 700 and 750 Btu/ft^3 (26.1–27.9 MJ/m^3). This gas will be injected into Pacific Gas and Electric's gas main, which runs across the landfill [61].

2. *Chemical Transformation*

Cellulose is a major component of certain solid wastes (paper, garbage, grass, straw, leaves, etc.) and therefore processes involving the chemistry of cellulose are being evaluated as possible candidates for resource recovery [62]. Cellulose, a polysaccharide $(C_6H_{10}O_5)_n$, can be degraded both thermally and chemically. The main products of thermal decomposition are carbon, carbon dioxide, and water. Some of the other products formed are acetic acid, acetone, formic acid, formaldehyde, furfural, ethane, carbon monoxide, and methane.

Hydrolysis of cellulose yields simple sugars by heating cellulosic wastes with an acid at elevated temperature. The sugars can then be fermented to produce ethyl alcohol, citric acid, animal fodder, and other useful substances. Under the right conditions, alkaline degradation of cellulose can result in complex reactions producing formic, acetic, glycolic, and lactic acids. Cellulose derivatives find use as varnishes, lacquers, adhesives, thickeners, coatings, film, emulsifiers, and a host of other uses.

High-pressure hydrogenation (around 4000 psig) of various carbohydrate compounds (such as cellulosic wastes) can yield an oil which is free of oxygen and water (as compared with pyrolysis oil) and contains aromatic and aliphatic hydrocarbons and tertiary amines.

Acetylation with acetic anhydride converts cellulosic wastes to cellulose acetate, an important raw material for film, yarn, plastics, and coatings.

A feasibility study by the Bureau of Mines has shown that the hydrogasification of the carbon content of municipal solid wastes converts it to methane [63]. Basically, carbonaceous waste is reacted with hydrogen to yield a product gas containing methane as follows:

$$C_{2.95}H_{4.4}O_{1.76}N_{0.03}S_{0.01}(H_2O)_{1.15} + 5.51H_2 \rightarrow 2.95CH_4 + 2.91H_2O + 0.03NH_3 + 0.01H_2S$$

(refuse—unseparated)

or

$$C_{3.57}H_{5.4}O_{2.14}N_{0.04}S_{0.01}(H_2O)_{1.12} + 6.66H_2 \rightarrow 3.57CH_4 + 3.26H_2O + 0.04NH_3 + 0.01H_2S$$

(refuse—metal and glass eliminated)

These processes and many other similar chemical conversion schemes have been successful on a laboratory scale. Commercial development will depend to a large extent on the economics of expansion and production.

3. *Oxidation*

Wet oxidation consists of processing the organic fraction of solid waste in an aqueous solution at high pressure (500–1000 psig) and temperature (200–260 °C) where partial oxidation occurs. It is estimated that for every 100 tons solid waste treated, 10 tons residue, 45 tons carbon dioxide and water, and 45 tons commercially useful chemicals can be obtained [64, 65].

Partial oxidation in which finely divided organic waste is partially combusted in an oxygen–nitrogen atmosphere (deficient in oxygen) results in a number of organic chemicals such as acetic acid, formic acid, formaldehyde, methanol, acetone, etc. [66].

IV. SELECTED CASE STUDIES

The two basic resources to be recovered from solid wastes are energy and materials [23–29]. These two goals are not necessarily in conflict for non-combustible materials such as glass, aluminum, and ferrous metals. The conflict arises in the recovery of the organic fraction that could be recycled for materials (such as fiber recovery) or burned for energy production.

At the present time, no large-scale system has yet been able to profitably recover materials from the organic fraction for recycling. The Franklin, Ohio, facility discussed below, sold reclaimed fiber for a short time. However, owing to market problems this operation was terminated, and the system is now advertised as another method of producing burnable fuel. The only reasonable method of fiber recovery seems to be source separation of both newsprint and corrugated cardboard (or home scrap from a paper plant, such as envelope cuttings). The fiber value of mixed paper (the type of material extracted from a recovery process) is quite low, whereas the energy value is much higher [21]. Most systems are thus extracting noncombustibles such as aluminum, steel and glass for materials, and burning the organics for fuel.

It is estimated that over 30 resource recovery systems greater than 1000 tons/day capacity will be in operation by 1982 [67], with over 150 municipalities presently showing keen interest in some form of resource recovery as an alternative to disposal [20].

Described below are some of the earliest systems that have had the benefit of recent analysis and evaluation. The first two systems were designed primarily for materials recovery, whereas the remainder were intended primarily as producers of fuel and energy.

All of these systems involve relatively complex capital-intensive technologies with reported or estimated capital investments ranging from $10,000 to about $30,000 per ton of daily processing capacity, depending on type of process, plant size, and other factors.

A. New Orleans, Louisiana

The New Orleans facility was built in response to the city's need to improve its solid waste disposal techniques, and the need for a facility to evaluate and demonstrate resource recovery processes. It is a joint venture between the city and a private firm (Waste Management, Inc.), and the technical and financial assistance of the National Center for Resource Recovery (NCRR).† The city provides the refuse to be processed as well as the landfill space for residues from the operation, Waste Management owns and operates the facility, and NCRR which developed the concept, monitors the operation as an agent of the city [68].

The plant is constructed modularly into two functional segments; a schematic of the operation is shown in Fig. 11. The first segment is the "reduction module" whose primary function is to reduce the size of the incoming refuse thus facilitating refuse recovery and converting the unrecoverable residue into an easily landfilled element. This module is capable of handling up to 750 tons/day (680 metric tons/day) of municipal refuse.

The second segment is the "recovery module" and consists of a series of separation processes for extracting marketable metals and glass to previously agreed-upon market specifications.

The process begins with front-end loaders pushing the refuse into a pit conveyor. The refuse then goes through a large trommel screen (45 ft long, 10.5 ft diameter, 4.75 in. diameter holes) which breaks open plastic and paper bags and sorts out most of the glass containers and metal cans as underflow material. The underflow also includes small light organic waste. The trommel overflow is sent to a shredder where it is reduced to a more manageable size.

The trommel underflow and the shredded materials are next con-

† NCRR is a private labor and industry-funded organization established as a means of encouraging the development and demonstration of resource recovery technology. It is based in Washington, D.C.

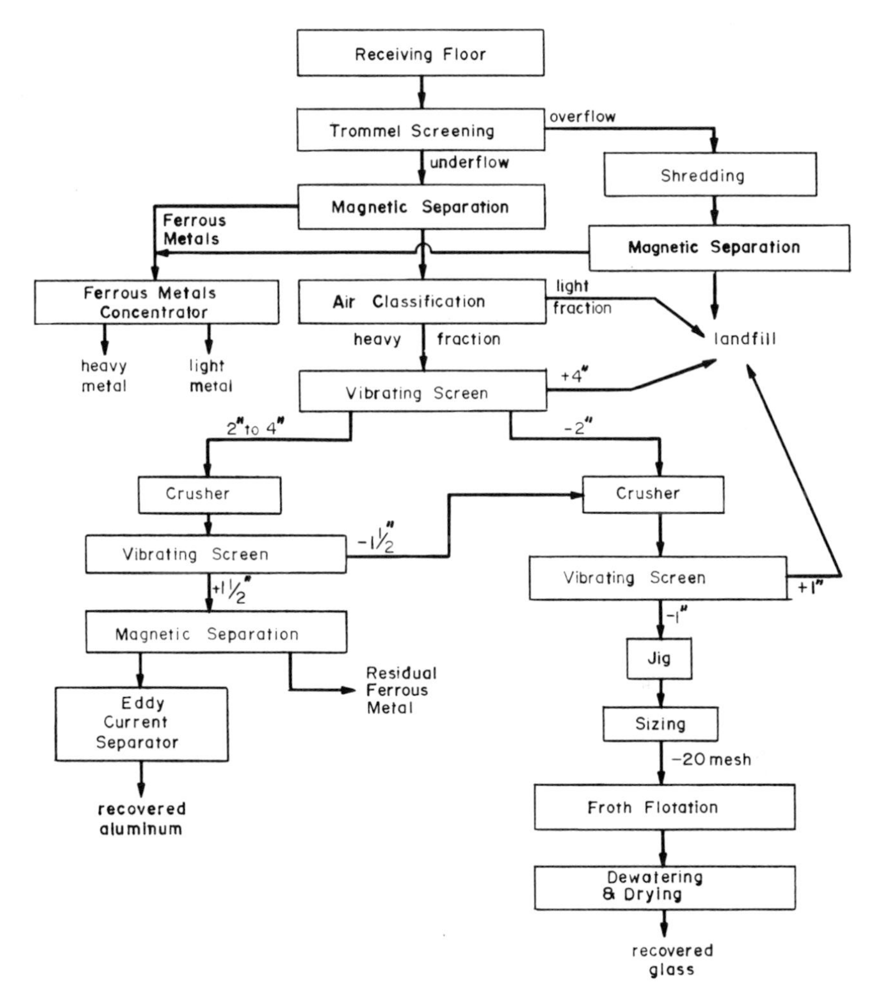

FIG. 11 New Orleans materials recovery system.

veyed separately under rotating magnetic drum separators. The recovered ferrous metals are sent to a ferrous metals concentrator in which two separate fractions of ferrous metal are recovered; a light metal consisting of cans and light-gage metals, and a heavy metal made up of casting, forgings and rolled stock.

The nonmagnetic residue from the shredded stream is diverted to a landfill loadout area for disposal whereas the nonmagnetic residue from the trommel underflow passes through an air classification step where the light organic material is separated and directed to landfill disposal and the heavy material is conveyed toward the aluminum and glass recovery systems.

The air-classified heavy material is subjected to a two-deck vibrating screen for separation into sizes larger than 4 in., smaller than 2 in., and between 2 and 4 in. The plus 4-in. material is directed to the landfill loadout; the minus 2-in. material enters the glass processing stream; and the 4 × 2 in. material is processed for aluminum.

Aluminum recovery consists of crushing and screening operations wherein any of the remaining glass (reduced to less than $1\frac{1}{2}$ in.) is recycled to the glass processing stream. The aluminum-rich stream passes over a rotating drum magnet to remove any residual ferrous metals and then enters an eddy current separator or "aluminum magnet" where an electromagnetic field and the resulting field in the conducting aluminum causes the aluminum to be repelled into a collecting chute.

The minus 2-in material stream from the vibrating screen area is funneled to a crusher that further reduces the size of the friable glass material. This is followed by a 1-in. vibrating screen where all materials larger than 1-in. are sent to the landfill loadout area, and the minus 1-in. material is fed into a minerals jig to separate light organic waste from the heavier (mostly glass) material.

The glass-rich stream from the jig undergoes a sizing operation that reduces the glass particles to a 20-mesh size by repeated screening and rod milling. This sized stream then enters a froth-flotation unit where the fine glass particles are floated off and the recovered glass slurry is dewatered and dried prior to storage.

Construction of the overall resource recovery facilities was completed in March 1978 and the plant has been processing around 650 tons/day (590 metric tons/day), which is nearly two-thirds of New Orlean's waste, including nearly all residential waste.

Ferrous metal recovery approximates 20 tons/day (18 metric tons/day) and aluminum recovery varies between 500 and 700 lb/day (225 and 315 kg/day). The glass recovery system is presently (early 1979) undergoing shakedown and modification. Modifications are also underway in the ferrous and aluminum recovery sections, in order to improve the quantity and quality of the recovered metals. NCRR is also actively pursuing an energy market for the light, organic fraction of the shredded waste which is currently being landfilled.

Negotiations with various private firms have elicited firm "letters of intent" for the sale of all recovered materials, namely aluminum, heavy and light ferrous metals, and color-mixed glass fines. Since the plant has not achieved full operation, the economics are not firmly established. The announced cost of the project is approximately $9 million.

The uniqueness of this venture is mainly in the level of materials recovery sophistication and the partnership between the city, the private firm, and NCRR. The model developed in New Orleans could well be applied in many other U.S. cities.

B. Franklin, Ohio

In 1967, the city of Franklin was faced with a frequently occurring problem namely that the landfill space was rapidly approaching its limits, and a solution had to be found. Working with a private local firm, the Black-Clawson Company, the city obtained an EPA demonstration grant for the construction of an integrated materials recovery system, including the disposal of sewage sludge from an adjacent wastewater treatment plant.

The plant began operation in mid-1971. The system was originally designed to reclaim ferrous metallics and fibrous material from the solid waste, and to provide for the ultimate disposal of the residual wastes that could not be reclaimed. A glass reclamation facility was added and made operational in early 1976.

The Franklin plant is located adjacent to a municipal wastewater treatment plant and the two facilities offer complementary service to each other. For example, the solid waste facility incinerates sewage sludge from the wastewater plant, whereas the latter treats wastewater from the solid waste plant [69, 70].

As illustrated in Fig. 12, the Franklin plant is comprised of three major systems: Hydrasposal,† Fibreclaim† (Fiber Recovery System), and Glass Recovery systems. The Hydrasposal system is further subdivided into pulping/separation and dewatering/incineration subsystems.

1. *Hydrasposal System*

The input to the Hydrasposal system is municipal refuse which is weighed and selectively screened for large unprocessable materials (tree trunks, refrigerators, tires, etc.) and then conveyed to a wet pulper. The wet pulper, which is in many respects similar to a kitchen sink disposal unit, consists of a 12 ft (3.7 m) diameter tube with a high-speed cutting blade in the bottom driven by a 300 hp (224 kW) motor. Process water is mixed with the solid waste in the pulper where all soft and brittle material is ground into a slurry. Large unbreakable material

† Hydrasposal and Fibreclaim are copyrighted trademarks for the Black-Clawson Company, Middletown, Ohio.

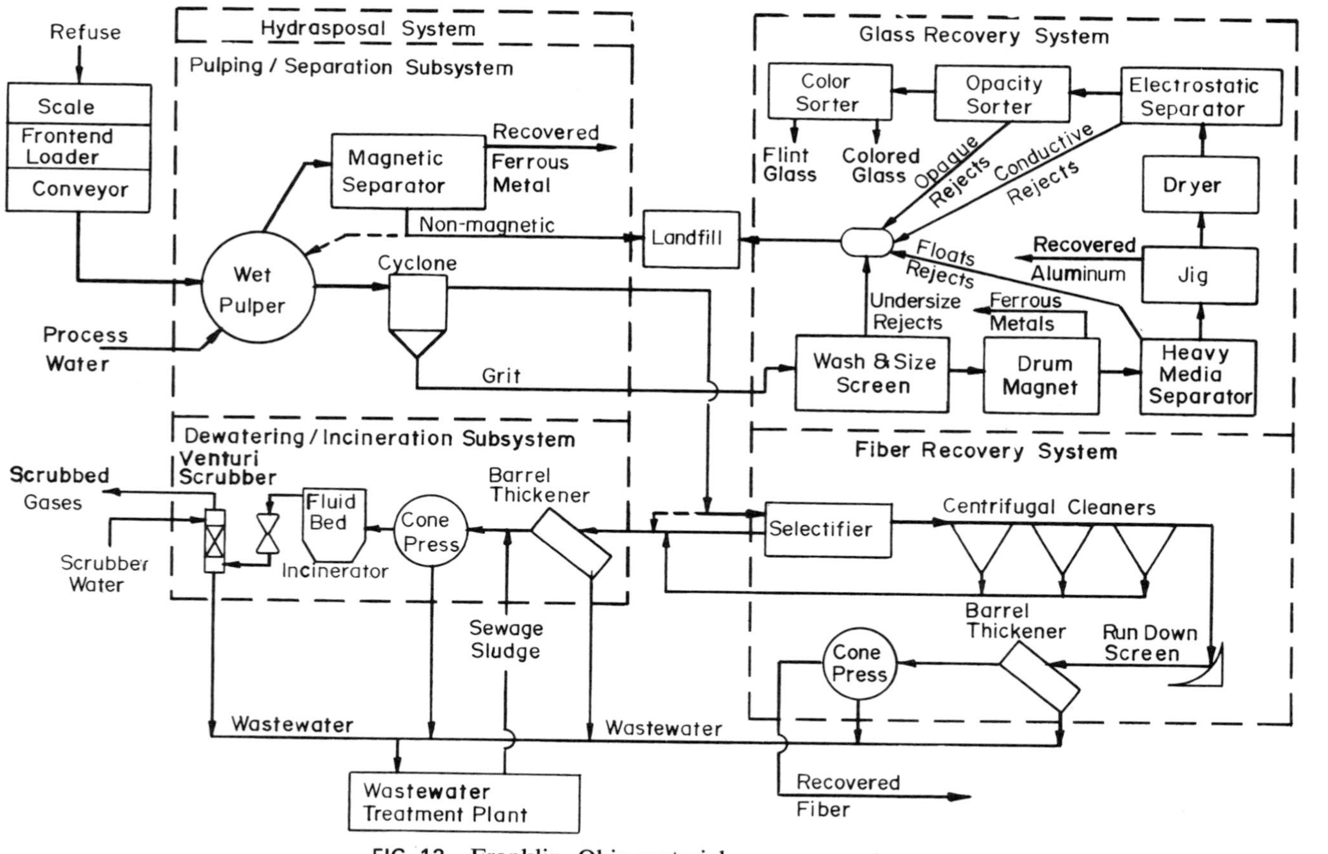

FIG. 12 Franklin, Ohio materials recovery system.

is thrown out of the pulper (by centrifugal force), washed, and fed to a continuous belted magnetic separator.

The magnetics obtained from the magnetic separator are combined with the magnetics obtained from the glass recovery system and sold. The nonmagnetics are either recycled to the wet pulper or landfilled.

The slurry, which contains the organic material, glass, small pieces of metal, and some aluminum, leaves the pulper through a grate below the blade. The grate consists of holes one inch (2.5 cm) in diameter. The liquid slurry is pumped to a liquid cyclone where the heavy fraction (grit) is separated from the organics, dewatered, and conveyed to the glass recovery system. The fibrous organic fraction is then directed either to the fiber recovery system via the selectifier tank, or to the dewatering/incineration (D/I) subsystem.

The D/I subsystem performs dewatering and subsequent incineration of materials obtained from any or all of three sources: (a) process stream from the liquid cyclone, (b) sewage sludge from the wastewater plant, (c) various streams from the fiber recovery system. These process streams are pumped to a barrel thickener which consists of three square-pitch screws rotating within perforated cylinders. The dewatered material is then mixed with sewage sludge and transferred to a cone press where a wheel rotating against an opposing screen presses water out through the screen. From the cone press the process stream is pneumatically fed to a fluidized-bed incinerator with a bed diameter of 21 ft (6.4 m). The process stream must contain at least 30% solids and the minimum calorific value must be 3000 Btu/lb (6.98 MJ/kg). The incinerator is rated at 1.0×10^6 Btu/min (1060 MJ/min) and the bed temperature is maintained within 1200 to 1400 °F (650–760 °C). The dewatering process is sufficiently efficient to allow the incineration to be self-sustaining, and no auxiliary fuel is necessary.

The combustion products are passed through a venturi scrubber for removal of particulate material; the effluent from the scrubber is pumped to the wastewater treatment plant.

2. *Fiber Recovery System*

Normal operation of the fiber recovery system recovers fibers from the lighter fraction leaving the liquid cyclone. The process stream is first diluted and passed through the selectifier screen. Here, materials larger than 0.062 in. (1.6 mm) in diameter are rejected to a rejects tank, and the process stream is directed to centrifugal cleaners for removal of grit, glass and other small, dense particles. The process stream slurry then flows down over 0.02 in. (0.51 mm) screens (rundown screens) where additional particles are separated. The rundown screens feed a barrel thickener followed by a cone press, both of which dewater the recovered fiber for subsequent sale to a nearby asphalt tile manufacturer.

The fiber recovery system can also be operated in a thickening mode when fiber is not being recovered. In this mode the process stream bypasses the centrifugal cleaners and mechanical thickeners, and is pumped to the D/I subsystem for ultimate incineration.

3. *Glass Recovery System*

This system was installed in 1974 and further modified in 1976.

The heavy gritty materials removed by the liquid cyclone form the feedstock to the glass recovery system. The first step is a washing and screening step that removes particles smaller than 0.25 in. (6.35 mm). The oversized material from the screen is passed to a drum magnet where magnetic materials are separated. The magnetically cleaned stream is then delivered to a heavy media separator where materials with a specific gravity less than 1.8 (mostly organics) float and are removed for landfill disposal. The heavier materials sink and are carried into a jig where aluminum is separated from glass and stones, and stored for market.

The glass-rich process stream is dewatered on a vibrating screen and dried in a rotary drum dryer. The dried material is conveyed to a high-tension electrostatic separator for the removal of any remaining conducting materials. The process stream is then passed on to an opacity sorter where glass is separated from the stream. The final step is color sorting which separates the glass into clear glass cullet and mixed colored glass cullet.

4. *Recovery Economics*

The major recovered materials at the Franklin facility are magnetic metals, aluminum, glass and paper fiber. For each ton of as-received refuse fed to the system the recovery fractions have been estimated as follows:

	lb (kg)/ton refuse
Magnetic metals	195 (88)
Aluminum	13 (6)
Glass	
flint	60 (27)
colored	35 (16)
Paper fiber	390 (177)

The recovered fiber product has a calorific value of about 7200 Btu/lb (16.7 MJ/kg). With the glass plant operating, approximately 15% of the refuse stream by weight, requires landfilling. Without the glass plant, 20.4% by weight, requires landfilling.

The Franklin plant was built at a capital cost of about $5 million and was designed to handle 150 tons/day. However, owing to an insufficient supply of refuse, the plant was generally operated at an average throughput of 35 tons/day. At this reduced operating capacity the plant was unable to enjoy any economy of scale and experienced an operating expense of $19.50/ton of refuse handled, versus income from the sale of recovered materials amounting to $6.60/ton of refuse and tipping or dumping fees of $8.50/ton of refuse. Thus resulting in a net operating cost of $4.40/ton of refuse. This does not take into account any amortization and interest charge for the capital expense, which could reflect unfavorably on the total operating cost.

The reduced operating capacity coupled with problems with the fiber market have caused the plant to cease operations indefinitely (March 1979). The Franklin project has however demonstrated the feasibility and reliability of the wet-pulping system, especially if it can be tied to the disposal of sewage sludge. The latter problem often costing $30–40 per ton of sludge handled.

Another significant advantage of the wet pulper is the elimination of the dust and explosion problem so frequently encountered in dry shredding operations.

Economic analyses of the Franklin operation have shown that this type of operation on a 500–1000 ton/day scale can be economically feasible (assuming stable markets are available for the recovered materials) with net operating costs, including amortized capital charges, around $7–9/ton of refuse handled. These operations could even be profitable ($3–4/ton of refuse handled) if the fiber recovery subsystem is eliminated entirely and the total organic stream from the liquid cyclone is dewatered and sold as a fuel product [69].

Based on this latter concept the town of Hempstead, New York, has constructed through the Black-Clawson Co., a wet pulping system to process municipal refuse into a fuel product now coming to be known as "wet refuse-derived fuel." The fuel will be burned on-site in specially designed boilers, and will produce steam for the generation of electricity. The plant built at a cost of $73 million is presently in a shakedown phase (early 1979), and will handle 2000 tons/day of municipal refuse when fully operational.

C. Saugus, Massachusetts

The Saugus project resulted from a joint venture RESCO (Refuse Energy Systems Company) between Wheelabrator-Frye Inc. and a local landfill operator who was facing imminent closure of the landfill by the community. The result was a waterwall incinerator facility for the combustion of unprocessed municipal refuse, with the resultant steam being sold to the nearby General Electric Company's Lynn River Works. The system can also recover ferrous metals and glassy aggregate from the incinerator residue [71].

The waterwall incinerator design is based on the Von Roll, Ltd., incinerators of Switzerland, who now have over 50 systems in operation worldwide, some of which are 20-yr old.

The basic system, shown schematically in Fig. 13, starts with a refuse dumping pit which receives refuse brought in by community and private haulers at a tipping fee of around \$14/ton. The pit has a holding capacity of 6700 tons (6080 metric tons). Bulky material such as furniture is transported by a crane to a 1200 hp (895 kw) fragmentizing hammermill to reduce the largest dimensions to a manageable size, and the fragments are discharged back into the pit.

Two refuse-fired steam generators, in parallel, are provided, each with a capacity to handle a maximum of 750 tons of refuse/day (680 metric tons/day).

The refuse from the pit is crane-fed to the boiler hoppers, where it falls through a chute to a reciprocating grate system consisting of three grates. The first grate is the drying grate from which the refuse tumbles to the middle grate where the actual burning takes place. Depending on the moisture content of the refuse, the fire can be extended over the third grate if desired. The speed of the refuse through the system can be varied through reciprocal action of the grates, thus assuring complete burnout without the need for auxiliary fuel.

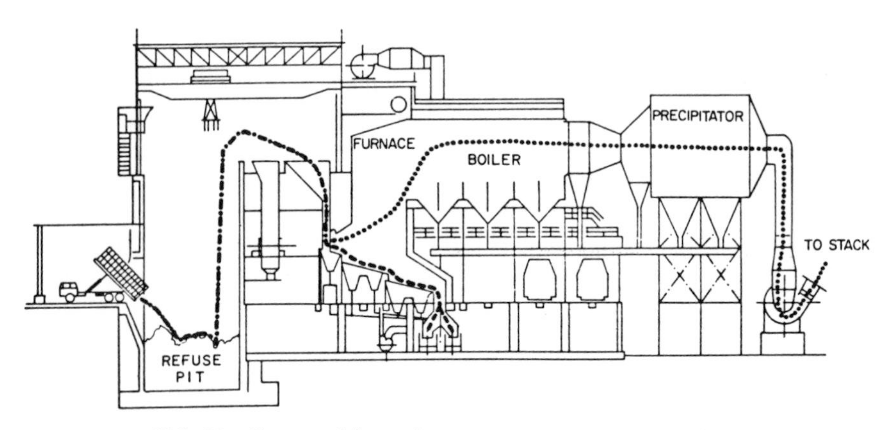

FIG. 13 Saugus, Massachusetts energy recovery system.

The hot combustion gases, which are in the range of 1650 °F (900 °C), together with furnace radiation, heat the water in the waterwalls of the boiler. Water for steam production comes from the municipal water supply and a condensate return line. The flue gases pass through two electrostatic precipitators designed to reduce particulate emissions to less than the required 0.05 g/std. ft^3 (0.12 g/std. m^3), and are subsequently discharged to the atmosphere through concrete stacks.

The unburned residue from the last grate along with fly ash collected by the precipitators, is water-quenched and passed by conveyor to a trommel screen (not shown in Fig. 13). The 8-ft. diameter rotating screen has 2-in. holes and separates smaller materials onto a vibrating conveyor that carries them past a drum magnet. The residue from the magnet is a glassy aggregate that is currently landfilled, whereas the ferrous metal recovered by the magnet is sold as low grade scrap. The bulky metal items that overflow the trommel are also sold to scrap dealers.

The Saugus plant has an average refuse capacity of 1500 tons/day (1360 metric tons/day) with a 1725 ton (1565 metric tons) peak capacity. The plant operates 24 h/day and 7 days/week although refuse is delivered to the refuse pit only 5 days a week. The plant has been operating at about 1050 tons of refuse a day (950 metric tons) and delivers superheated steam to the General Electric Company at 625 psig (430 kPa) and 825 °F (441 °C), at an average rate of 285,000 lb/h (130,000 kg/h) and a peak delivery capability of 350,000 lb/h (159,000 kg/h).

Approximately 7–8% of the incoming refuse is recovered as low grade ferrous scrap for sale to scrap dealers. An experimental program is underway to utilize the glass aggregate (10–12% of incoming refuse) with asphalt in road construction.

The plant experienced air emission problems in the summer of 1978 which were traced to black carbon flakes being emitted owing to the inability of the flakes to hold an electrical charge in the precipitators. A test program and experimentation subsequently resolved the problem. The plant has also had superheater and grate corrosion problems which have been solved by using special alloys.

The Saugus facility was built at a total capital cost of around $50 million, financed by a combination of $30 million in tax-free solid waste disposal revenue bonds and $20 million invested by the joint partnership. It represents the largest privately financed resource recovery facility in the United States.

Operating since October 1975, the plant had processed nearly one million tons of municipal refuse by early 1979, with nearly 5.5 billion pounds of steam sold to the General Electric Plant. Also, not a single truckload of refuse has been turned away from the plant during that period.

The Saugus project stands as a model of how direct high temperature combustion of unprocessed municipal refuse may be achieved to generate steam and/or electricity, and may act as a guide to other communities interested in establishing a refuse disposal service via a cooperative effort with private enterprise.

D. St. Louis, Missouri

The St. Louis project differs from the Saugus operation in that in the former the municipal refuse is processed and prepared before feeding into existing coal-fired power plants, without any need for expensive modifications or retrofitting.

The original studies leading to the St. Louis system were conducted by the engineering firm of Horner & Shifrin, Inc., for the City of St. Louis with the cooperation of Union Electric Company and partial funding by EPA [72]. These studies provided enough data to proceed with full-scale operation.

The system in St. Louis consists of the processing facility and the power plant. The processing facility, shown schematically in Fig. 14, is capable of producing 300 tons (270 metric tons/day) of supplemental fuel during one shift. The shredder and conveyor assembly has a rated capacity of 45 tons (40 metric tons) of refuse per hour [73].

The refuse is pushed from the receiving floor to a conveyor allowing constant feeding rates and visual inspection. Elevation is gained by a belt conveyor and dropped into a hammermill. This shredder is 60 in. (1.5 m) in diameter and 80 in. (2 m) long, and connected to a 1250 hp (930 kW) 900 rpm motor. Of the shredded refuse, discharged from 2 × 3 in. (5 × 7.5 cm) grates, 98% is less than 1 in. (2.5 cm) in size. The density of the shredded refuse varies from 4 to 12 lb/ft^3 (64–192 kg/m^3).

The shredded refuse is fed to an air classifier where about 70–80% by weight is collected as the light fraction. The heavy fraction goes to magnetic separation of the ferrous materials. The light fraction drops into a storage bin. Because of the high bridging tendency of shredded refuse, the storage bin has a greater cross-sectional area at the bottom than at the top. Auger screws are used to unload the bin from the bottom into compactors for the trip to the power plant.

The second phase of this process, not illustrated in Fig. 14, is the firing facility. The shredded refuse is conveyed by a pneumatic feeder from a receiving bin to a surge bin, which is again constructed so as to have the larger cross-section on the bottom and thus eliminate bridging.

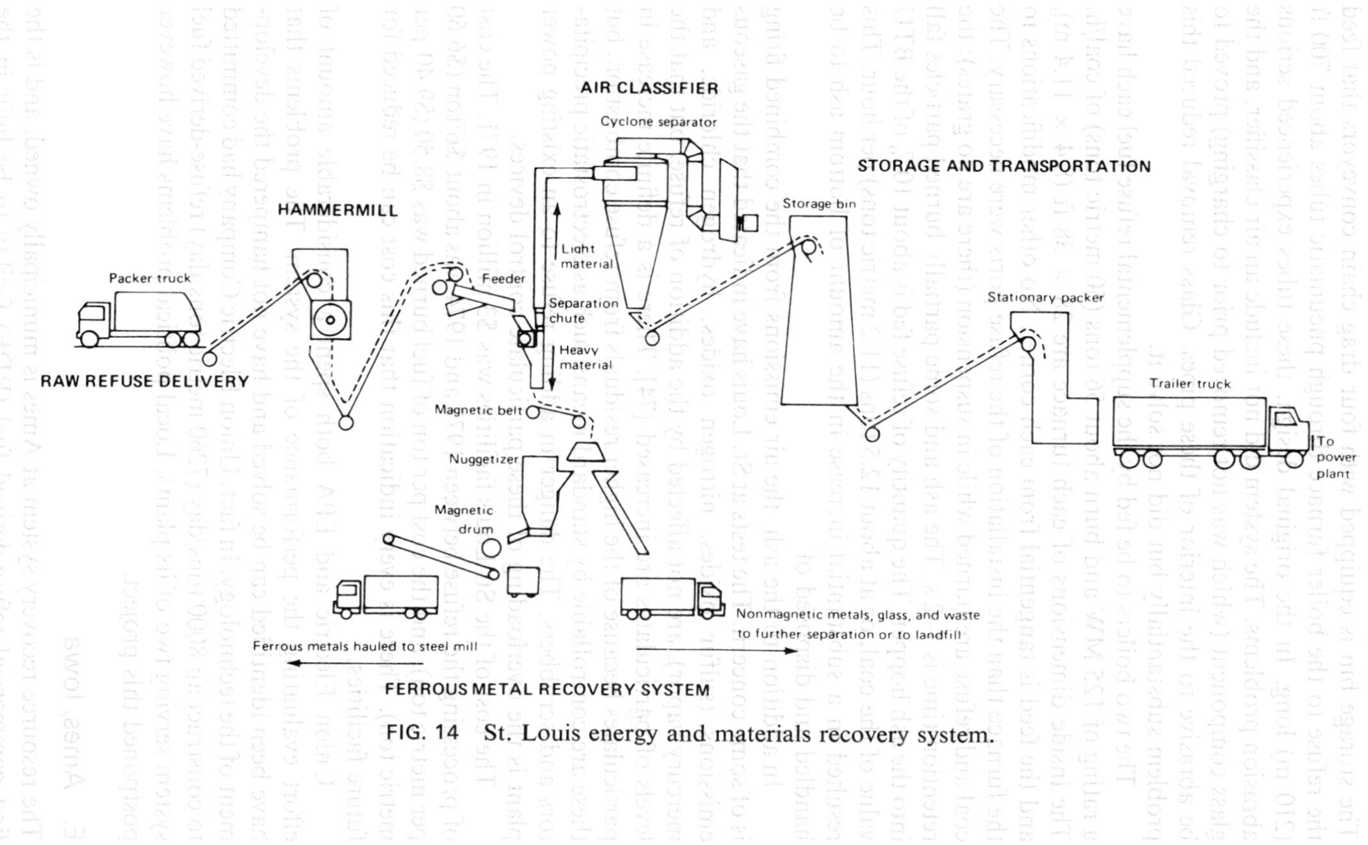

FIG. 14 St. Louis energy and materials recovery system.

The storage bin is equipped with four drag chain conveyors that feed the refuse to the boiler furnace through pneumatic tubes about 700 ft (210 m) long. In the original design, these tubes experienced serious abrasion problems. The system did not include an air classifier, and the glass component (which was not removed prior to charging) proved to be abrasive to the interior of these pipes. Glass removal reduced this problem substantially but did not solve it.

The two boilers to be fed by the supplemental refuse fuel each have a rating of 125 MW and burn about 56 tons (50 metric tons) of coal/h. The inside dimensions of each furnace are 28 × 38 ft (8.4 × 11.4 m), and the feed is tangential from each corner. No other modifications to the furnace than the installation of the refuse ports were necessary. The coal and refuse are burned while in suspension (there are no grates); the retention time is 1–2 s. The ash and some partially burned particles fall into the ash hopper. The quantity of refuse fed is about 10% of the BTU value of the coal, or at about 12.5 tons (11.2 metric tons) per hour. This resulted in a substantial increase in the amount of bottom ash to be handled and disposed of.

In addition to the ash, the air emissions from the combined firing is of some concern. The tests at St. Louis have indicated that the gaseous emissions (sulfur oxides, nitrogen oxides, hydrogen chloride, and mercury vapor), are not affected by the addition of refuse, but that the levels of particulates are increased [74]. There is a definite increase in particulates because of the higher residuals from refuse combustion, but these are controllable by standard means such as electrostatic precipitators and scrubbers. The danger in adding refuse to an existing power plant is the overloading of these particulate control devices.

The cost of the St. Louis facilities was $3 million in 1971. The cost of processing the refuse between 1972 and 1974 was about $6/ton ($6.60 per metric ton), and the cost per ton of fuel burned was $8.50 ($9.40 per metric ton). There is every indication that this cost can be reduced for future facilities.

Union Electric and EPA both spent a considerable amount of effort evaluating the performance of the system. The problems that have been identified can be solved and have not hampered the development of the technology. In fact, Union Electric Company had committed to construct an 8000 tons/day (7260 metric tons/day) refuse-derived fuel system serving two of its plants. Local political problems have however postponed this project.

E. Ames, Iowa

The resource recovery system at Ames is municipally owned, and is the first commercial refuse-derived fuel (RDF) facility to be built in the

U.S. The plant, patterned after the St. Louis demonstration project, produces RDF which is burned as supplemental fuel in a city-owned power plant. The power plant includes a 33 MW suspension-fired boiler and two 20 MW spreader-stoker (grate-equipped) boilers.

The chief motivation behind the consideration for building this facility was a familiar one—the landfill used by Ames and its environs was rapidly approaching capacity limits. A feasibility study indicated that the city could successfully operate a resource recovery system, based on the supplementary fuel concept, as an economical and environmentally acceptable alternative to landfilling. The project was initiated in June 1973; field construction of the plant, along with modifications of the existing boilers to handle RDF, began in April 1974; the plant was started up in September 1975 and has been operational since November of that year.

The plant has a maximum refuse-handling capacity of 50 tons/h (45 metric tons/h), and typically operates at about 30 tons/h (27 metric tons/h) for 8–12 h per day depending on the refuse supply. It accepts all general household and commercial refuse delivered by both commercial and private haulers. The plant schematic is shown in Fig. 15.

The wastes are moved from the tipping floor into the processing equipment by a front-end loader which pushes the material onto a variable speed conveyor. The conveyor transfers the waste to the first-stage shredder where it is reduced to an average 7 in. (18 cm) size. This shredder has 9 × 9 in. (23 cm) grate openings and 48 hammers that weigh approximately 150 pounds (68 kg) each. The shredder rotates at 690 rpm and is driven by a 1000 hp (746 kW) motor.

Upon leaving the shredder, the fragmented waste stream is subjected to a magnetic separator whereby approximately 90% or more of the ferrous metal is extracted. The ferrous extracts are deposited on a conveyor belt and transported to storage for subsequent transport and sale.

The remainder of the material, less the larger ferrous materials, moves to a second-stage shredder which reduces the material to an average 2 in. (5 cm) size. This shredder has a drive unit identical to the first-stage shredder, but has grate openings that are 3 in. × 5 in. (716 × 12.7 cm) and hammers that weigh only 45 pounds (20 kg).

The finely shredded material next passes to an air classifier where it is subjected to an air stream of approximately 25,000 ft^3/min (708 m^3/min) in a baffled chamber. Here the lighter particles are carried upward into the high velocity stream whereas the heavier particles fall downward and out of the chamber, thus providing the separation of "heavies" from "lights."

The light material stream leaving the top of the classifier enters a

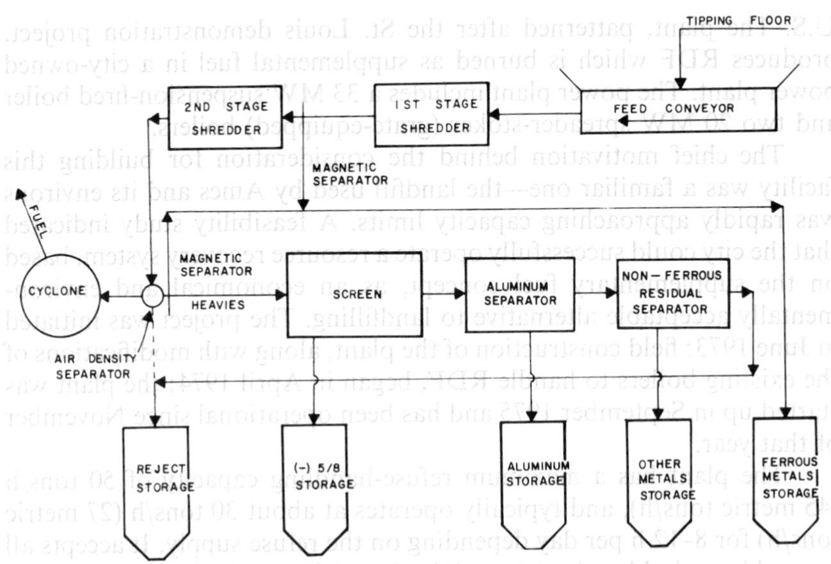

FIG. 15 Ames, Iowa energy and materials recovery system.

cyclone, is separated from the air stream, and drops to a screw feeder RDF to be fed to a pneumatic pipeline. The RDF is pneumatically delivered to a conical-type storage bin located about 600 ft (183 m) away at the power plant facility. The bin is 84 ft (26 m) in diameter and provides storage of approximately 550 tons (499 metric tons) of RDF. The RDF drops through grates in the bin floor to four separate drag conveyor lines. These conveyors meter the RDF into four separate airlock feeders which inject it into four separate pneumatic conveying lines that feed directly into the combustion zones of the power plant boilers about 300 ft (90 m) away.

The heavy materials which drop out of the air classifier are conveyed to a magnetic separator where more ferrous materials are separated and transferred to storage. The remainder of the materials are transported to the aluminum recovery portion of the plant. First, a 48 in. (107 cm) diameter trommel screen separates a minus $\frac{5}{8}$-in. fraction, which is mostly glass and grit. The larger materials leaving the trommel are subjected to an eddy current separator where mostly aluminum is recovered, and the residual material is transported to the rejects storage for subsequent delivery to the landfill.

The aluminum recovery system has produced minimal amounts of product due to operating problems and the fact that the refuse fed to the system contains very little aluminum.

Ferrous metals are recovered at the rate of 10–12 tons/day, which amounts to 5–6% of the incoming refuse. These are sold at 47½% of the Chicago bundle price to Vulcan Materials Company in Gary, Indiana.

The refuse derived fuel is by far the largest recycled fraction, averaging about 85% of the incoming refuse. This fraction is mostly paper, plastics, and wood, and has an average heat content of 5200 Btu/lb (12.1 MJ/kg). The RDF is used primarily in the suspension-fired boiler, where it is burned with pulverized coal and provides between 10 and 20% of the total fuel requirements of the boiler. The spreader-stoker boilers can accept the RDF as supplementary fuel whenever the suspension-fired boiler is out of service for maintenance and inspection [75].

In addition to the above recovery functions, the plant provides a facility whereby anyone wishing to dispose of used oil may bring it to the plant in a separate container and this oil is placed in a 10,000 gallon (37,850 L) underground storage tank for later resale. The plant also has wood chipping equipment which processes logs, branches and scrap lumber that are brought in; wood chips are sold to local buyers.

The plant also has the ability to bale paper or cardboard. Bundled newsprint is accepted at the plant, and this material is resold as recycled paper or to companies processing materials for insulation.

The total reject materials sent to the landfill have averaged about 8.5% of the incoming refuse.

The economics of the Ames solid waste recovery system are presented in Table 11. These are figures for 1977 operations and indicate a net operating cost of between 11 and 12 dollars/ton of refuse handled.

Several process modifications to the plant have been required during 1978. These have included retrofitting the plant with dust collection equipment, and the suspension-fired boiler with dump grates. Also, two disk screens were installed between the first and second stage shredders. These serve to separate the glass and grit from the pulverized waste at an early stage; a very desirable step since these materials are very abrasive and have, in the past, caused severe abrasion of pneumatic lines and other mechanical devices subjected to them. Also, the early screening results in reduced power consumption in the secondary shredder.

The Ames facility has, since startup, consistently processed the surrounding communities' waste, and has produced fuel that is regularly burned in the city power plants' boilers.

F. Baltimore, Maryland

The Baltimore plant is based on the Landgard process, which is centered on the pyrolysis of shredded municipal waste and the subsequent in situ

Table 11
Ames Solid Wastes Recovery System Data[a]

Capital Cost:		
Construction & Engineering	$5.6 million	
Misc. Equipment and Start-up	0.5 million	
Land and Land Improvements	0.1 million	
Total	$6.2 million	
Salable Products: (2-year Average)	% of Total	
Refuse-derived fuel	84.3	
Ferrous metals	6.6	
Miscellaneous: wood chips, aluminum, baled paper, and used oil	Approx. 0.6	
Non-salable to landfill	Approx. 8.5%	
1977 Operations:		
Expenses: operation & maintenance	$582,382	
bond payments	465,352	
		$1,047,734
or $21.66 per ton (48,381 tons)		
Revenues:		
Fuel (RDF)	$353,325	
Metals	105,571	
Gate fees	24,964	
Wood chips	2,752	
Miscellaneous	12,014	
		$498,626
or $10.31 per ton		
1977 Shared deficit of operations		$549,108
or $11.35 per ton		

[a] From Chantland [76].

combustion of the pyrolysis products. Partially funded by the EPA and the Maryland Environmental Services, the plant was designed and constructed as a turn key operation by Monsanto Enviro Chem Systems Inc. for the City of Baltimore. The plant, scaled up from a 35 ton/day (32 metric tons/day) pilot plant operated by Monsanto at the time, was designed to receive, shred and pyrolyze 1000 tons/day (907 metric tons/day) of unsorted municipal refuse and to recover energy (in the form of steam), magnetic metals, glassy aggregate, and char. The plant is shown in Fig. 16.

Processing starts on a receiving floor where bulldozers push the refuse onto two conveyors that lead to two separate shredders, each 73 in.

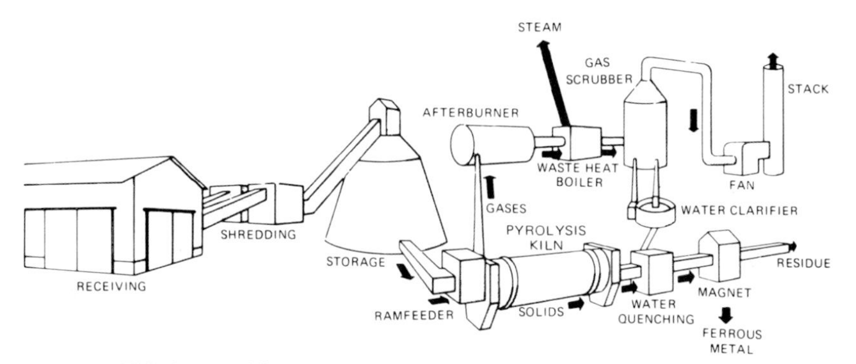

FIG. 16 Baltimore energy and materials recovery system.

(1.85 m) in diameter and 99 in. (2.5 m) long, driven by 900 hp (670 kW) motors.

The shredded refuse is stored in a conical bin with a 2000 ton (1814 metric tons) capacity. A drag-line bucket is used to undercut the refuse pile and move it out of the bottom of the bin to a kiln feed conveyor. The conveyor discharges to two ram feeders, which extrude the refuse through stainless steel tubes into the pyrolysis kiln. As the refuse tumbles down the inclined kiln, it is dried, volatilized, and partially combusted before being discharged at the lower end into a water quench bath.

The residue from the quench bath is removed by a drag conveyor and transferred to a flotation unit where the light char and ash float and are conveyed to a dewatering operation prior to storage of the char. The heavier materials (metals and glass) sink, and are conveyed to a belt magnet where the ferrous metals are separated for sale as scrap, and the glassy aggregate is transferred to a storage pile for use in asphalt road construction.

Fuel oil burners and air fans are located at the discharge end of the kiln to provide a flow of hot gases and combustion air countercurrent to the refuse flow. Air is provided at 40% of the stoichiometric combustion requirements, and approximately 7 gallons (27 L) of fuel oil per ton of refuse are consumed.

The offgases resulting from pyrolysis are partially combusted in the kiln, and provide the heat necessary to sustain the pyrolysis reactions at a temperature of 1200 °F (650 °C) for the offgases, and 2000 °F (1090 °C) for the solids.

The gaseous pyrolysis products exit the kiln at the feed end and proceed through a crossover duct where air is added to complete the combustion of the gases and any carbon and soot entrained in the gas stream. The gases enter an afterburner tangentially, and the resulting cyclonic flow pattern causes molten fly ash particles to be thrown to the vessel side walls from where they flow to a slag tap hole in the bottom of the afterburner.

Quench air is added to the afterburner exit gases to cool them below the ash fusion temperature. The gases then flow through two parallel waste heat boiler/economizer assemblies that generate about 200,000 lb (90,700 kg) steam/h at 100–260 psia (690–1790 kPa), for sale to a nearby utility.

Finally, the flue gases are drawn through a wet gas scrubber by an induced draft fan (the fan produces sufficient suction to draw the gases through the entire system), and are discharged through a humidifier to the atmosphere.

1. *Plant Performance*

The Baltimore plant has not operated as designed. Since startup, in January 1974, the plant had been plagued with numerous equipment malfunctions and shutdowns.

Many of the malfunctions were attributed to the rotary kiln performance, or rather nonperformance, and its effects on the local and downstream equipment and operations.

When the kiln for the Baltimore plant was designed from the prototype, the geometric scaling did not account for the combined aerodynamic and thermodynamic changes during the scaleup. As a consequence, the kiln process was unstable, and the resulting instability often caused kiln temperatures to exceed design temperatures by as much as 630 °F (330 °C). Furthermore, kiln temperatures would often fluctuate by as much as 300 °F (150 °C) over a 20 minute period, making the process virtually impossible to control. Consequently, the residue discharged from the kiln ranged in quality from fused slag balls 4 ft (1.2 m) in diameter to slightly burned paper. Such extremes in residue quality caused the discharge conveyor to be overloaded and the kiln to be shutdown on many occasions.

In addition to the high and fluctuating kiln temperatures, faulty refractory installation techniques and the differential thermal expansions in the kiln shell caused the refractory to spall and fall out.

The excessive kiln temperatures caused refuse metals to volatilize into aerosols. Along with these aerosols, large amounts of particulates were also introduced into the gas stream. In the kiln, fines within the shredded refuse were entrained by the countercurrent gas flow as they contacted refuse falling from the feeder tubes and also the refuse tumbling down the inclined kiln. The wet scrubber was unable to efficiently remove the submicron metal particles as well as the other particulates from the gas stream, and thus could not meet the state's particulate emissions standards. In addition, the induced draft fan being downstream of the scrubber was subjected to corrosive acids condensed in the flue gases. Also, solids accumulation on the impellers caused the fan to vibrate excessively. This caused weekly fan shutdowns to permit sandblasting and rebalancing.

In addition to kiln-related problems there were other operational problems. For example, the afterburner in the Baltimore plant was operated in a slagging mode as opposed to the nonslagging mode of the prototype. This variation in design did not anticipate nor provide for problems with the molten slag. As the molten slag passed through the tap hole at the bottom of the afterburner, it would be chilled by quench water thus causing it to solidify. The plugging of the tap hole at the bottom of the afterburner caused extensive downtime. To maintain slag flow, the temperature in the afterburner had to be increased by about 500 °F (260 °C) to 2500 °F (1370 °C), and this in turn caused frequent refractory failures in the afterburner.

Large variations in the heterogeneous refuse composition, density and moisture content, coupled with conservative design of the material handling equipment led to further problems and shutdowns. For example, the accumulated refuse in the conical bin would often cause a densified mass intertwined with rags and wire. Frequently, the refuse was so compacted it had to be removed manually. Compacted waste in the ram feeders frequently became so jammed that the waste could not be extruded into the kiln until the system was shutdown and the waste cleared away. Large agglomerations of fused slag jammed both the kiln and the afterburner residue conveyors to the point where the process had to be shutdown.

As a result of the various equipment difficulties and process deficiencies, Monsanto had difficulty meeting throughput capacity guarantees specified in a $4 million performance bond with the city of Baltimore (80% availability during a continuous 60 day test run). Furthermore, the plant failed to meet the Maryland particulates standard of 0.03 g/std. ft^3 (0. 07 g/std. m^3). In February 1977, Monsanto withdrew from the project thus forfeiting the $4 million performance bond, and recommended that the city convert the plant into a conventional incinerator. However, the city believed that the system had sufficient technical merit to warrant further investment to improve the system reliability.

2. *Plant Modifications*

Since the Baltimore plant had not operated as designed, it had to be extensively field-modified.

To improve the control and reliability of the kiln process, an air bustle was installed in the kiln firehood to uniformly distribute the incoming combustion air across the firing end of the kiln. The kiln refractory was replaced with upgraded materials and better installation techniques were used. Also, vessel skin cooling was introduced along with expansion slots inserted at the vessel ends. After these modifications, the kiln operated quite satisfactorily with no further downtime owing to refractory failure. However, the refuse metal volatilization in the kiln still continued.

An explosion suppression system was installed on the shredders, and the shredders themselves were vented to the atmosphere. Also, various improvements were made to the materials handling equipment.

The above modifications were performed prior to Monsanto's withdrawal from the project in early 1977. Following Monsanto's exit, the city assumed responsibility for the plant and continued operating at rates of 450–600 tons/day (410–545 metric tons/day).

Even though plant performance had improved substantially following the above modifications, no sustained operation was experienced owing to frequent shutdowns resulting from various other malfunctions. Consequently, the plant was shutdown in March 1978 to permit numerous extensive modifications to the facility. The future plant configuration will be changed appreciably as a result of these modifications.

Owing to its severe equipment wear and refuse retrieval difficulties, the use of the bin storage and recovery unit will be discontinued. The storage site will be reworked to permit access by a skiploader for assured retrieval of the shredded refuse. The original slagging afterburner will be used as a duct, and a new nonslagging afterburner will be placed in series with it. The wet scrubber, induced draft fan, and dehumidifier will be replaced with two dry electrostatic precipitators, two new induced draft fans, and a 220 ft (67 m) stack. The entire residue separation system (recovery of char, metals and glassy aggregate) will be eliminated owing to its questionable economics and the many modifications that would be necessary to make it functional.

While these modifications have a fair probability of being successful, it is likely that they will reduce somewhat the thermal efficiency of the plant. However, if the modifications substantially improve the plant reliability, the plant should prove to be an economical means of disposing of municipal waste [77]. The plant is expected to be on-line again by mid-1979.

3. *Summary*

The thermal efficiency of the Baltimore plant was approximately 50% for an average refuse feed rate of 30 tons/h (27 metric tons/h). The capital outlay for the plant has been $22 million. During the limited plant operation from the startup in January 1974 to the shutdown for major modifications in March 1978, the annual operating and maintenance cost was $3 million, and the annual steam revenue was $1 million. The net operating cost based on historical data was $58.20/ton (64.15/metric ton) of refuse processed. However, if the annual throughput of 74,000 tons (67,000 metric tons) could be substantially increased to the design level of 300,000 tons (272,000 metric tons) by optimizing the plant operation, operating costs could be reduced to $7.10/ton (7.80/metric ton) of refuse processed [78].

G. Summary

As an approach to resource recovery, mixed municipal refuse processing is shown to offer some attractive features. In addition to integrating easily into conventional waste collection and transfer operations, most such systems divert very large fractions of the total waste input, leaving a nonmarketable residue for landfilling of no more than 25% by weight, or 10% by volume.

The mixed-waste processing systems considered in the previous case studies have illustrated the application of technology in various novel configurations, and can be deemed successful demonstrations of commercial-scale resource recovery. Even though in most cases the processes required various alterations, such technical modifications are not entirely unusual in order to achieve effective operation in state-of-the-art demonstrations. In the case of the Baltimore project, the modifications were extensive, costly, and satisfactory results were often elusive. Nevertheless, the project provided invaluable experience in the use of innovative equipment and techniques, and the rotary kiln has been demonstrated to be an excellent primary reaction vessel.

The developmental and demonstration work presently underway should do much in the next few years to reduce uncertainties regarding technical performance and reliability of many of the proposed systems (see Table 7). The principal long-term questions relate more to issues of economic feasibility and the extent to which the new technologies can be made to compete with conventional land disposal methods on the one hand, and virgin material and fossil fuel supply sources on the other.

The subject of resource recovery economics will be discussed in the next section.

V. THE ECONOMICS OF RESOURCE RECOVERY

In the preceding sections we have considered the esthetic and environmental rationale for resource recovery (i.e., reduced raw material consumption and reduced waste disposal), have examined the composition and quantities of municipal solid waste that suggest it is indeed an "urban ore" worthy of harvesting, and have been apprised of the fact that many unit operations familiar in chemical, mechanical, and minerals processing engineering can be, and already have been applied to solid waste processing and resource recovery (as described in the case studies). Nevertheless, the unmistakable fact remains that large-scale and widespread recycling and resource recovery will occur only when the economics justify it. This means that the resource recovery costs to the community must at least be competitive with the more traditional means of solid waste disposal such as incineration and landfill.

The first step in the economic analysis of resource recovery is to determine the capital and operating costs of the technology to be installed. However, at present such data are hard to obtain, and whatever data do become available are too scant and inconsistent to base an economic analysis on. It is, however, possible to estimate realistic capital and operating costs and byproduct (recovered resources) revenues, and to make a critical economic assessment based on such estimates. One such assessment has been presented in a recent study and will be discussed at length below.

A. Cost–Benefit Analysis of Resource Recovery

A study conducted by Franklin Associates [3] presents an overview of the costs and benefits of resource recovery for the 150 largest standard metropolitan statistical areas (SMSAs) in the United States, using 1985 as a target year for development potential (including SMSAs of over 200,000 population). This view encompasses 62.4% of the nation's municipal solid waste generation. The basis of this cost–benefit analysis was as follows: All capital investment, operating costs, and revenues from recovered resources were based on 1974 economic data. The population basis, and waste quantity and composition were projected to 1985 on the basis that facilities planned and constructed in 1975 would anticipate conditions ten years hence. Thus the potential recovery of materials and energy is related to a base year of 1985. Finally, the

Table 12
Basic Municipal Solid Waste Generation Factors, 1975–1990 (Projected)[a]

Year	Location	Municipal solid waste generated[a] (1000 tons)[b]	Total U.S. population[c] (thousand)	Waste generation		
				tons per person per year	lb per person per year[d]	lb per person per day[e]
1975						
	USA	140,000	215,000	0.651	1,300	3.57
	SMSAs	96,800	134,300	0.721	1,440	3.95
1980						
	USA	160,000	224,000	0.714	1,430	3.91
	SMSAs	110,400	139,800	0.790	1,580	4.33
1985						
	USA	180,000	233,500	0.771	1,540	4.22
	SMSAs	124,400	145,800	0.853	1,705	4.67
1990						
	USA	200,000	243,000	0.823	1,645	4.51
	SMSAs	138,100	151,800	0.910	1,820	4.99

[a] From Franklin [4], Table 11.

[b] 1 ton = 0.9072 metric ton.

[c] From Franklin Associates, Ltd., based on U.S. Bureau of Census, U.S. Department of Commerce, Series E Population Projections.

[d] SMSAs basic generation rates were calculated on 10.7% above total U.S. average waste generation, based upon Ref. 4, Appendix 2, p. 8.

[e] Based on 365 days per year.

Note: The SMSA population was calculated each year on the basis of percentage increase in total U.S. population, with the 1970 census as the base year.

costs and benefits were contrasted with the costs of conventional disposal techniques such as landfill and incineration.

The calculated results of this study were derived from national average data for waste generation and composition based upon the studies in Ref. 4, 79, and 80. Only municipal waste was included in the study. The data are summarized in Table 12.

The economic (revenue) potential of municipal waste was determined as follows: First, the 1985 waste composition was estimated (third column, Table 13). Then the recoverable fractions were identified, i.e., ferrous metal, aluminum, glass, and paper. The mixed combustibles are assumed to be an energy source directly as fuel, converted to steam, or pyrolysis oil or gas. Since current techniques cannot achieve 100% recovery, recovery efficiency factors were assigned as follows: ferrous metal, 0.85; aluminum, 0.62; glass, 0.65; recyclable paper, 0.23. The energy content was calculated at the "higher heating value" of waste. This was assumed at 4500 Btu/lb (10.47 MJ/kg) or 5000 Btu/lb (11.63 MJ/kg) for the combustible fraction of wastes. Finally, the potential value of a recovered product to an industrial purchaser was determined. These product values were estimated on the basis of the prevailing (1974) prices and adjusted to reflect an allowance for freight. The resulting revenues are shown in Table 13. The metals and glass are worth $4.57 per input ton of waste. The composite value of municipal waste is $7.74 per input ton, if the combustible fraction is used as fuel supplement to coal-fired boilers. Conversion of the combustible (organic) fraction to pyrolysis oil or gas gives the waste a value of $13.32 per input ton. If the combustible fraction is converted to steam, the value of the composite waste rises to $19.57 per input ton.

The economic parameters of the resource recovery facilities depend on the configuration of these facilities. The Franklin study estimated that the most prevalent configuration (for 1985) would be that of materials recovery (ferrous metal, aluminum, and glass) with the shredded organic fraction being used as a fuel supplement to coal fired boilers. The next most prevalent configuration would be materials recovery with the organic fraction being converted to steam. The third most prevalent configuration was estimated to be materials recovery followed by pyrolysis. Tables 14–19 present itemized capital investment and operating costs (in 1974 dollars) for 1000 tons/day resource recovery systems encompassing the above three configurations. To determine the total capital investment for a given resource recovery configuration it is necessary to add the values in Tables 14 and 16 (materials recovery and conversion to steam) and Tables 14 and 18 (materials recovery and pyrolysis); Table 14 stands alone if the energy product is shredded fuel

Table 13
Recoverable Fractions in Mixed Municipal Waste, 1985[a]
(from 100,000 tons of Waste)

Waste component	Quantity generated	Percent of waste	Recoverable quantity	Percent recoverable	Value ($/ton)[b]	Total value ($)	Value ($/input ton)	Value adjusted for freight ($/ton)	Adjusted value	Value adjusted ($/input ton)
Ferrous metal	8,500	8.5	7,200	7.2	35	252,000	2.52	30	216,000	2.16
Aluminum	800	0.8	500	0.5	300	150,000	1.50	300	150,000	1.50
Glass	9,300	9.3	6,100	6.1	21	128,100	1.28	15	91,500	0.91
							5.30			4.57
Paper[f]	32,000	32.0	7,500	7.5	20	150,000	1.50	20	150,000	1.50
Other combustibles[f]	47,200	47.2	66,700	66.7	5[c]	333,500	3.33	2.50[c]	166,750	1.67[c]
Subtotal	97,800	97.8	88,000	88.0		937,500	10.13		774,250	7.74
Other inorganic and nonrecoverable	2,200	2.2	12,000[g]	12.0[g]						
Total	100,000	100.0	100,000	100.0		937,500	10.13[d]		774,250	7.74[e]
							14.05[d]			13.32[e]
							20.30[d]			19.57[e]

[a] From Franklin [3].

[b] FOB recovery plant.

[c] Shredded combustible refuse for use in coal fired boilers. Pyrolysis oil and gas is valued at about \$8.75 per total input ton (\$1.75/10^6 Btu × 5 million Btu recoverable per ton); steam is valued at \$15.00 per input ton or \$2.50 per 1000 lb of steam and 6000 lb/ton of refuse.

[d] If the system is a pyrolysis unit, the income is \$5.30/ton plus \$8.75 = 14.05/input ton (see footnote [c]); if the system is a steam generating facility, the income is \$5.30 plus \$15.00 = 20.30/input ton.

[e] If pyrolysis; income = \$13.32/input ton. If steam: income = \$19.57/ton input. For pyrolysis and steam, it was assumed that paper recovery does not take place.

[f] This tonnage and percentage is expressed on the basis of raw refuse generated, i.e., without deduction of moisture content of about 25% of as-received tonnage. Likewise, the Btu values are on the same basis.

[g] This is unrecoverable and includes ferrous metal, glass, aluminum, and ash content of organics that are also unrecoverable.

Note: This waste is on a forecast composition for 1985, not the current composition. The rather high aluminum content is based upon a substantial growth in the demand for beverage containers over present use.

supplement. Similarly the operating costs are determined by adding Tables 15 and 17 (materials recovery and conversion to steam); Tables 15 and 19 (materials recovery and pyrolysis); Table 15 stands alone when the energy product is shredded fuel supplement.

A number of assumptions were necessary in developing the costs shown in Tables 14–19. The recovery facilities were assumed to be municipally owned and operated. Tax-free municipal bonds can be sold at about 7% annual rate. Municipalities also have an advantage over private owners with respect to property and income taxes. A 1000 ton/day capacity facility was used as the basis for the economic analysis. After capital investment was estimated appropriate scale factors were used to estimate costs for other plant sizes. The operating schedule was assumed to be 260 days/yr and 24 h/day. Capital investment is divided

Table 14
Capital Requirements in Dollars for 1000 t/day Shredding and Materials Recovery Plant (Shredded Organic Fraction Used as Fuel Supplement to Coal-Fired Boilers)[a]

Investment	Throughput rate, TPY[c]	
	260,000	360,000
Amortized		
Engineering, R&D (12% × FPI)	1,416,000	1,416,000
Plant start-up (2 months DOE[b])	222,000	356,000
Total	1,638,000	1,772,000
Fixed plant (FPI)[d]		
Receiving and outloading facilities	1,500,000	
Materials handling and storage	1,600,000	
Processing plant	5,000,000	
Controls, instrumentation, and electrical	200,000	
Miscellaneous equipment	800,000	
Ferrous metal recovery	700,000	
Aluminum and glass recovery	2,000,000	
Total	11,800,000	11,800,000
Recoverable		
Land (7.5 acres at $50,000)	375,000	375,000
Working capital (3 months DOE)	333,000	533,000
Total	708,000	908,000
Total capital requirement	14,146,000	14,480,000

[a] From Franklin [4].
[b] Direct operating expenses.
[c] Tons per year
[d] Fixed plant investment.
Note: Abbreviations apply to all Tables.

into (a) amortized investment, (b) fixed investment, and (c) recoverable investment.

Amortized investment includes engineering costs, research and development expenditures, and allowance for a period of plant startup. The capital requirements in this category are not depreciable and cannot be written off as operating expenses during the year in which they are incurred. The assumption here is that they will be charged off over a 5-yr period, with interest computed at an annual rate of 7%.

Fixed investment represents the direct costs of plant facilities; these

Table 15

Annual Cost of Operation in Dollars for 1000 t/day Shredding and Materials Recovery Plant (Shredded Organic Fraction Used as Fuel Supplement to Coal-Fired Boilers)[a]

Expenses		Throughput rate, TPY	
		260,000	360,000
Direct operating expenses			
Operating labor	36 at 10,000	360,000	
Maintenance labor	18 at 12,000	216,000	
Direct supervision	9 at 14,000	126,000	
Total labor		702,000	
Payroll loadings at 30%		211,000	
Total labor plus loadings		913,000	
Other direct expenses		420,000	
Total DOE		1,333,000	2,133,000
Fixed plant overhead			
Plant superintendent(s)	1 at 20,000	20,000	
Engineers	1 at 16,000	16,000	
Laboratory technicians	0.5 at 12,000	6,000	
Total overhead labor		42,000	
Payroll loadings at 30%		13,000	
Total overhead labor plus loading		55,000	88,000
Other fixed expense			
Capital related at 0.01 × total investment		141,000	145,000
Operating related at 0.05 × total DOE		67,000	107,000
Total other fixed expense		208,000	252,000
Total fixed plant overhead		263,000	340,000
Capital charges			
Amortized investment		400,000	432,000
Fixed plant investment at 11.4%		1,345,000	1,345,000
Recoverable investment at 7.0%		50,000	64,000
Total capital charges		2,095,000	1,841,000
Total annual cost of operation		3,691,000	4,314,000

[a] From Franklin [4].

were, in turn, placed into four broad categories: (a) waste handling, preparation, and storage equipment and facilities, including, wherever, appropriate, the equipment and machinery needed for crushing, grinding, or otherwise sizing the raw wastes for further processing; (b) facilities for waste conversion, whether accomplished by combustion, mechanical, or other processes; (c) resource recovery processes, involving in some cases the physical or mechanical sorting of marketable waste components, and in others the recovery and processing or utilization of combustion products; and (c) auxiliary and support facilities covering structures, utilities, environmental control equipment,† and all other plant requirements. In converting total fixed investment to its equivalent annual cost, a rate of 11.4% was applied (capital recovery at 5% interest over 20 yr or 9.4%/yr) plus insurance, administrative, and general expenses at 2.0%/yr of fixed investment.

Recoverable investment encompasses those requirements for

Table 16
Capital Requirements, in Dollars for 1000 t/day Materials Recovery and Steam Generating Plant[b]

Investment	Throughput rate, TPY	
	260,000	360,000
Amortized		
Engineering, R&D (12% × FPI)	1,680,000	1,680,000
Plant startup (2 months DOE)	248,000	397,000
Total	1,928,000	2,077,000
Fixed plant (FPI)		
Boiler unit	6,000,000	
Installation	2,400,000	
Environmental control systems	1,400,000	
Auxiliary and support facilities	3,200,000	
Steam distribution	1,000,000	
Total	14,000,000	14,000,000
Recoverable		
Land (5 acres at $50,000)	250,000	250,000
Working capital (3 months DOE)	372,000	595,000
Total	622,000	845,000
Total capital requirement	16,550,000	16,922,000

[a] Tons per day.
[b] From Franklin [4].

† This includes landfilling of the 12% unrecoverable fraction (see footnote *g* to Table 13).

capital that, on completion of the project, will be returned to the project's owners. In this case, the recoverable investment tied up in the project includes land and sufficient working capital to cover the system's costs of operation for a 3-month period.

The operating costs of each system includes (a) direct operating costs, i.e., labor, materials, supplies, utilities, variable overhead and other expenses related directly to the amount of waste material handled; (b) fixed costs, i.e., those incurred with time rather than with the scale of plant operation, and covering the costs of maintaining and administering the overall operation; and (c) capital charges, i.e., those directly attributable to the amount of capital tied up in the facility. These

Table 17

Annual Cost of Operation in Dollars for 1000 t/day Materials Recovery and Steam Generation Plant[a]

Expenses		Throughput rate, TPY	
		260,000	360,000
Direct Operating			
Operating labor	30 at 10,000	300,000	
Maintenance labor	15 at 12,000	180,000	
Direct supervision	9 at 14,000	126,000	
Total labor		606,000	
Payroll loadings at 30%		182,000	
Total labor plus loadings		788,000	
Other direct expenses		700,000	
Total DOE		1,488,000	2,381,000
Fixed plant overhead			
Plant superintendent(s)	1 at 20,000	20,000	
Engineers	2 at 16,000	32,000	
Total overhead labor		52,000	
Payroll loadings at 30%		16,000	
Total overhead labor plus loading		68,000	109,000
Other fixed expense			
Capital related at 0.01 × total investment		167,000	168,000
Operating related at 0.05 × total DOE		74,000	119,000
Total other fixed expense		241,000	287,000
Total fixed plant overhead		309,000	396,000
Capital investment			
Amortized at 24.4%		470,000	507,000
Fixed plant at 11.4%		1,596,000	1,596,000
Recoverable at 7.0%		44,000	59,000
Total		2,110,000	2,162,000
Total annual cost of operation		3,907,000	5,050,000

[a] From Franklin [4].

capital charges, in turn, cover capital recovery on amortized investment (depreciation and interest); insurance and administrative charges applied to fixed plant; and interest on the recoverable investment.

Operating costs for each system were developed in the same way as fixed investment, by determining the direct labor, materials, and utility requirements associated with each major system function. Total direct operating costs for a particular system then were found by totaling the separate costs for the appropriate functions of processes required in the system's operation.

The capital investment requirements for three different recovery configurations and seven different facility sizes (capacities) are shown in Table 20; also shown are the related operating costs (all costs in 1974 dollars). As can be seen, the capital requirements range from a low of \$8.8 million for a 500 tons/day facility to a high of \$66.8 million for a 3000 tons/day facility. Furthermore, the operating costs vary from \$2.2 million/yr to \$15.5 million/yr.

The revenue values developed in Table 13 were applied to the annual throughput of each plant and the net operating costs in dollars and dollars per ton of input waste were determined. Note that the larger the system, the lower the net operating cost.

Table 18
Capital Requirements in Dollars for 1000 t/day Materials Recovery and Pyrolysis System[a]

	Throughput rate, TPY	
Investment	260,000	360,000
Amortized		
Engineering, R&D	1,200,000	1,200,000
Plant startup	188,000	300,000
Total	1,388,000	1,500,000
Fixed Plant (FPI)		
Pyrolysis units (installed)	4,700,000	
Energy recovery system	1,500,000	
Environmental controls	1,400,000	
Auxiliary and support facilities	2,400,000	
Total	10,000,000	10,000,000
Recoverable		
Land	250,000	250,000
Working capital	282,000	451,000
Total	532,000	701,000
Total capital requirement	11,920,000	12,201,000

[a] From Franklin [4].

In order to compare the costs of resource recovery with conventional solid waste disposal methods, a composite average disposal cost (Table 21) was developed. This cost of $7.10 (1974 dollars) was close to the landfill cost that most cities face if they acquire new sites away from the central city.

The Franklin study furthermore estimates the number of potential 1985 resource recovery facility installations by geographical region and facility sizes. First, the approximate population base required to support various size facilities, from 500 to 5000 tons/day, was estimated with

Table 19
Annual Cost of Operation in Dollars for 1000 t/day Materials Recovery Pyrolysis System[a]

Expenses		Throughput rate, TPY	
		260,000	360,000
Direct Operating			
Operating labor	22 at 10,000	220,000	
Maintenance labor	11 at 12,000	132,000	
Direct supervision	6 at 14,000	84,000	
Total labor		436,000	
Payroll loadings at 30%		131,000	
Total labor plus loadings		567,000	
Other direct expenses		560,000	
Total DOE		1,127,000	1,803,000
Fixed plant overhead			
Plant superintendent(s)	1 at 20,000	20,000	
Engineers	2 at 16,000	32,000	
Laboratory technicians	2 at 12,000	24,000	
Total overhead labor		76,000	
Payroll loadings at 30%		23,000	
Total overhead labor plus loading		99,000	158,000
Other fixed expense			
Capital related at 0.01 × total investment		119,000	122,000
Operating related at 0.05 × total DOE		56,000	90,000
Total other fixed expense		175,000	212,000
Total fixed plant overhead		274,000	370,000
Capital investment			
Amortized at 24.4%		339,000	366,000
Fixed plant at 11.4%		1,140,000	1,140,000
Recoverable at 7.0%		37,000	49,000
Total		1,516,000	1,555,000
Total annual cost of operation		2,917,000	3,728,000

[a] From Franklin [4].

Table 20
Investment Requirements for Basic Resource Recovery Options by System Capacity and Recovery Configuration Alternatives[a]

Size facility TPD and TPY throughput[b]	Description	Basic configuration and investment ($1000)				Annual cost of operation ($1000)	Operating revenue ($1000)	Net operating cost ($1000)	Net operating cost ($/input t)	Number of employees required
		Transfer stations	Materials separation and shred	Energy recovery process	Total investment					
500 TPD 130,000 TPY	Shredding, materials recovery, and fuel prep.	0	8,800		8,800	2,200	1,006	1,194	9.18	44
1,000 TPD 260,000 TPY	Shredding, materials recovery, and fuel prep.	0	14,200		14,200	3,700	2,012	1,688	6.49	65
1,500 TPD 390,000 TPY	Shredding, materials recovery, and fuel prep.	0	18,800		18,800	4,500	3,018	1,482	3.80	86
2,000 TPD 520,000 TPY	Shredding, materials recovery, and fuel prep.	1,800	23,000		24,800	5,900	4,024	1,876	3.61	114
3,000 TPD 780,000 TPY	Shredding, materials recovery, and fuel prep.	1,800	30,500		32,300	7,500	6,036	1,464	1.88	135
4,000 TPD 1,040,000 TPY	Shredding, materials recovery, and fuel prep.	3,600	37,200		40,800	10,200	8,050	2,150	2.07	162

5,000 TPD 1,300,000 TPY	Shredding, materials recovery, and fuel prep.	5,400	43,700		49,100	11,800	10,062	1,738	1.34	183
1,000 TPD 260,000 TPY	Shredding, materials recovery, and steam generation	0	14,200	16,600	30,800	7,600	5,088	2,512	9.66	122
2,000 TPD 520,000 TPY	Shredding, materials recovery, and steam generation	1,800	23,000	27,100	51,900	11,900	10,176	1,724	3.32	192
3,000 TPD 780,000 TPY	Shredding, materials recovery, and steam generation	1,800	30,500	34,500	66,800	15,500	15,265	235	0.30	243
1,000 TPD 260,000 TPY	Shredding, materials recovery, and pyrolysis	0	14,200	11,900	26,100	6,400	3,463	2,937	11.30	107
2,000 TPD 520,000 TPY	Shredding, materials recovery, and pyrolysis	1,800	23,000	19,200	44,000	10,600	6,926	3,674	7.07	182
100,000 TPY	Sanitary landfill	0			1,800	400 to 700	0	300	4.00 to 7.00	9

[a] From Franklin [3].

[b] All systems are based upon 260 days per year operation (5 days/week) 24 h/day; however, all systems can be operated more days as required without further capital investment.

Table 21
Calculation of Composite Disposal Costs Using Conventional Incineration, Remote Landfill, and Close-in Landfills in Large Metropolitan Areas [a]

Type of disposal	Fraction of waste	Average cost ($/ton)	Composite cost of disposal ($/ton)
Incineration	0.20	12.75	2.55
Landfill Remote	0.45	7 [b]	3.15
Close by	0.35	4 [c]	1.40
Total	1.00	—	7.10

[a] From Franklin [3].
[b] Consists of: transfer station operation at $1.55/t, haul costs as $2.50/t, and landfill operation at $3.00/t. Assumes a haul distance of 20 miles from transfer station.
[c] Assumes land is already owned by agency disposing of waste; no transfer station and location convenient to packer truck routes.

the facility size based on the operating rate of 24 h/day at 260 and 300 days/yr. The data projected for 1985 are shown in Table 22. Next a profile was drawn of the 150 largest SMSAs in the United States. This profile included the 1985 population, waste processable each year.

Table 22
Population Base and Operating Characteristics by Resource Recovery Facilities Capacity [a]

Rated system capacity, TPD	Waste throughput at 260 DPY [c]	SMSA population base servable 1985 [b]	Waste throughput at 300 DPY	SMSA population base servable 1985 [b]
500	130,000	152,000	150,000	176,000
750	195,000	228,000	225,000	264,000
1,000	260,000	305,000	300,000	352,000
1,500	390,000	457,000	450,000	528,000
2,000	520,000	610,000	600,000	703,000
3,000	780,000	915,000	900,000	1,055,000
4,000	1,040,000	1,220,000	1,200,000	1,407,000
5,000	1,300,000	1,525,000	1,500,000	1,758,000

[a] From Franklin [3].
[b] Calculated at a waste generation rate in SMSAs of 4.67 lb/per person per day in 1985 or 0.853 t per person per year.
[c] DPY = days per year.

Table 23
Summary of Potential 1985 Resource Recovery Installations by Region and Size[a]

Region	Population (million)	Waste available for processing: 10^6 TPD	Waste available for processing: lb/capita/day[a]	Plants, TPD: 500	1000	2000	3000	4000	5000	Total plants	Capacity: 10^3 TPD	Capacity: 10^6 TPY	Capacity: Average 10^3 TPD plant	Capacity: Daily per capita (lb)
New England	8.00	6.14	4.20	6	6	2		2		16	21.0	5.46	1.3	3.7
Middle Atlantic	34.42	26.39	4.20	3	12	6	4	3	9	37	94.5	24.6	2.5	3.9
South Atlantic	18.83	14.47	4.21	10	15	9		2	1	37	51.0	13.3	1.4	3.9
East South Central	5.31	4.09	4.22	7	5	3				15	14.5	3.77	0.97	3.9
West South Central	10.82	8.35	4.22	3	9	6	1	1		20	29.5	7.67	1.5	3.9
East North Central	30.05	22.94	4.18	8	17	9	1	4	5	44	83.0	21.6	1.9	3.9
West North Central	8.30	6.39	4.22	1	5	2	3	1		12	22.5	5.85	1.9	3.9
Mountain	4.48	3.45	4.22		5	4				9	13.0	3.38	1.4	4.1
Pacific (including Hawaii)	25.60	19.64	4.20	7	13	6	3	2	5	36	70.5	18.3	2.0	3.9
Total	145.81	111.86	4.20	45	87	47	12	15	20	226				
Total capacity (10^3 TPD)				22.5	87	94	36	60	100		399.5	103.9	1.8	3.9
Percent total plants				19.9	38.5	20.8	5.3	6.6	8.8	99.9				
Percent total capacity				5.6	21.8	23.5	9.0	15.0	25.0	99.9				

[a] From Franklin [3].

the number of facilities (by size) that could be installed in each metropolitan area, and tonnage that could be annually processed for recovery. The details for each metropolitan area would be too lengthy to reproduce here (e.g., the St. Louis area with a population in 1985 of 2.7 million people would generate 2.07 million tons of processable municipal solid waste annually, and could support one 3000 tons/day plant and one 4000 tons/day plant; similarly, the Boston area with 3.14 million people in 1985 would generate 2.4 million tons of municipal waste annually, and could support two 4000 tons/day plants and one 1000 tons/day plant). However, a regional summary is shown in Table 23. These projections estimate that the 150 largest SMSAs would serve 145.8 million people in 1985; would be able to process 112 million tons solid waste annually, and would require a total of 226 waste processing plants.

Based on all the above data, estimates, and projections, a national picture of resource recovery was drawn. The results are presented in Tables 24–27, and represent one particular configuration profile out of several that could be developed. These results indicate that 100% of the projected development of resource recovery facilities (226 facilities) would require that \$6.3 billion be invested and that \$1.5 billion be expended annually on operating expenses; the revenue potential would

Table

Summary of Resource Recovery Potential in 1985 in the 150 Largest

System option	Type of resource recovery facility	Capacity (TPD)	Number in place	Capital invest-ment (\$1000)	New employ-ment	Annual operating cost (\$1000)	Annual revenue (\$1000)
1	Materials and fuel	500	45	396,000	1,980	99,000	45,270
2	Materials and fuel	1,000	40	568,000	2,600	148,000	80,480
3	Materials and fuel	2,000	20	496,000	2,280	118,000	80,480
4	Materials and fuel	3,000	6	193,800	810	45,000	36,216
5	Materials and fuel	4,000	15	612,000	2,430	153,000	120,750
6	Materials and fuel	5,000	20	982,000	3,660	236,000	201,240
	subtotal		146	3,247,800	13,760	799,000	564,436
7	Materials and steam	1,000	28	865,200	3,415	215,600	142,464
8	Materials and steam	2,000	12	622,800	2,304	142,800	122,112
9	Materials and steam	3,000	6	400,800	1,458	83,000	91,590
	Subtotal		46	1,888,800	7,177	451,400	356,166
10	Materials and pyrolysis	1,000	19	493,900	2,033	121,600	65,797
11	Materials and pyrolysis	2,000	15	660,000	2,730	159,000	103,890
	Subtotal		34	1,155,900	4,763	280,600	169,687
	Total		226	6,292,500	25,700	1,531,000	1,090,289
	Product revenue summary (\$1000)						1,090,289
	Composite conventional disposal Costs, landfill and incineration					737,477	0

[a] From Franklin [3].
[b] It was assumed that paper for recycling would be recovered only from materials and shredded fuel systems; not from
[c] The steam and pyrolysis systems are assumed to be operated on a 7 day/week schedule, on a throughput of 5-day

be $1.09 billion and the composite net operating cost would be $4.24 per ton of input waste which compares very favorably against the disposal cost of $7.10 per ton. Even if 25% of the possible recovery facilities were built and operated (Table 25) $1.66 billion in capital investment would be required along with an annual expenditure of $403 million for operating costs. The revenue potential would be $288 million and the net operating cost would be $4.18 per input ton of waste, which also compares favorably against the alternative of $7.10 per ton for disposal.

This study, although hypothetical, is quite reasonable and is a useful exercise for highlighting the various factors that bear on the economics of resource recovery. The most obvious finding of the above economic analysis is that resource recovery systems are not self-sustaining economic operations, i.e., they do not recover revenue sufficient to offset total costs but instead show a net cost of operation. This is true of the individual systems considered in Table 20 and also the nationwide configurations shown in Tables 24 and 25. However, where incineration, remote landfill, or other conventional waste disposal is necessary, resource recovery offers an economically viable alternative. All of the resource recovery systems analyzed show operating costs (see Table 20)

24

Metropolitan Areas (Based Upon 100% Implementation of Maximum Possible)[a]

Net annual operating cost ($1000)	Net annual operating cost ($/ton)	Annual waste throughput (1000 tons)	Annual recovery by type of product (1000 tons)					
			Ferrous	Aluminum	Glass	Paper[b]	Combustible	
53,730	9.18	5,850	421	29	357	439	3,902	
67,520	6.49	10,400	749	52	634	780	6,937	
37,520	3.61	10,400	749	52	634	780	6,937	
8,784	1.88	4,680	337	23	285	351	3,121	
32,250	2.07	15,600	1,123	78	952	1,170	10,405	
34,760	1.34	26,000	1,872	130	1,586	1,950	17,342	
234,564	3.22	72,930	5,251	364	4,448	5,470	48,644	
73,136	10.05[c]	7,280	524	36	444		5,402	
20,688	3.32[c]	6,240	449	31	381		4,630	
1,410	0.30	4,680	337	23	286		3,473	
95,234	5.23	18,200	1,310	80	1,111		13,505	
55,803	11.30[c]	4,940	356	25	301		3,665	
55,110	7.07[c]	7,800	562	39	476		5,787	
110,913	8.71	12,740	918	64	777		9,452	
440,711	4.24	103,870	7,479	508	6,336	5,470	71,601	
			224,359	155,745	94,522	109,395	121,793	fuel
							273,000	steam
							111,475	pyrolysis
							506,268	
737,477	7.10	103,870						

steam or pyrolysis systems.
volume. Increasing the throughput on these systems would lower the per ton operating cost significantly.

that are lower than conventional incineration ($12–15 per input ton of waste), several have net costs low enough to compete with the composite waste disposal cost ($7.10 per input ton of waste), and some even compete with landfill costs ($4–7 per input ton of waste).

The cost figures indicate that most resource recovery systems are capital intensive, i.e., a large capital investment is required for each system. Therefore, the fixed costs of operation are quite high in relation to total costs. These systems should be operated at or near capacity to minimize unit costs and to maximize salable product output. In addition, the systems show economies of scale, so that the larger the system, the more attractive the net operating cost. However, it should also be borne in mind that the raw material of resource recovery systems is mixed municipal wastes, and a plant must be sized to serve its raw materials supply. Thus even though economies of scale favor the larger plant size the actual plant size is dictated by the resource base, i.e., by the population to be served (city or regional area), the rate and type of waste generated, and the rate of waste collection and delivery to the central processing plant.

The Franklin study has addressed the 1985 resource recovery

Table
Summary of Resource Recovery Potential in
(Based Upon 25% Implementation

System option	Type of resource recovery facility	Capacity (TPD)	Number in place	Capital Invest-ment ($1000)	New employ-ment	Annual operating cost ($1000)	Annual revenue ($1000)
1	Materials and fuel	500	12	105,600	528	26,409	12,072
2	Materials and fuel	1,000	10	142,000	650	37,000	20,120
3	Materials and fuel	2,000	5	124,000	570	29,500	20,120
4	Materials and fuel	3,000	2	64,600	270	15,000	12,072
5	Materials and fuel	4,000	4	163,200	648	40,800	32,200
6	Materials and fuel	5,000	5	245,500	915	59,000	50,310
	Subtotal		38	844,900	3,581	207,700	146,894
7	Materials and steam	1,000	7	216,300	354	53,900	35,616
8	Materials and steam	2,000	3	155,700	576	35,700	30,528
9	Materials and steam	3,000	2	133,600	286	31,000	30,530
	Subtotal		12	505,600	1,716	120,600	96,674
10	Materials and pyrolysis	1,000	5	130,500	535	32,000	17,315
11	Materials and pyrolysis	2,000	4	176,000	728	42,400	27,704
	Subtotal		9	306,500	1,263	74,400	45,019
	Total		59	1,657,000	6,560	402,700	288,587
	Product revenue summary ($1000)						288,587
	Composite conventional disposal Costs, landfill and incineration					193,840	0

[a] From Franklin [3].
[b] It was assumed that paper for recycling would be recovered only from materials and shredded fuel systems; not from
[c] The steam and pyrolysis systems are assumed to be operated on a 7 day/week schedule, on a throughput of 5-day

potential on a broad national scale in terms of SMSAs, whereas in reality resource recovery is governed by local political jurisdictional boundaries. Many areas could be expected to fragment by political jurisdiction within an SMSA and, therefore would support a smaller plant than is indicated by the total population within the SMSA itself. Thus, few 5000 tons/day plants may be justifiable from a strictly plant economic viewpoint, and 500–1000 tons/day plants may be the principal sizes to be expected in practice. Consequently the economics of these "lower to middle range" plant sizes are most important. On the other hand, a 5000 tons/day plant providing materials recovery and auxiliary shredded fuel, processes ten times as much waste as a 500 ton/day plant, with only 5.6 times as much investment and at about 15% of the net operating cost (see Table 20). Thus it can be safely suggested that resource recovery should be undertaken on a "grand scale" with much cooperation between local governments, otherwise only the very largest cities would appear attractive for resource recovery plants.

Perhaps the most critical economic factor is the marketability of the products, i.e., the recovered resources. All of the resource recovery techniques lead to products that must compete with established

25

1985 in the 150 Largest Metropolitan Areas[a]

of Maximum Possible)

Net annual operating cost ($1000)	Net annual operating cost ($/ton)	Annual waste throughput (1000 tons)	Annual recovery by type of product (1000 tons)					
			Ferrous	Aluminum	Glass	Paper[b]	Combustible	
14,328	9.18	1,560	112	8	95	117	1,041	
16,880	6.49	2,600	187	13	159	195	1,734	
9,380	3.61	2,600	187	13	159	195	1,734	
2,928	1.88	1,560	112	8	95	117	1,041	
8,600	2.07	4,160	300	21	254	312	2,775	
8,690	1.34	6,500	468	32	396	487	4,335	
60,805	3.20	18,980	1,366	95	1,158	1,423	12,660	
18,284	10.05[c]	1,820	131	9	111		1,351	
5,172	3.32[c]	1,560	112	8	95		1,158	
470	0.30	1,560	112	8	95		1,158	
23,926	4.84	4,940	355	25	301		3,667	
14,685	11.30[c]	1,300	94	6	79		965	
14,696	7.07[c]	2,080	150	10	127		1,543	
29,381	8.69	3,380	244	16	206		2,508	
114,113	4.18	27,300	1,965	136	1,665	1,423	18,835	
			58,968	40,936	24,841	28,470	31,697	fuel
							74,100	steam
							29,575	pyrolysis
							135,372	total
193,840	7.10	27,300						

steam or pyrolysis systems.
volume. Increasing the throughput on these systems would lower the per ton operating cost significantly.

Table 26

Summary of Potential Resource Recovery Costs and Benefits in 150 Major Metropolitan Areas, 1985[a]

Extent of development	Number of recovery facilities	Annual throughput (1000 tons)	Total capital investment ($1000)	New employment	Annual operating cost ($1000)	Annual revenue ($1000)	Net annual cost ($1000)	Net annual cost ($/ton)	Population served (1000)
Full	226	103,870[b]	6,292,500	25,700	1,531,000	1,090,289	440,711	4.24	145,810
25%	59	27,300[c]	1,657,000	6,560	402,700	288,587	114,113	4.18	36,500

[a] From Franklin [3].
[b] This is 57.7% of total waste generated in the United States in 1985.
[c] This is 15.2% of total waste generated in the United States in 1985.

Table 27

Annual Recovery of Products and Value of Recovered Products in 150 Major Metropolitan Areas, 1985[a]

Extent of Development	Ferrous metal	Aluminum	Glass	Paper	Combustible	Total	Unrecovered
Full							
Tons (1000)	7,479	508	6,336	5,470	71,601	91,394	12,476
Dollars	224,359	155,745	94,522	109,395	506,268	1,090,298	
25%							
Tons (1000)	1,965	136	1,665	1,423	18,835	24,024	3,276
Dollars (1000)	58,968	40,936	24,841	28,470	135,372	288,587	

[a] From Franklin [3].

commodities directly or indirectly in the marketplace. Each product has a unique relation to its potential market, depending on quality, unit price, and quantity of recovery. Materials recovered for recycling are often relatively valuable but only if a market exists for the grade and quality of product derived from the recovery system. In the Franklin study it was assumed that recoverable resources would be obtained in the quantities and values summarized in Table 13. In order to realistically qualify the various resource recovery configurations in light of the highly variable and risky market conditions that are likely to prevail for recovered products and commodities, the sensitivity of each recovery configurations' economics to changes in the estimated resource revenues would be advisable. Such a sensitivity analysis can serve as a measure of the relative risk of resource recovery and may often emphasize the necessity for a thorough market study prior to undertaking any resource recovery project [81].

It is worth noting that local governments are often hesitant to experiment with new or unproved technology since this represents a radical departure from traditional waste management practices and introduces a "high risk" of taxpayer funds. Thus, in order to introduce technically and economically viable resource recovery systems, local governments will be required to adopt relatively sophisticated technology and competitive marketing skills.

In summary then, it must be emphasized that the optimistic results of the Franklin study depend on very stringent conditions requiring government, industry, and the public to form an alliance on a national scale. The public must be willing to support the investment; government and industry must be willing to make the shift from a waste-oriented disposal operation to a market-oriented industrial resource operation; and the manufacturing industry must be willing to purchase and utilize recycled materials. Furthermore, customers must be willing to buy all of the recovered products at the minimum price stipulated, and the recovery processes must meet the customers specifications for the products.

Only from the simultaneous fulfillment of these requirements will the benefits appear to outweigh the costs, and resource recovery will then emerge attractive not only from a resource conservation and environmental viewpoint (minimizing X_{RM} and X_D as discussed in the introductory section of this chapter) but also from an economic viewpoint.

B. Looking Ahead

Clearly, resource recovery is at a stage that causes one to look forward with cautious optimism. Until recently, most resource recovery technologies have not been able to compete economically with sanitary landfill or incineration. Now, however, certain economic factors have come into play to change this picture. For example, mandatory air pollution controls have made incineration a very costly disposal means; available landfill sites are becoming more and more scarce; the progressive increases in petroleum and natural gas prices have provided a market and substantial credit for solid-waste derived fuel. Such economic facts certainly brighten the prospects for resource recovery. However, to ensure long-term success, maximum attention must be directed to creation of financial incentives, correction of economic inequities, creation of end product markets, and improvement of technologies [82].

Financial incentives can be provided through low-interest loans and subsidies to producers and users of materials from municipal waste. Recycling subsidies payable to whomever recovers waste materials, would increase not only the frequency of voluntary recycling drives on the part of charitable institutions but also permanently establish a growing structure of firms and individuals engaged in the collecting of waste materials. Tax incentives have already been developed whereby the Internal Revenue Service permits the use of Industrial Revenue Bonds for the purpose of building recycling plants. These provide lower-cost capital because of their tax-free nature, and at the same time provide industry with tax advantages through accelerated depreciation and investment tax credit. Such incentive couples municipal funding with operation and management by responsible private entrepreneurs. In most states, the municipal and county authorities are constrained by law from negotiating contracts that extend beyond the term of the incumbent administration. Private enterprise cannot afford the risk of investing capital unless afforded the protection of a sufficiently long contract to recover its investment and earn a profit. This dilemma will require special enabling legislation by the states.

Several economic inequities currently favor virgin production. For example, virgin timber profits are treated as capital gains for tax purposes, an advantage that is not available to waste paper. Another example is the practice of assessing solid-waste management costs against municipalities and their general tax funds rather than against those who generate the wastes in proportion to volume. This favors the waste maker and discourages recycling of paper and other materials. Discriminatory transportation rates are a significant factor in restraining recycling. Freight rates for shipping materials to be recycled are as much as 50% higher than rates for comparable virgin or primary materials. Such inequities necessitate a reexamination and change in tax policies and freight rate structures with an eye toward encouraging recycling rather than impeding it.

The issue of market creation especially applies to local, state, and national procurement policies which in many cases discriminate against recycled material. Specifications that call for virgin or new materials should be changed and rewritten based on performance standards desired. Government agencies must take the lead in stimulating the demand for products made with recycled materials. In this regard, the General Services Administration (GSA) has already set an example: Two-thirds of the paper purchased must contain 3–100% recycled fiber. GSA also requires substantial recycled glass or mineral fiber in bat, blanket, board, and solid block insulation. At the same time, GSA removed the "virgin only" specification from most of its material purchases.

In the case of technology improvement, the federal government should accelerate research and development, and technology transfer. The EPA has funded several technology demonstrations, and the Bureau of Mines has initiated technologies for incinerator residue processing, raw refuse processing, pyrolysis, and hydrogenation. Some of these technologies need to be led into large-scale demonstrations and subsequently to full commercialization. In addition, tax policies can be changed to provide a basis for expanding research and development activities by industrial firms capable of recovering resources and materials from wastes.

The four areas of concern, namely financing, inequity correction, market enhancement, and technology improvement, continue to encourage initiative especially at the state level. As of May 1976, 12 states had grant or loan programs for construction of resource recovery systems by municipalities, 20 were planning statewide systems or regulating resource recovery activities through guidelines, and six had authority to create agencies to operate resource recovery facilities (see Table 28). These numbers are expected to further improve under the impetus of the Resource Conservation and Recovery Act of 1976.

The prospects for long-range progress in recycling and resource recovery are promising in view of the basic economic pressures associated with the costs of virgin materials, land disposal, and energy. Furthermore, the removal of financial and economic impediments to expanded recycling coupled with improved recovery technology will enable resource recovery to provide an attractive alternative to traditional disposal methods.

For economic and political reasons recycling is destined to become an increasingly important phase of economic activity. It represents the

Table 28
Summary of State Involvement in Resource Recovery, 1976[a]

States with grant or loan authority (12)	States involved in planning or regulation (20)[b]	States with operating authority (6)
California	California	Connecticut
Connecticut	Connecticut	Florida
Florida	Florida	Maryland
Illinois	Hawaii	Michigan
Maryland	Illinois	Rhode Island
Massachusetts	G-Maryland	Wisconsin
Minnesota	G-Massachusetts	
New York	Michigan	
Ohio	Minnesota	
Pennsylvania	Montana	
Tennessee	New York	
Washington	North Carolina	
	Ohio	
	Oregon	
	Pennsylvania	
	G-Rhode Island	
	South Dakota	
	Vermont	
	Washington	
	Wisconsin	

[a] From McEwen [83].
[b] G—EPA implementation grant.

solution to both the problems of increasing solid waste and its attendant demands on land, and those associated with the growing scarcity of natural resources.

REFERENCES

1. A. J. Teller, "*Ecosystem Technology: Theory and practice*," *AIChE Monogr. Ser.* **70** (9) (1974).
2. Office of Solid Waste Management Programs, *Third Report to Congress—Resource Recovery and Waste Reduction*, EPA, SW-161, Washington, 1975.
3. W. E. Franklin, *Potential for Resource Recovery in the United States*, Franklin Associates, Ltd., for ALCOA, Pittsburgh, Pa., May, 1975.
4. W. E. Franklin, *Baseline Forecasts for Resource Recovery, 1972 to 1990*, Midwest Research Institute, for Office of Solid Waste Management Programs, EPA, Washington, Mar., 1975.
5. C. Stern, et al., *Impacts of Beverage Container Legislation on Connecticut and a Review of the Experience in Oregon, Vermont, and Washington State*, Dept. of Agr. Econ., Univ. of Conn., Storrs, Conn., Mar., 1975.

6. Applied Decision Systems and Decision Making Information, Inc., "A study of the effectiveness and impact of the Oregon minimum deposit law," prepared for the State of Oregon Dept. of Transportation, Highway Division, Oct., 1974.
7. J. G. Abert and M. J. Zussman, *AIChE J.* **18** (6), 1089 (1972).
8. N. L. Drobny, H. E. Hull, and R. F. Testin, *Recovery and Utilization of Municipal Solid Waste*, EPA, SW-10c, Washington, 1971.
9. J. G. Abert, H. Alter, and J. F. Bernheisel, *Science* **183** (4125), 1052 (1974).
10. *Decision-makers Guide in Solid Waste Management*, EPA-SWMP, SW-500, 1976.
11. R. C. Ziegler, et al., *Environmental Impacts of Virgin and Recycled Steel and Aluminum*, Calspan Corp., prepared for EPA-OSWMP, Feb., 1974.
12. E. J. Ostrowski, "The bright outlook for recycling ferrous scrap from solid waste," in *Energy and Resource Recovery from Industrial and Municipal Solid Wastes, AIChE Symp. Ser.* **73** (162), 93 (1977).
13. L. C. Blayden, "The chemistry of recycling aluminium," in *Energy and Resource Recovery from Industrial and Municipal Solid Wastes, AIChE Symp. Ser.* **73** (162), 85 (1977).
14. "Montgomery county, a county executive report to the people," *Montgomery County Sentinel*, Sept. 23, 1971.
15. "Hercules wins Delaware contract," *Solid Waste Report*, Silver Springs, Md., Oct. 19, 1970.
16. News release, *New High-Volume Fuel Source Available through Composting Would Ease Energy Crisis*, Cobey-Ecco Co., Crestline, Ohio, Feb., 1974.
17. Rust Engineering Co., *Engineering Services for Urban Forest Products Facility*, Birmingham, Alabama, 1971.
18. M. H. Stanczyk and P. M. Sullivan, "Physical and chemical benefication of metal and mineral values contained in incineration residues," *AIME preprint* No. 69-B-54, Feb., 1969.
19. W. J. Campbell, *Environ. Sci. Technol.* **10** (5), 436 (1976).
20. Office of Solid Waste, *Fourth Report to Congress—Resource Recovery and Waste Management*, EPA, SW-600, Washington, 1977.
21. E. J. Farkas, *Ind. Eng. Chem. Fundament.* **16** (1), 40 (1977).
22. M. Maaghoul, "Fuels from municipal refuse," presented at the 69th Annual Meeting, AIChE, Chicago, Ill., Nov., 1976.
23. L. McEwen and S. Levy, *Waste Age* **8** (2), 42 (1977).
24. "Resource Recovery Round-up," *Solid Waste Systems* **6** (5) 25 (1977).
25. "Resource Recovery Round-up," *Resource Recovery & Energy Rev.* **4** (3), 18 (1977).
26. J. H. Flandreau, *Resource Recovery & Energy Rev.* **4** (1), 16 (1977).
27. J. G. Abert, *Waste Age* **8** (3), 30 (1977).
28. C. A. Ballard, *Waste Age* **8** (3), 58 (1977).
29. N. J. Weinstein and R. F. Toro, *Municipal-scale thermal Processing of Solid Wastes*, U.S. Dept. of Commerce, NTIS No. PB 263–396/4WP, 348P, 1977.
30. M. L. Smith, *Waste Age* **4** (5), 15 (1973).
31. *Shredding Cuts Space Requirements by 70%, Solid Waste Management* **19** (2), 32 (1976).
32. R. DeZeeuw, E. B. Haney, and R. B. Wenger, *Solid Waste Management* **19** (4), 22 (1976).
33. P. H. McGauhey, *Waste Age* **6** (7), 2 (1975).

34. A. O. Chantland, *APWA Reporter* **18** Dec., 1974.
35. See various publications from Institute for Scrap Iron and Steel Inc., 1729 H. Street, N.W., Washington, D.C. 20006.
36. A. M. Gaudin, *Principles of Mineral Dressing*, McGraw-Hill, New York, 1939.
37. G. J. Trezek and G. Savage, *Waste Age* **6** (7), 9 (1975).
38. G. J. Trezek, *Size Reduction in Solid Waste Processing*, College of Engineering, Univ. of California, Berkeley, Ca., 1973.
39. O. T. Zimmerman and I. Lavine, *Chemical Engineering Laboratory Equipment*, Industrial Research Service, Dover, N.H., 1943.
40. R. G. Zalosh, et al., *Assessment of Explosion Hazards in Refuse Shredders*, ERDA, Contract No. E (49-1)–3737, Washington, D.C., Apr., 1976.
41. A. F. Taggart, *Handbook of Mineral Dressing*, Wiley, New York, 1974.
42. National Center for Resource Recovery, *Materials Recovery System*, Washington, D.C., 1972.
43. R. D. McChesney and V. R. Degner, "Metal and glass recovery from municipal solid waste," in *Energy and Resource Recovery from Industrial and Municipal Solid Wastes, AIChE Symp. Ser.* **73** (162), 77 (1977).
44. J. A. Campbell, "Electromagnetic Separation of Aluminum and Nonferrous Metals," 103rd Annual Meeting, AIME, Dallas, Feb., 1974.
45. R. D. McChesney and V. R. Degner, "Hydraulics, Heavy Media and Froth Flotation Processes Applied to the Recovery of Metals and Glass from Municipal Waste Streams," 78th National Meeting, AIChE, Salt Lake City, Aug., 1974.
46. R. H. Perry, et al., *Chemical Engineers Handbook*, Section 21, p. 73, 4th Edition, McGraw-Hill, New York, 1963.
47. National Center for Resource Recovery, *Materials Recovery System, Engineering Feasibility Study*, Washington, 1972.
48. J. A. Scher, H. N. Myrik, and R. B. Seymor, "Chemistry of pyrolysis of Organic Solid Wastes," in *Industrial Solid Waste Management*, University of Houston, 1970.
49. R. C. Bailie and D. M. Doner, "Energy and Resource Recovery from Industrial and Municipal Solid Wastes," *AIChE Symp. Ser.* **73** (162), 102 (1977).
50. J. L. Kuester and L. Lutes, *Environ. Sci. Technol.* **10** (4), 339 (1976).
51. S. J. Levy, *A Review of the Status of Pyrolysis as a Means of Recovering Energy from Municipal Solid Waste*, 3rd United States–Japan Conference on Solid Waste Management, Tokyo, Japan, 1976.
52. J. Jones, *Chemical Eng.* **85** (1), 87 (1978).
53. D. L. Klass, *Wastes and Biomass as Energy Resources: An Overview*, Institute of Gas Technology, Chicago, Illinois, Jan., 1976.
54. R. A. Kormanik, "A Resume of the Anaerobic Digestion Process," *Water and Sewage Works*, Annual Reference Number, 1968.
55. E. C. Clausen, O. C. Sitton, and J. L. Gaddy, *Chemical Engineering Progr.* **73** (1), 71 (1977).
56. G. M. Adams, et al., *Total Energy Concept at the Joint Water Pollution Control Plant*, International Conference on Water Pollution, WPCF, Anaheim, Cal., Oct., 1978.
57. S. J. Hitte, *Anaerobic Digestion of Solid Waste and Sewage Sludge to Methane*, EPA-SW159, Washington, July, 1975.

58. J. T. Pfeffer, "Reclamation of Energy from Organic Refuse—Final Report," Grant No. EPA-R-80076, Office of Research and Monitoring EPA; National Environmental Research Center, Cincinnati, Ohio, Apr., 1973.
59. Waste Management Inc., "Title I Preliminary Engineering for A.S.E.F. Solid Waste to Methane Gas," Contract No. E (11-1)-2770, Energy Research and Development Administration, Washington, Jan., 1976.
60. F. R. Dair and R. E. Schwegler, *Waste Age* **5** (2), 6 (1974).
61. R. W. Headrick and C. Southard, "Mountain View Landfill Gas Processing System Project," Pacific Coast Gas Assn., Operating Section, Distribution Conference, S. Lake Tahoe, Nev., Apr., 1978.
62. R. R. Grover, J. F. Barbour, and V. H. Freed, *AIChE Symp. Ser.* **68** (122), 86 (1972).
63. H. F. Feldmann, *AIChE Symp. Ser.* **68** (122), 94 (1972).
64. C. G. Golueke, *Abstracts, Excerpts, and Reviews of the Solid Waste Literature*, Vol. IV, USEPA, SERL Report No. 71–72, Univ. of California, Berkeley, 1971.
65. W. J. Boegley, Jr., W. L. Griffith, and W. E. Clark, *The Development of a Wet Oxidation Process for Municipal Refuse*, U.S. Dept. of HUD, ORNL-HUD-15, UC-41-Health and Safety, 1971.
66. W. W. Shuster, *Partial Oxidation of Solid Organic Wastes*, U.S. Public Health Service, SW-7rq, Washington, 1970.
67. C. G. Ganotis and R. E. Hopper, *Environ. Sci. Technol.* **10** (5), 425 (1976).
68. *New Orleans Resource Recovery Facility Implementation Study*, National Center for Resource Recovery, Washington, 1977.
69. Systems Technology Corporation, *A Technical, Environmental and Economic Evaluation of the Wet Processing System for the Recovery and Disposal of Municipal Solid Waste*, EPA-SW109c, Washington, 1975.
70. Systems Technology Corporation, *A Technical, Environmental and Economic Evaluation of the Glass Recovery Plant at Franklin, Ohio*, EPA-SW146C, Washington, 1977.
71. W. K. MacAdam and S. E. Standrod, "Design and operational considerations of a plant extracting energy from solid waste for industrial use," *Proceedings of ASME Industrial Power Conference*, Pittsburgh, Pa., May, 1975.
72. Horner and Shifrin, Inc., *Study of Refuse Supplementary Fuel for Power Plants*, for City of St. Louis, 1975.
73. Horner and Shifrin, Inc., *Energy Recovery from Waste*, EPA-SW36di, Washington, 1972.
74. Midwest Research Institute, *St. Louis Union Electric Refuse Firing Demonstration, Air Pollution Test Report*, National Technical Information Service, Springfield, Va., 1974.
75. H. D. Funk and S. H. Russell, "Operating Experience of the Ames Solid Waste Recovery Plant," in *Energy and Resource Recovery from Industrial and Municipal Solid Waste, AIChE Symp. Ser.* **73** (162), 52 (1977).
76. A. O. Chantland, *The Ames Experience*, presented at the conference on Utilization of Wood Wastes, Madison, Wi., Oct., 1978.
77. A. J. Helmstetter and R. A. Haverland, *An Evaluation of the Resource Recovery Demonstration Project, Baltimore, Maryland*, EPA, SW-719, Washington, D.C., September, 1978.
78. R. A. Haverland and D. B. Sussman, *Baltimore, a Lesson in Resource Recovery*, EPA, SW-712, Washington, D.C., July, 1978.

79. R. A. Lowe, et al., *Energy Conservation Through Improved Solid Waste Management*, EPA-OSWMP, SW-125, Washington, D.C., 1974.
80. W. E. Franklin, et al., *Alternative Strategies and Plans for Effective Solid Waste Management and Resource Recovery in the Twin Cities Metropolitan Area*, Metropolitan Council of the Twin Cities Area, St. Paul, Minn., Feb., 1975.
81. Midwest Research Institute, *Resource Recovery—The State of Technology*, prepared for Council on Environmental Quality, Feb., 1973.
82. J. Boyd, *Environ. Sci. Technol.* **10** (5), 422 (1976).
83. L. B. McEwen, Jr., *A Nationwide Survey of Waste Reduction and Resource Recovery Activities*, EPA, SW-142, Washington, D.C., 1977.

7

Solid Waste Systems Planning

Jarir S. Dajani

Department of Civil Engineering, Stanford University, Stanford, California.

Dennis Warner

Gannett, Fleming, Corddry, and Carpenter, Inc., Harrisburg, Pennsylvania.

I. INTRODUCTION

Preceding chapters have discussed the different unit operations involved in solid waste disposal and resource recovery systems. In order to provide the best possible service to society, these systems must be efficiently integrated within the environment in which they must function. Such integration requires that the components of the systems be selected and combined in order to achieve their intended purpose and contribute to overall environmental improvement. This chapter reviews some of the basic planning methodologies currently used in the solid waste disposal and resource recovery fields and indicates how they may be used to develop efficient, integrated systems. Since attention has focused on these methodologies only recently, much of the following discussion reflects the direction in which planners and decision makers are moving rather than the current state of planning techniques.

A solid waste disposal and resource recovery system is composed of three basic functions:

1. A collection and transportation function, which covers such items as containers, trucks, collection crews, and district routing and scheduling strategies.
2. A processing and disposal function, which covers transfer stations, incinerators, resource recovery facilities, and landfills.
3. Auxiliary services, such as management, planning, administration, marketing, and the enactment and enforcement of ordinances and legislation.

It has been estimated that about 80% of the cost of overall solid waste management is spent on collection and transportation. Although it can be expected that this proportion might drop as more capital intensive resource and energy recovery facilities are built, it will always remain a significant part of the total cost. It also is the portion that is the most labor intensive and thus most subject to inflationary pressures. Because of this, the planning and design of collection systems lie at the core of solid waste management planning and, indeed, are often the actual determinants of the number, size, location, type of processing, and ultimate disposal facilities developed.

Solid waste management represents the fifth largest public expense in the United States after education, highways, welfare, and public safety. It involves a national bill of 5 billion dollars, of which 4 billion are needed to pay for collection. Thus, overall cost figures clearly show that collection is the most significant component of the solid waste disposal effort. For example, a community of 500,000 people that spends about $3.5 million annually for solid waste collection and about $1.0 million for disposal is paying about $7.00 per capita per year for collection alone. Furthermore, there usually is employed one collector or driver for every 600 persons and one truck for every 2000 persons. On a national basis, this represents 350,000 employees and almost 100,000 trucks.

These costs can be reduced in a number of ways, the most obvious being an increase in collection efficiency by improved truck utilization and operation. Other possibilities include the reduction of residential wastes at the source utilizing home compactors or residential composting. Kitchen grinders and garbage disposal units reduce waste volumes, although the wastes are simply diverted from the solid waste collection system to the sewerage system. A more recent method employs the modular integrated utility system, which produces electricity by burning solid wastes at a central location near the collection system. Costs can be reduced through (a) legislative action limiting the use of objectionable materials, (b) special taxes designed to render the use of certain packaging materials uneconomic, and (c) public education aimed at reduction and change of the community waste stream. Furthermore, the economic and technical feasibilities of new collection facilities, such as pneumatic and slurry systems, should be explored.

II. PLANNING METHODOLOGY

A. Questions and Decisions

In planning solid waste collection, processing, and disposal systems, a number of basic questions are first raised:

1. What geographic area should be planned for?
2. What is to be collected and in what type of containers?
3. Who will collect it?
4. How often?
5. Where?
6. In what type of vehicle?
7. What routing, crew size, and schedules should be used?
8. What type of processing and disposal facilities (including material recovery and energy conversion) should be included in the system plan?
9. How many transfer stations (intermediate facilities) are needed, what should their capacity be, and where should they be located?
10. How many incinerators, landfills, or other types of processing and final disposal facilities are needed, at what capacity, and where should they be located?
11. To which set of intermediate and final facilities should the waste generated in a given location be sent?

Figure 1 outlines the steps involved in the collection, transport, processing, and ultimate disposal of residential solid wastes. The first question, concerning the geographic area is usually determined on the basis of existing political and taxing jurisdictions. There may be, of course, considerable merit in combining a number of these jurisdictions into a single district. Such an arrangement may generate "economies of scale" in large processing and disposal facilities by operating at lower costs per ton of input wastes than those of smaller facilities. It should be

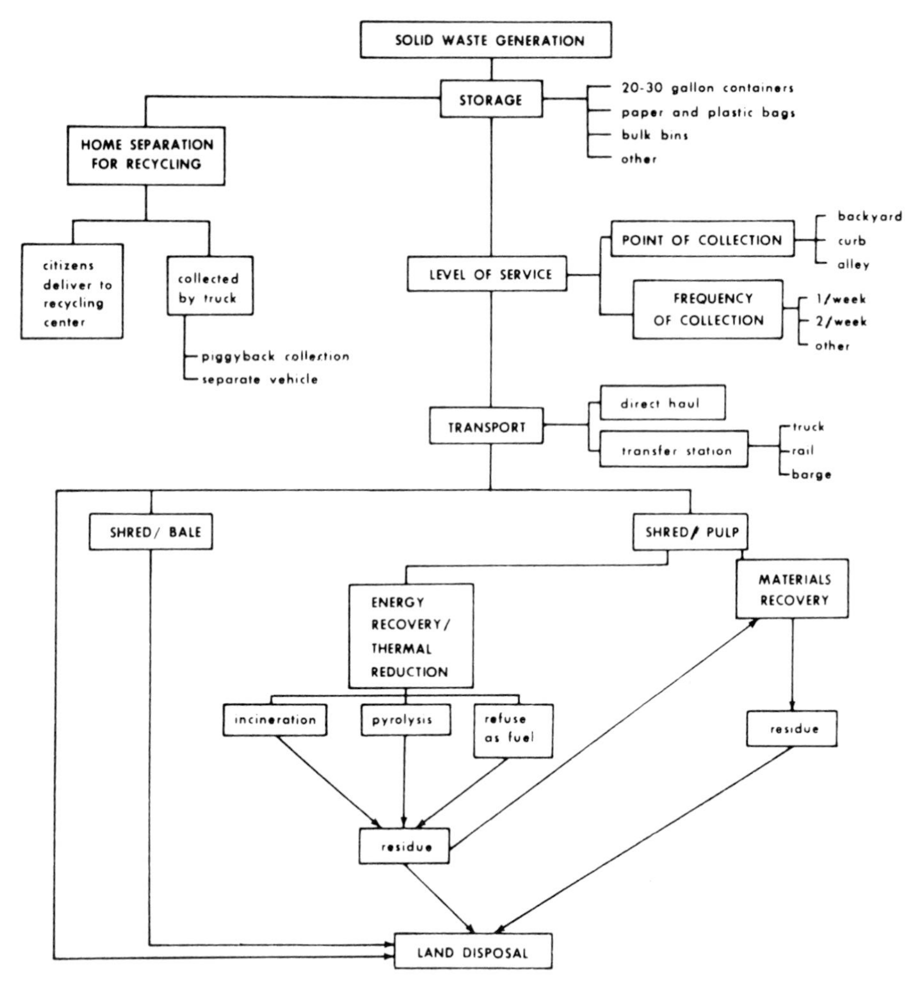

FIG. 1 Decision options in solid waste systems planning. From Ref. 6.

remembered, however, that these larger facilities sometimes involve larger unit costs for collection and transportation. Tradeoffs do exist between large and small systems, and it is possible to determine a minimal, or at least a localized optimal, cost system. Intergovernmental cooperation between adjacent jurisdictions should be sought whenever the savings resulting from large-scale transfer, intermediate handling, and final disposal facilities outweigh the additional costs incurred as a result of collecting the waste generated from a larger geographic area.

Solid waste systems planning is a complex process involving considerations of technology, economics, finance, management, and public policy. There is no deterministic technique that incorporates all of these issues into a rigorous analytical pattern. Comprehensive planning remains very much an "art" subject to the imagination and breadth of knowledge of the planner. At present, effective planning of solid waste systems may be more dependent upon logic and an intuitive balance of planning inputs than upon the ability to command sophisticated analytical techniques. This is not to deny the power of such techniques, which have become increasingly important in recent years. It only serves to stress the fact that the primary skill required in comprehensive planning is the ability to discriminate among the broad universe of possible planning activities in order to choose those elements essential to the success of the plan. Once the critical elements of a given plan have been identified, the specific tools and techniques of analysis can be utilized in developing an acceptable plan solution.

In the following section, the general planning process in the area of solid waste systems is discussed in terms of planning purposes and objectives, plan components, and ultimate implementations. The final two sections describe specific techniques for obtaining answers to the basic questions raised earlier in this chapter. The first six of these questions usually are addressed in the planning of street collection systems. This activity generally involves short-range planning of a labor-intensive activity. In contrast, the last four questions raise issues related to the long-range planning of an often highly capital intensive activity. Although the short-range day-to-day decisions relate to an activity that consumes four-fifths of solid waste management expenditures, long-haul processing and disposal can be expected to consume an increasingly larger proportion of the total bill, as cost and scarcity of urban land increases, and more sophisticated processing and recovery facilities are adopted. Short- and long-range planning are not mutually exclusive, especially since long-range decisions can be expected to have a significant bearing on the cost and efficiency of the daily collection activity.

B. The Planning Process

According to the U.S. Environmental Protection Agency (EPA), planning "is the conscious process for achieving proposed objectives which rationally and fully considers any likely contingencies and alternatives" [1]. In effect, this process should be viewed as a controlled, flexible activity subject to continual revision. It has a formal structure, which can be described as a systematic method for recognizing a problem, establishing objectives, collecting relevant data, generating alternative solutions, choosing a recommended solution, and, finally, evaluating the ultimate success of the solution.

In the most basic sense, planning is an aid in the decision-making process [2]. It must be realized, however, that planning is not a necessary prerequisite to decision making, for decisions will be made by public officials, political leaders, and businessmen even in the absence of planning. Thus, the burden borne by the planner includes not only the job of determining a solution to perceived problems, but also the task of presenting this solution in a manner that is both understandable and acceptable to the decision makers. This double burden is particularly evident in the field of solid waste systems planning.

The term "planning" has been used to denote a wide variety of activities ranging from a simple physical design to a complex social arrangement. A spectrum of "planning" activities, consisting of "part design" at one end and "regional plan" at the other, can be ranked according to an increasing degree of inherent complexity, as follows:

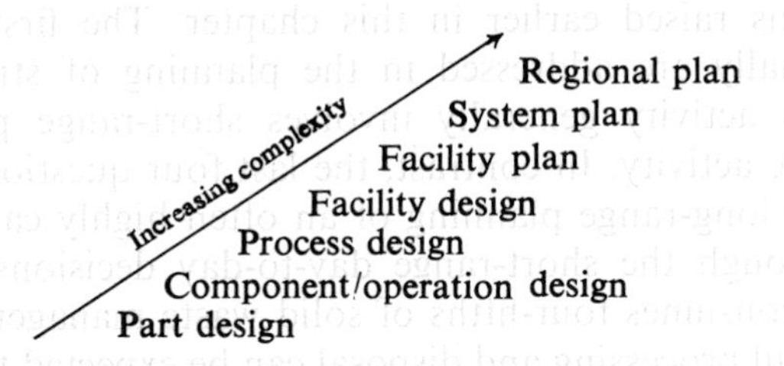

Only the aspects of facility, system, and regional plans are considered here. Each of these plans, and especially the one at the regional level, deals more with the interrelationships between subject areas than with the internal details of a given subject. At its highest level, comprehensive planning involves numerous issues of law, transportation, government policy, housing, site design, and management that require coordination by the planner [31]. The interrelationships between subject areas must be understood and observed in plan formulation or the recommended solution will be based upon unrealistic premises.

In general, the greater the geographic area encompassed within a plan, the more general and less specific are its objectives. They may be narrowly technical in scope, as in performance criteria for collection vehicles, or broadly policy oriented, as in overall state goals for solid waste regionalization. In the entire planning process, the formulation of objectives is probably the most critical and yet least understood aspect. Engineers whose training is directed toward technical design process, tend to impose upon the planning process a narrow range of conceptual models and, when faced with a planning problem, to immediately initiate data collection and analysis suitable to these few models. A comprehensive solid-waste systems plan, however, requires more than just the derivation of the most apparent solutions. It requires a determination of the entire problem, a preliminary generation of a wide range of potential solutions, the formulation of specific objectives, and finally the collection of specific data for analysis. Objectives may be broad or narrow, but they are essential to the conduct of the planning process. As such, they must be formulated early and be explicit.

The EPA requires all comprehensive solid waste management plans to have the following six objectives [1]:

1. Adopt a sound planning process.
2. Establish a management system to implement optimum proposals for storage, collection, transportation, disposal, processing, and resource recovery.
3. Apply organization theory, financial management, cost control, and the management sciences to solid waste management.
4. Coordinate solid waste plans with other governmental agencies and further intergovernmental cooperation.
5. Integrate the solid waste plan with comprehensive area-wide plans.
6. Develop programs with the greatest promise for the ultimate solution of solid waste problems, such as the conservation of resources and protection of the environment.

A comprehensive solid-waste management plan [1] should (a) act as an internal policy guideline; (b) provide the public with a framework of standards for assessing the implementation; (c) provide for integrated management of various solid waste activities; (d) establish procedures for achieving actual design and operation; and (e) promote legislative improvements within the jurisdiction. Successfully serving all of the above functions requires consideration of a broad range of planning inputs. Space limitations prevent a full discussion of the critical planning inputs; however, they can be classified in terms of technical, financial, institutional, and restrictive issues.

Technical issues involve the engineering analysis and design of physical facilities and system operations associated with collection, transportation, processing, and disposal. For the solid waste engineer, these activities traditionally are treated as deterministic problems contained within a fixed set of assumptions regarding costs, organizations, and policies. More often than not, however, little or no consideration is given to the more realistic dynamic aspects of managerial theory, legislative changes, and institutional constraints. Since effective comprehensive planning deals more with the interrelationships between these various subject areas than with the specific details of any one area, the engineer must consider the issues affecting systems development. The following paragraphs will outline some of the critical subject areas requiring attention.

Financial issues include economic aspects, which can be defined as the expenditure of valued resources to achieve certain system objectives, and financial aspects, which can be defined as the cash flows occurring within a system. Resources are limited in quantity and have some market value, or opportunity cost. In solid waste systems, as in most public services, the resources include labor, management, equipment, physical facilities, and money. An efficient utilization of system resources implies either minimum resource use for a given set of system outputs or maximum system outputs for a given level of resource inputs. The techniques of economic analysis, such as cost-benefit or cost-effectiveness analysis, are often used in the evaluation of alternative plans [3].

The financial aspects deal less with intrinsic system worth and more with the ability of a system to generate sufficient monetary flows to remain solvent. Among the various ways of financing solid waste systems are tax revenues, user charges, general obligation bonds, revenue bonds, corporate bonds, and private leasing arrangements [4, 5]. Bond issues require sophisticated financial administration, and consequently most communities prefer to finance ordinary solid waste collection and disposal out of general tax revenues. The recent development of capital-intensive resource recovery facilities, however, has been dominated by private investment, although usually with government assistance, and greater attention now is paid to the financing modes involving corporate bonds and lease arrangements. In all cases, capital and recurrent costs usually involve different sources of finance. Capital costs are lump-sum investments for new equipment and facilities, whereas recurrent costs are continuous expenditures necessary to meet the on-going operation and maintenance expenses of the system. Capital finance is obtained normally through federal grants or borrowing; recurrent finance is obtained through user charges, fees, or tax revenues.

Institutional issues include the types of organizational structures within which solid waste operations are conducted, the role of management, and the establishment of administrative guidelines. Solid waste systems may be managed by cities, counties, councils of government, or even state agencies. They also may be managed within the private sector as a privately owned, publicly regulated utility or as a purely private business undertaking. Sometimes an overall solid waste system is divided between public and private sectors, such as privately operated collection services and publicly operated disposal facilities. A critical issue of current concern is the extent to which small systems should be integrated into larger regional systems in order to obtain greater efficiencies of operation and improved levels of service. Manpower planning and industrial relations also are growing in importance as the tendency for municipal employees to unionize and strike for improved conditions of employment grows in all parts of the country. Under such circumstances, solid waste systems engineers and planners in the public sector must learn to take into account problems of management and administration that formerly were of concern only in the private sector [6].

Restrictive issues refer to the constraining effects of institutions, policies, and legislation upon solid waste systems. Institutional constraints include limitations upon the authority of system managers to change their methods of operation or to expand their areas of service. The limitations differ somewhat between public and private systems. The former are more sensitive to the political processes of public elections and jurisdictional boundaries, whereas the latter are more affected by franchise agreements and internal financial resources. In addition, the motivating policies and objectives are likely to be vastly different between public and private systems. Publicly operated solid waste systems usually are concerned mainly with minimizing costs and maximizing system reliability. Private systems, on the other hand, tend to be concerned primarily with maximizing the profits of system operation. Planning forecasts, therefore, should be qualified by the purposes for which the systems are operated.

Policies can also have indirect effects, as in the case of industrial attitudes toward the appearance, convenience, and cost of packaging materials. Private industry traditionally has placed great emphasis on the marketing aspects of packaging and little on the disposal aspects. These marketing policies have flourished because of widespread public indifference to the problems of limited resources and because of the national orientation towards maximum production and consumption. As a result, the major indirect effect of these attitudes and policies has been a great increase in the generation of solid wastes. Because of the importance of such indirect effects, comprehensive solid waste systems planning must attempt to take into account both organizational policies and prevailing community values.

Legislative constraints include the laws and regulations that control and guide the planning and management of solid waste systems. The Solid Waste Disposal Act of 1965 and its subsequent resource recovery-oriented amendments have established the basic federal policies for solid waste management [7]. Most of the states have followed the lead of the federal government. During the early 1970s, many states enacted new solid waste management statutes, some of them for the first time. Typical provisions were general adherence to improved environmental quality, strong injunctions against open refuse fires and open dumps, moderate encouragement of regional solid waste management efforts, and relatively ineffectual support for resource recovery and conservation. However, not all states acted. Some, most notably New York and Washington, not only enacted new legislation, but also established major funding programs to support implementation of the acts [8, 9]. Despite these progressive trends, legislative controls are not always beneficial. Local zoning restrictions on collection routes and disposal sites can severely hamper an otherwise useful solid waste plan. Similarly, resource recovery tends to be discouraged by many legislative acts and administrative rulings that provide preferential freight rates, tax depletion allowances, and packaging requirements for virgin materials over recovered materials [10, 11]. However, all of these restrictive issues are dynamic and subject to change. Nevertheless, the actual presence of such constraints, as well as the possibility of their changing for better or worse in the future, calls attention to their importance in solid waste systems planning.

The development of comprehensive solid waste systems plans incorporating the full range of above issues is no easy task, and it is unreasonable to expect the individual planner or engineer to have a firm grasp of all such issues. Resources available for planning purposes always are limited and frequently are considered (by the planners) to be insufficient. The development of an appropriate solid waste systems plan, therefore, involves initial identification of the critical planning inputs essential to plan success. From the almost endless array of possible plan activities, the planner must choose those key inputs allowed by available resources of time, staff, and funds. Figure 2 presents a model of the major activities included in the development of a solid-waste systems plan. In addition, the model links plan formulation with subsequent implementation, which in turn is followed by systems monitoring and evaluation. These latter activities provide information and insights that should be included as inputs in new planning activities.

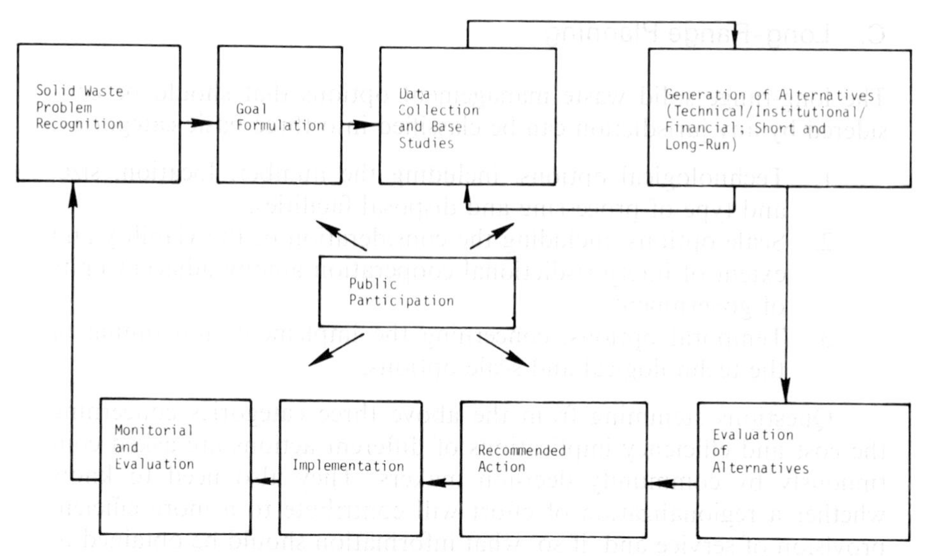

FIG. 2 Solid waste systems planning model.

A starting point is the recognition of a solid waste problem. This may be initiated by the engineer or planner, by the general public, or by some evaluation mechanism within the solid waste system itself. The next step is to formulate basic goals for the solution of the problem. The goals lead to an iterative process involving data collection and the generation of alternatives. Within each alternative, the basic technical, financial, and institutional issues discussed earlier should be considered. Once an appropriate range of alternative solutions has been formulated, various analytical techniques are available to evaluate the alternative and to select one or more recommended courses of action. Comprehensive planning must go beyond data collection and analysis to include specific recommendations whenever appropriate.

Continuing with the model, the next general step in systems planning is implementation, which in turn is followed by monitoring of system performance and overall evaluation of success. The results of the monitoring and evaluation steps logically feed directly back into the planning process by calling attention to new problems requiring solution. Accompanying the entire process of planning, implementation, and evaluation should be a steady input of public participation in the form of public hearings, citizens' committees, open meetings, and wide publicity. Public participation should be encouraged, not only because of increasing legal requirements for such inputs, but also because of the important benefits to be obtained from the open discussion of proposed solid waste systems affecting the general public.

C. Long-Range Planning

The long-range solid waste management options that should be considered by any jurisdiction can be classified into three basic categories:

1. Technological options, including the number, location, size, and type of processing and disposal facilities.
2. Scale options, including the consideration of the viability and extent of interjurisdictional cooperation among adjacent units of government.
3. Temporal options, concerning the implementation timing of the technological and scale options.

Questions stemming from the above three categories concerning the cost and efficiency implications of different actions are asked continuously by community decision makers. They also need to know whether a regionalization of effort will contribute to a more efficient provision of service and, if so, what information should be obtained to assist them in setting up future interjurisdictional cooperation. Decision makers must be able to determine whether and when the introduction of resource recovery plants will contribute to cost reduction and overall system efficiency. New landfills must be planned and old landfills eventually closed, and it is essential, in an economic sense, to be able to forecast the need for such actions. And last, but not least, the community as a whole needs to obtain a better understanding of the actual costs of environmental protection and citizen satisfaction in situations involving controversial actions.

Decisions relating to issues usually found within the long-range planning function can be made in a variety of ways. At one extreme is intuitive judgment based on personal experience and a liking or disliking of the status quo. A more analytical approach would attempt to enumerate a range of possible future courses of action and assess the costs and benefits associated with each in order to develop a plan most suitable for the jurisdiction in question. The most rigorous of the "rational" planning methods is relatively new and utilizes the power inherent in mathematical programming and optimization theory to select the most efficient combination of technological, scale, and temporal options open to the community. A variety of such models have been developed within the last decade, examples of which are given in Refs. 12–18. All are concerned with balancing the economies of scale resulting from larger processing and disposal facilities against the additional costs of transportation needed to supply these facilities with the waste generated in the region.

The EPA has recently completed the initial development work on a computerized model that has the capacity of performing this long-range planning function [19]. Termed "WRAP" for Waste Resources Allocation Program, this model produces outputs that provide a dynamic profile of regional solid-waste management solutions in terms of the location, size, and type of facilities needed and the specific waste generating areas supplying each of these facilities at various times. It is not the intention here to delve into the mathematical details of problem formulations and algorithmic solutions, but it is instructive to present a linear programming formulation of a typical long-range planning problem. The formulation presented below provides the basic generic form of most of the more complicated optimization models. This simplified example is of the static variety and presents the optimal solution at one specific future time. A dynamic model would include different time periods as additional variables in the optimization problem.

Consider the area shown in Fig. 3, which has been divided into a number of waste generation zones representing areas of similar population, land use, and waste generation characteristics. Each zone is assigned a centroid, which is assumed to be the point at which all wastes are generated within the zone. The zones are numbered $i = 1, 2, \ldots, m$ and the waste generated in each zone is designated W_i (in tons).

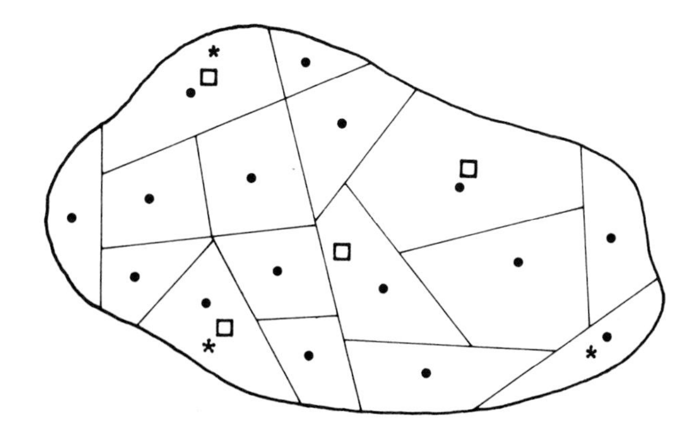

FIG. 3 Hypothetical solid waste planning problem.

Intermediate processing and final disposal sites available in the general area are selected for use. The former sites include transfer stations and resource recovery facilities, whereas the latter denote landfill locations. The final disposal sites are designated $j = 1, 2, \ldots, n$ and the intermediate processing sites are numbered $k = 1, 2, \ldots, p$. Whenever one processing location has the potential for a number of alternative processes, each combination of mutually exclusive processes is given a different number, although they may be contenders for the same site. Materials flows, thus, are denoted by W_{ij} for flows between sources and disposal sites and by W_{ik} for flows between sources and processing facilities. Flows between processing and disposal facilities are denoted by W_{kj}. Transportation costs in dollars per ton between these three sets of facilities are designated by c_{ij}, c_{ik}, and c_{kj}, respectively. Similarly, the

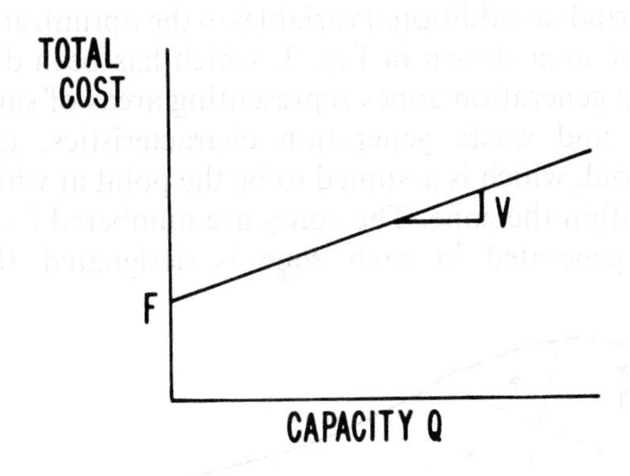

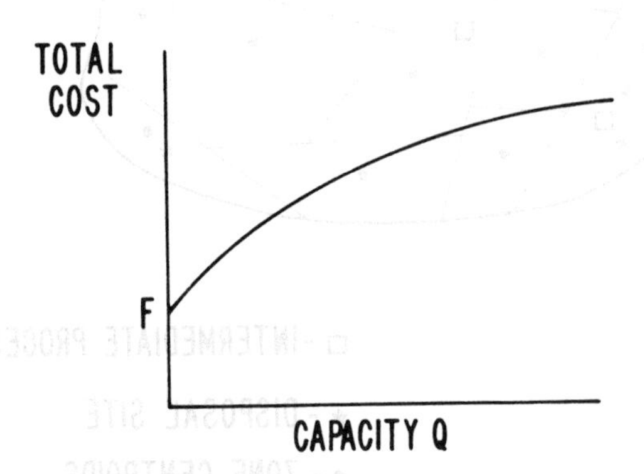

FIG. 4 Cost functions for disposal and processing facilities: (a) Linear cost function; (b) Cost function with economies of scale.

capacities of disposal and processing facilities are denoted by Q_j and Q_k, respectively. Since processing facilities will result in some reduction of the total amount of waste, it is necessary to define an indicator of the magnitude of reduction for each process being contemplated. This indicator is defined as the ratio of the weight of the outputs of the process to the weight of the inputs and is denoted by α_k.

The final input required for the development of the model is that of facility cost data. Such costs usually are given in terms of two components, fixed and variable costs. The simplest (though not very realistic) assumption is that of a linear cost function, as shown in Fig. 4a. Most processing and disposal facilities, however, have a cost function similar to that shown in Fig. 4b, which represents declining incremental costs with increasing capacity. This indicates the existence of economies of scale, which is another way of saying that the cost per unit of capacity decreases as the size of the facility increases. Such functions usually are approximated by a number of straight line segments. For the purposes of this example, a linear relationship similar to the one shown in Fig. 4a is assumed. Each processing or disposal facility will have a fixed charge needed to set up the facility, F_j or F_k, and a variable cost that is dependent on the actual size of the facility, V_j or V_k.

The above inputs provide all the necessary data for formulating and solving a simple optimization problem. A solution algorithm selects that combination of size, location, and waste assignment which will handle the wastes of the region at the lowest possible cost. In a simple case involving two zones, one disposal site, and two possible intermediate transfer process options, all waste routing options would be as shown in Fig. 5. Each arrow represents a possible flow of raw or processed waste, and the optimization algorithm selects that combination of waste flows which results in the lowest total system cost. The objective function, thus, takes the following form:

$$\left(\sum_{i=1}^{m}\sum_{j=1}^{n} W_{ij}c_{ij} + \sum_{i=1}^{m}\sum_{k=1}^{p} W_{ik}c_{ik} + \sum_{j=1}^{n}\sum_{k=1}^{p} W_{kj}c_{kj}\right)$$

transportation costs

$$+ \left(\sum_{j=1}^{n} F_j y_j + \sum_{k=1}^{p} F_k y_k\right)$$

fixed charges

$$+ \left(\sum_{i=1}^{m}\sum_{j=1}^{n} W_{ij}V_j + \sum_{i=1}^{m}\sum_{k=1}^{p} W_{ik}V_k + \sum_{k=1}^{p}\sum_{j=1}^{n} W_{kj}V_j\right)$$

variable costs

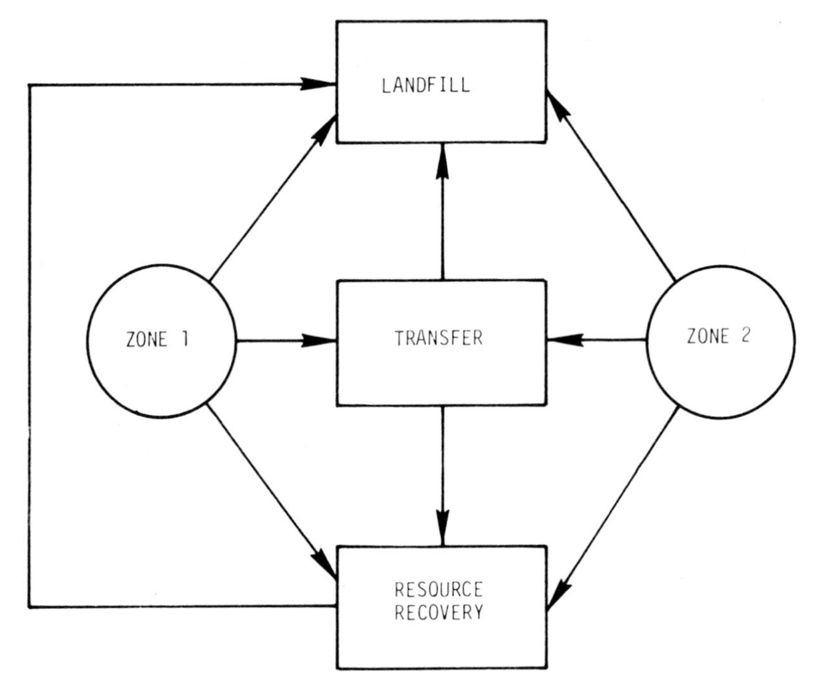

FIG. 5 Options in a simple hypothetical case.

This objective function must be minimized subject to the following constraints:

1. All wastes generated in a given zone are processed and disposed of:

$$\sum_{j=1}^{n} W_{ij} + \sum_{k=1}^{p} W_{ik} = W_i \qquad \text{for } i = 1, 2, \ldots, m$$

2. Flows entering an intermediate processing facility are equal to those leaving it:

$$\sum_{i=1}^{m} \alpha_k W_{ik} = \sum_{j=1}^{n} W_{kj} \qquad \text{for } k = 1, \ldots, p$$

3. Inputs to each intermediate facility do not exceed the capacity of that particular facility:

$$\sum_{i=1}^{n} W_{ik} \leqslant Q_k y_k \qquad \text{for } k = 1, \ldots, p$$

where y_k is an integer variable that has a value of 1 if the facility is selected in the solution and a value of 0 if it is not.

4. Flows to a given disposal site do not exceed the ultimate capacity of that site (Q_j^*):

$$\sum_{i=1}^{m} W_{ij} + \sum_{k=1}^{p} W_{kj} \leqslant Q_j^* y_j \qquad \text{for } j = 1, \ldots, n$$

5. The usual mathematical restrictions for linear programming formulations apply:

$$W_{ij}, W_{ik} \geqslant 0 \qquad \text{and} \qquad y_k, y_j = (0, 1)$$

It should be stressed at this point, that the above example represents a static linear formulation of the problem, with all the constraining effects that it implies. It has been presented for conceptual purposes. The addition of a time component is possible through the addition of terms to the objective function, such that the same function is repeated for each additional intermediate period to be considered. This will, in turn, require the addition of constraint equations representing the relationship that must be forced to exist between subsequent time periods.

D. Short-Range Planning

Short-range planning deals with decisions that do not involve capital-intensive undertakings and cover actions able to be implemented within a short period of time. The area of solid waste collection frequently involves issues of short-range planning. An extensive discussion of solid waste collection practice has been published by the American Public Works Association and should be read by any engineer or planner concerned with the planning problems in this field [20].

The initial question facing the planner is "What to collect?" A recent survey of solid waste activities in the United States indicates that prior to collection about one-third of the population separates its waste into garbage, which is generally putrescible organic matter, and rubbish, which is generally nonputrescible and inorganic [20]. In this case, the source separation is made by the generators of the solid wastes. More than half of the population, however, lives in communities where no waste separation is required. In general, systems that collect unseparated garbage and rubbish are more common in rural areas, whereas separate collection operations are more common in urban areas. The survey also shows that a variety of different collection arrangements are available.

A community can either operate its own collection system or leave that responsibility to private enterprise operating under contract to the city. A third, but less common form of collection is one in which private companies are engaged by individual homeowners. The first option is selected by two-fifths of the communities across the country. Overall, more than half of all household wastes are collected by public collectors, about a third by private collectors, and a tenth are disposed of by individuals themselves. In the case of industrial and commercial wastes, about two-thirds of the national total is collected by private contractors. Another one-third of the industrial wastes and one-eighth of the commercial wastes are privately collected by the industries and commercial interests themselves.

In theory, the direct collection of solid wastes by a public agency ensures an acceptable level of service because of continuing public scrutiny. Similarly, the collection of wastes by private contractors theoretically also results in adequate service because of the ability of consumers to take their business elsewhere in the marketplace. In practice, however, the individual (or community) often is faced with a single unresponsive collection system. A municipal collection system tends to suffer from problems of inefficiency and political interference typical of municipal operations. Contracting the collection function to private enterprise may alleviate these disadvantages, but may bring on the problems characteristic of private operations. Among these are the risk of contract default, excessive rates, and the lack of flexibility in the quality of service offered. The third alternative, in which homeowners directly contract with a private firm, is the only one available in many rural areas. It has the disadvantage of lacking the mechanism of public control that would ensure an adequate, economical, and sanitary level of service.

The frequency of collection varies significantly from one location to another. About half of the collection systems in the United States operate on a once-per-week schedule, one-third collect twice weekly, and two-thirds of the systems that require the separation of rubbish and garbage have only a weekly pickup. In general, there is a nationwide shift from twice-weekly collection schedules to once a week.

A number of methods are used to transport the waste from premises to collection vehicles. At one extreme, residents may be required by city ordinances to place their waste containers at the curb or alley for pickup, which is the most economical method from the standpoint of the city. It is not the most desirable from the residents' point of view, however, because it requires them to do part of the work. Furthermore, it can result in unsightly street appearances and, possibly, additional street cleaning costs. The alternative of the city or contracting firm collecting the containers from the back door can be about 50% more expensive. In some urban areas the collector takes an empty tub or basket to the back door, empties the container into it, and carries the wastes to the truck. This method relieves residents of the burden of setting out their containers and has few esthetic or sanitary drawbacks. Nevertheless, it does result in time-consuming vehicle stops, inefficient equipment use, and expensive service.

The choice of the collection vehicle has, of course, significant effects on the overall efficiency of the collection operation. The most common types of vehicles are trucks with capacities of between 12.2 and 18.4 m^3 (16 and 24 yd^3). They usually are equipped with compactor bodies that are either manually or mechanically loaded and are manned by a crew of two or three, including the driver. Openbodied dump trucks are sometimes used for the collection of bulky items. A national trend toward increasing the mechanization of the collection process is evident through a number of recent innovations. These include the use of detachable containers, suction-trucks, or even trucks equipped with small pick-up scooters in order to save walking time in low-density areas.

A major management decision facing solid-waste collection agencies is that of allocating routes to collection crews. One approach is to allocate a definite daily task to each crew, thus dividing the community into areas that each provide about one day of work for each crew. Thus, a crew will be assigned between three to six collection routes per week, depending on the frequency of collection. Such an approach allows residents to know the collection days and provides the crew with an incentive to finish early. Labor problems may arise, however, if the schedule cannot be maintained. The rigidity of the approach makes it susceptible to breakdowns and interruptions. An alternative way of allocating definite tasks to crews includes assignments on a weekly basis, in which a crew begins on Monday and continues each morning at the ending point of the previous day until the assignment is completed. It sometimes is desirable to have a standby crew available to assist the regular crews with unusually high loads or to provide a central overflow route that could be collected by any of a number of crews serving surrounding routes. Both of these techniques are designed to add flexibility to the collection effort.

The economics of overall solid waste collection, processing, and disposal are a function of all of the above considerations, together with others that depend on the nature of the community and the selected methods of disposal. Population density is a significant cost factor. A community having fewer than 10 dwelling units per acre can expect to pay up to 50% more in collection costs per ton than a community that is developed at twice that density. Land use and zoning regulations have a definite effect on the distribution of activities in the community and, consequently, upon the distribution of waste generation and the distances to be traveled by collection vehicles.

One output of the long-range planning process is the assignment of waste generation areas to processing and disposal facilities of given characteristics. In order to translate this information into a short-range working plan for collection, processing, and disposal, the planner needs to take two further steps. Each waste generation area must be divided into specific collection routes and the actual path to be followed by each collection vehicle must be determined. The first step is referred to as districting and route balancing, the second is called microrouting. These steps have been described in detail by the Office of Solid Waste Management Programs (OSWMP) of EPA [21, 22]. A summary of the techniques suggested in these two documents for use by local planners is presented below.

Districting and route balancing is the process of determining the optimum number of services that constitutes a fair day's work and of dividing the collection task among the crews to give them equal work loads. It can be used for a variety of purposes, such as estimating the number of trucks and men needed for a system, evaluating crew performance or system changes, and assessing the costs of changes in the levels of collection service. It is accomplished by simulating the total effort required for performing the collection effort on a truck-by-truck basis.

The simulation starts by determining the number of waste-generating units to be serviced per truck load $\boldsymbol{N}$,

$$\boldsymbol{N} = x_1 x_2 / x_3$$

where x_1 is the vehicle capacity in m^3, x_2 is the vehicle load density in kg/m^3, and x_3 is the unit service load in kg per unit. The next step is to determine the number of loads that can be completed in a working day through an analysis of the time budget of a truck and its crew. The time, h, spent by each collection truck and crew during a given day can be broken down into several components, where a is the travel time from garage to the beginning of the route, b is the collection time on the route, c_1 is the travel time from the route to the disposal or processing site, c_2 is the travel time from the disposal or processing site back to the route, d is the waiting time spent at the site, e is the travel time from the disposal site back to the garage at the end of the day, f is the time spent on official breaks, g is the time lost to breakdowns and unforeseen events, and n is the number of loads per day for a given truck. For a truck that carries n loads per day, the total hours Y in a working day are equal to the following:

$$Y = a + b + n(c_1 + c_2 + d) - c_2 + e + f + g$$

By subtracting the overall nonproductive time spent traveling to and from work and on rest periods and breakdowns $(a + e + f + g - c_2)$ from the total time available, Y, the available time remaining for actual collection and disposal $[b + n(c_1 + c_2 + d)]$ can be determined.

All of the time values in the above equation can be estimated from actual operating experience. The collection time b is equal to ts, where t is the estimated collection time per service unit and s is the total number of units to be serviced per day. The value of s, of course, is equal to the number of service units per truck load N times the number of loads n. Inserting these values into the above equation and rearranging the variables results in the number of loads that a given truck can collect per day:

$$n = \frac{Y - (a + e + f + g - c_2)}{tN + (c_1 + c_2 + d)}$$

When considering the entire collection network, the total number of trucks T required for the task can be found from the following equation

$$T = Sf/sw$$

where S is the total number of service units, f is the weekly collection frequency, w is the number of work days per week, and s is the average number of units serviced per truck per day. After an adequate number of services per truck per day s has been determined, a waste generation area can be districted, i.e., divided into equal work load sections according to the day of the week. An additional step might include dividing each daily work load section into specific truck routes on the basis of the same route balancing procedure described above.

The next step in planning short-range solid waste collection is that of microrouting, which is the determination of the best path to be followed by each collection vehicle and crew on the basis of the existing street pattern in the waste generation area. In order to determine the most efficient path, it is necessary to minimize nonproductive time spent on repeated travel routes, on travel through streets with no service requirements, and on delays resulting from U-turns, left turns, and congested streets. An intuitive technique for the development of reasonably efficient routings is that of allowing the drivers to select their own routes with no guidance from management. This method works well when drivers know their areas and are interested in maximizing collection efficiency. The opposite approach is to mathematically describe the collection system by determining quantitative relationships between its various components. Although a variety of mathematical programming techniques have been utilized in this manner to obtain theoretically optimal routings, these techniques are highly complex and have a low probability of being accepted by local planners.

Another approach to microrouting, which is neither theoretical nor intuitive, but instead heuristic, has been developed by EPA [22]. A heuristic approach applies human intelligence, experience, common sense, and certain rules of thumb to develop an acceptable, but not necessarily optimal, solution to a problem. Such approaches have been found to be particularly useful in situations in which rigorous mathematical solutions have been found to be impossible, infeasible, or impractical. The EPA heuristic routing method involves a simple manual solution to the path selection process on the basis of the following rules:

1. Routes should not be fragmented or overlapping. Each route should be compact, consisting of street segments clustered in the same geographical area (see Fig. 6).
2. Total collection plus haul times should be reasonably constant for each route in the community (equalized work loads).
3. The collection route should be started as close to the garage or motor pool as possible, taking into account heavily traveled and one-way streets (see rules 4 and 5).

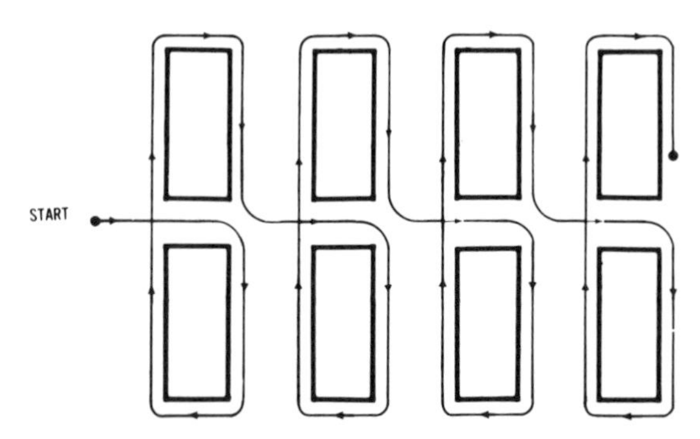

FIG. 6 Specific routing pattern for one-way street, one-side-of-the-street collection. From Ref. 22.

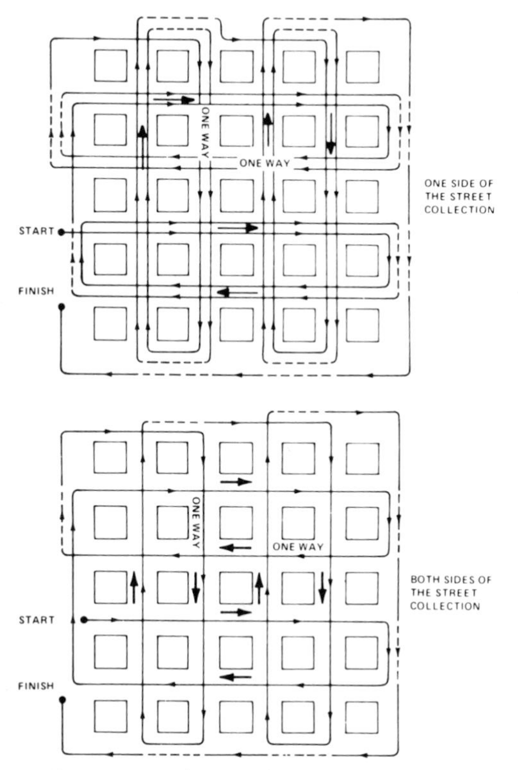

FIG. 7 Specific routing patterns for multiple one-way streets. From Ref. 22.

4. Heavily traveled streets should not be collected during rush hours.
5. In the case of one-way streets, it is best to start the route near the upper end of the street, working down through the looping process (see Fig. 7).
6. Services on dead end streets can be considered as services on the street segment that they intersect, since they can only be collected by passing down the street segment. To keep left turns at a minimum, collect the dead end streets when they are to the right of the truck. They must be collected by walking down, backing down, or making a U-turn.
7. When practical, steep hills should be collected on both sides of the street while vehicle is moving downhill for safety, ease, speed of collection, and wear on vehicle, and to conserve on gas and oil.
8. Higher elevations should be at the start of the route.
9. For collection from one side of the street at a time, it is

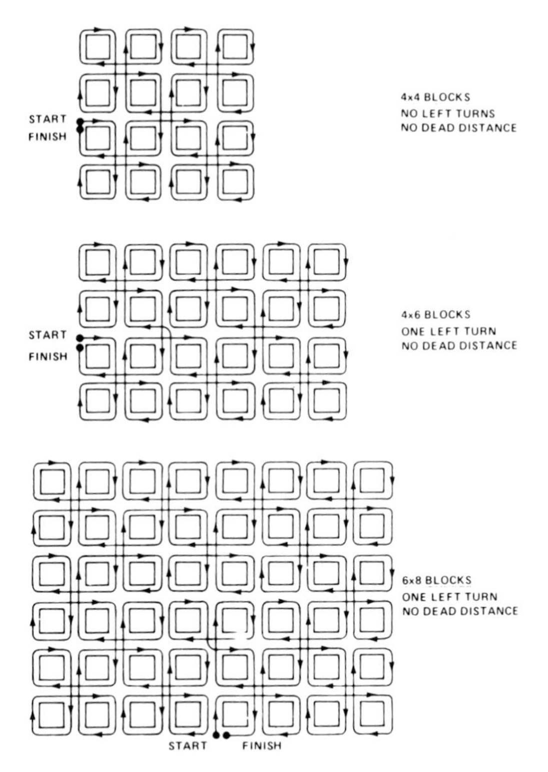

FIG. 8 Combinations of the four-block pattern, one-side-of-the-street collection. From Ref. 22.

generally best to route with many clockwise turns around blocks (see Fig. 8).

Heuristic rules 8 and 9 emphasize the development of a series of clockwise loops in order to minimize left turns, which are generally more difficult and time-consuming than right turns. In addition, right turns are safer, and especially so in the case of right-hand drive vehicles.

10. For collection from both sides of the street at the same time, it is generally best to route with long, straight paths across the grid before looping clockwise (see Fig. 9).
11. For certain block configurations within the route, specific routing patterns should be applied (see Fig. 10).

In order to use the heuristic approach outlined above, the planner needs maps of the waste generation area under study and information on the number and type of service units on each side of each street

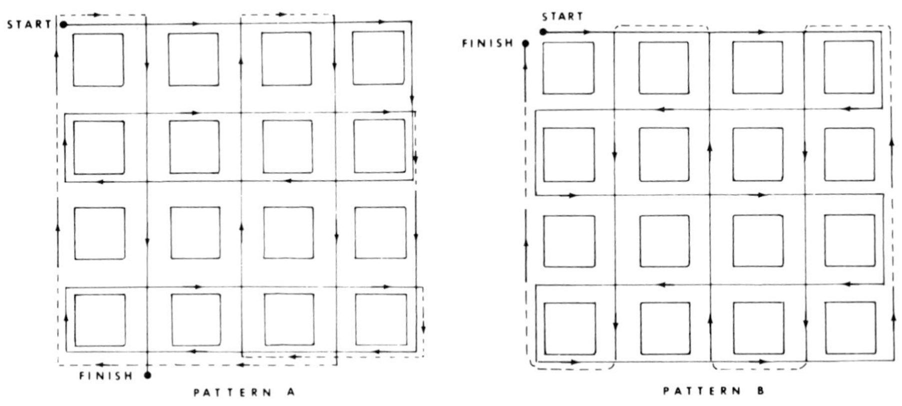

FIG. 9 Specific routing patterns for both-sides-of-the-street collection. From Ref. 22.

segment. He also must know which streets are one-way or dead end, the level of congestion in these streets, and the locations to which users are required to deliver wastes. The application of the rules-of-thumb listed above results in characteristic paths in given urban development

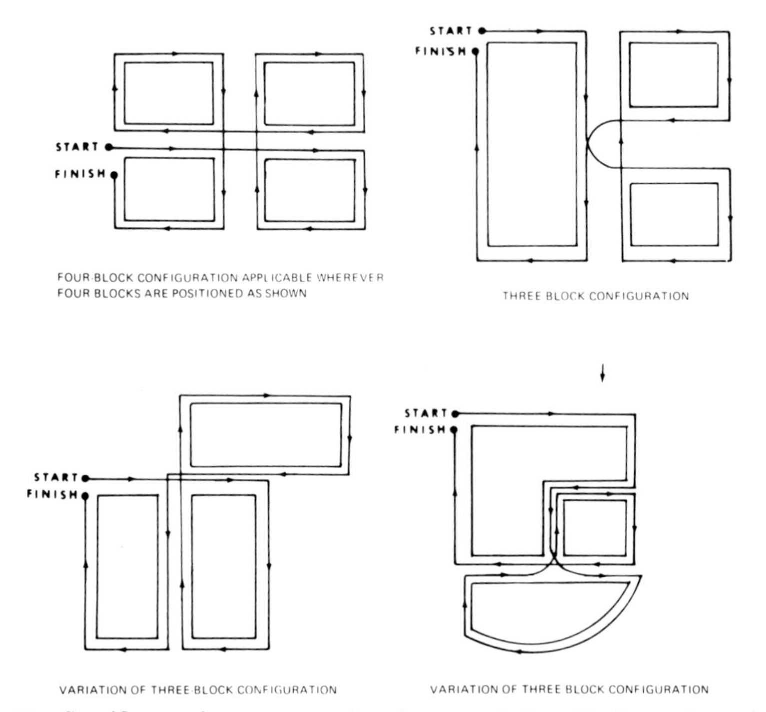

FIG. 10 Specific routing patterns for three- and four-block configurations. From Ref. 22.

patterns. Typical solutions for specific block and street configurations are shown in Figs. 6–10. Of course, other street configurations result in different patterns as a result of the application of the heuristic rules given above.

More sophisticated, but less practical, microrouting techniques are also available. These are based on the utilization of operations research models for obtaining the route that results in minimum mileage for a single vehicle or for a number of vehicles, so that all parts of a network are covered. The two basic variations of this problem and formulation are one that attempts to cover all nodes in a network, thus picking up waste from a set of fixed points, and another that attempts to cover all arcs in a network by assuring that each road is traveled at least once and that the waste is picked up along that route. The former is known as the "traveling salesman" problem, the latter as the "Chinese postman" problem. Since continuous route collection problems are the ones most commonly encountered in urban areas, the general formulation of the "Chinese postman" problem is given below:

$$\text{Minimize} \qquad \sum_{i=1}^{N} \sum_{j=1}^{N} C_{ij} X_{ij}$$

$$\text{Subject to} \quad \sum_{k=1}^{N} X_{ki} - \sum_{k=1}^{N} X_{ik} = 0 \qquad \text{for } i = 1, \ldots, N$$

$$X_{ij} + X_{ji} \geqslant 1 \qquad \text{for all arcs } (i, j) \epsilon A$$

$$X_{ji} \quad \text{and} \quad X_{ij} \geqslant 0 \qquad \text{(integers)}$$

where N is the number of nodes in the network; A is the set of all arcs in the network; X_{ij} is the number of times the arc from node i to node j is traversed; and C_{ij} is the length of the arc from node i to node j.

The objective function in this integer programming formulation minimizes the total travel distance required to cover each route in a road network at least once. This latter restriction is ensured by the constraints that the summation of X_{ij} and X_{ji} is equal to or larger than unity, and that each of them individually is zero or a positive integer. The first constraint equation is a node balance equation, which ensures that a vehicle arriving at a node must also depart from that same node. A number of computer algorithms have been developed for the solution of this and other routing problems. Reviews of the different approaches used and their corresponding solution algorithms can be found in Refs. 13 and 23–25.

III. EVALUATION METHODOLOGY

A. Purposes and Objectives

The ability to evaluate the productivity of solid waste systems is a necessary prerequisite for the proper planning of actions pertaining to those systems. In order for management to take appropriate actions pertaining to long- or short-range planning and resource allocation, it must have an adequate data base and an adequate mechanism for providing the necessary inputs for decision making on a continuous basis [32, 33]. Such mechanisms are currently not available since the output of a public service function is a multidimensional array of both quantitative and qualitative descriptors affecting different clusters of people at different times. The definition of these outputs is not an easy matter, nor is their assessment.

It is not uncommon to find two communities of equal size, only a few miles apart, allocating inordinately different amounts of resources per capita to solid waste collection. It has become a standard argument that private collection may be more efficient and less costly than public collection. In either case, it is not clear whether the more expensive undertaking provides an overall service that is better or worse than its "more efficient" counterpart. Neither is there a conceptual framework, a data base, or an adequate mechanism for providing the answer to such complex problems. Typically, the data available for decision making have been operational, such as the cost per ton collected or disposed. Such data, however, do not provide any insight into the extent to which a service is achieving its proposed goals. Only during the last few years have such questions been addressed. The concept of productivity is meaningless unless it is measured in terms of system *efficiency* and *effectiveness*.

Evaluation implies measurement and the ability to compare. The results of the evaluation process become, in turn, inputs into the planning process, so that a plan generates a condition warranting evaluation, which then provides feedback to guide future planning. In this manner, planning and evaluation can be seen as two poles of an iterative field of action, with each taking outputs from and providing inputs to the other. The task facing the solid waste engineer is to develop a systems plan that provides for relevant, yet practical, measurement of objectives. This interaction of planning, evaluation, objectives, and measurement is basic to comprehensive solid waste systems management.

B. The Evaluation Process

Much of the difficulty in the evaluation of solid waste systems stems from a lack of a conceptual framework and from the confusion in the use of terminology and definitions. The confusion, in turn, can be partially attributed to the fact that the direct output of a public system is rarely the final, desired, and intended purpose of the system. Issues are further complicated by the existence of a plethora of both desirable and undesirable unexpected consequences that are often sufficiently significant to warrant inclusion in the evaluation process. To minimize these problems, the evaluation process must be based upon a model, or models, that clearly differentiates between the types of system consequences, and that indicates the interrelationships between inputs and outputs. Such models should incorporate considerations of the efficiency, effectiveness, and ultimate social impacts of solid waste systems.

The initial consequences of a public system occur at the efficiency level involving the direct outputs of the production process measured in such terms as the number of tons collected, housing units served, or vehicle miles driven. The efficiency of the process can be defined as a ratio of outputs to the input resources, which results in such measures as tons per vehicle mile or tons per man day. These direct outputs, however, are not necessarily desired for their own sake, but rather as an intermediate input into the production of a more complex good, the quality of service. It is difficult to compare efficiency measures across geographic boundaries because of variations in local conditions affecting the level of efficiency in the utilization of resources. The key to the measurement of performance of a public system, therefore, is the effectiveness of the system.

Effectiveness is a compound overall measure of the different components that make up the quality of service. A system has a measurable level of effectiveness only when it is used and its effects are perceived by the community. Public systems generally affect both the users of the service and the community being served, but to different degrees. Since effectiveness is influenced by both expectations and subsequent perceived satisfaction, the degree of effectiveness of a solid waste plan varies from community to community and neighborhood to neighborhood. It is better, therefore, to concentrate on the process by which effectiveness measures can be developed in a given community than on the identification of values which have universal utility.

In addition to the outputs of solid waste systems at the efficiency and effectiveness levels, a third level of outputs is largely beyond the control of the planners and system managers. The consequences of a

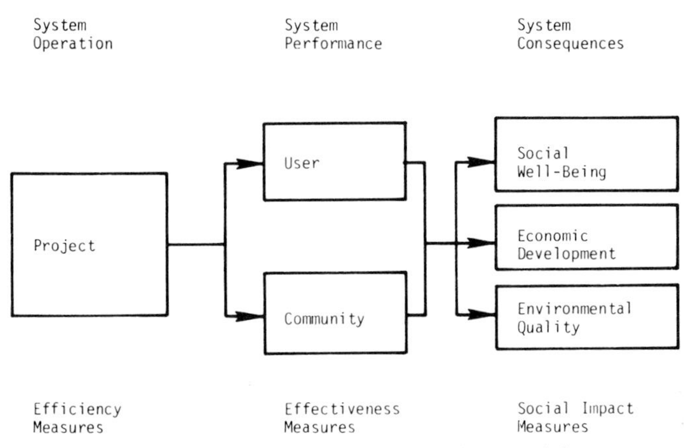

FIG. 11 Public systems evaluation model.

solid waste plan do not stop with the operation of a system. Because of the interrelationships between technology, social structures, public policies, and human values, solid waste systems have far-reaching effects upon other aspects of society, regardless whether the planner is aware of them or not. These impacts are classified into social well-being, economic development, and environmental quality. They are the secondary and indirect consequences of the system upon the community and region. In the broadest sense, however, these impacts encompass the ultimate purposes of all public works planning, that is, the improvement of the general welfare of society as a whole.

Figure 11 illustrates a generalized form of the above three-level model of system outputs. This Public Systems Evaluation Model has been found to be applicable for the assessment of a wide variety of public services, including waste water disposal and water supply [26], urban mass transit [27], statewide resource recovery efforts, and solid waste collection [28]. Since the model distinguishes different levels of outputs, it also can be used for the assessment of productivity. The efficiency level contains only physical inputs and outputs under the direct control of the system manager. These primary responses can be viewed as the basic "costs" of the system, and an efficient system will minimize these costs. The effectiveness measures also are primarily physical in nature, but they require interaction, or usage, by the public for the responses to occur. These secondary responses can be viewed as basic "benefits" of the system. It is the responsibility of the system manager to maximize these benefits. Since both the efficiency and effectiveness measures are defined in physical terms, they can be readily quantified and evaluated. This is true not only for solid waste systems, but for most other public services. Thus, the model allows a straightforward definition of efficiency and effectiveness that clearly separates these immediate and quantifiable aspects from the indirect and often unquantifiable impacts that occur in the areas of social well-being, economic development, and environmental quality.

Within the specific field of solid waste systems, the Public Systems Evaluation Model can be utilized in the following manner: The efficiency level comprises the operation of a solid wastes system and is determined by technical production functions under the control of the system manager. As indicated earlier, measurements of system operation are scaled in terms of physical units of weight, volume, distance, time, energy, etc. This level reflects the capabilities, or the technical potential, of the system and is not dependent upon specific project usage by the community. Project operation and technical outputs at the efficiency level are the primary concern of the design engineer.

At the effectiveness level, the efficiency outputs interact with the users and the general community. System performance, thus, is based upon the usage of system outputs by the public. For solid waste systems, typical areas of concern include street cleanliness, public cooperation, worker productivity, and policy implementation. Effectiveness measures often are stated in per capita terms, such as collection costs, waste generation, and system complaints per capita. In all cases, the measurement must reflect some degree of interaction between the solid waste system and the public. If a system is not utilized or underutilized by the public, measurements of this level should show a low degree of effectiveness. This interaction between system outputs and the users of the outputs is independent of the theoretical efficiency of the system, and it is conceivable that a system with a high efficiency rating could have a low effectiveness rating, and vice versa. The outputs that occur at the effectiveness level usually are the primary concern of physical planners and of engineers interested in overall system performance.

The social impact level includes the ultimate consequences of solid waste systems. Outputs at this level may involve changes in social well-being (public health, social opportunities, community attitudes), economic development (employment opportunities, new business formation), and environmental quality (open spaces, recreational facilities, community esthetics). There may be no discernible link between the initial solid waste system input and the ultimate social impacts. It would be erroneous, for example, to claim direct causality between an improved solid waste system and a subsequent increase in industrial employment. Furthermore, it could be difficult, if not impossible, to show even indirect linkages between system input and employment output. Too many factors interfere with a definite tracing of even indirect causality. Nevertheless, it is not unreasonable to contend that an improved solid waste system is one of several critical inputs contributing to the ultimate impact of increased employment. Because the major problems currently hampering evaluation at the social impact level are a poor conceptual understanding of impact linkages and a lack of relevant measurement variables, further research is urgently needed. Overall, the project-related consequences that occur at the social impact level are primarily the concern of policy makers.

Each of the three levels in the Public Systems Evaluation Model can be rated in terms of general dimensions relevant to solid waste systems and the needs of the systems planner. These dimensions include the time effect necessary for a given system output to occur, the degree of causality linking the system input to output of interest, the present state of knowledge concerning the input–output relationships, and the ease of measuring the output in the field. Over the three levels, the effects of these evaluation dimensions can be characterized as follows [26]:

	Efficiency *Effectiveness* *Impact*
Time effect	(increasing) ⟶
Degree of causality	(decreasing) ⟶
State of knowledge	(decreasing) ⟶
Ease of measurement	(decreasing) ⟶

At present, the state of the art in the evaluation of public systems in general, and solid waste systems in particular, suffers from a lack of consensus regarding evaluation concepts, measurement techniques, and overall methodology. There is great need for further research into the cause and effect relationships between inputs and outputs and for better methods of identifying and measuring specific evaluation variables. Until such time as evaluation is seen as an integral part of the overall planning process, the development of solid waste systems will be hampered by an inability to take full advantage of past experiences and actions.

C. Application

Figure 12 demonstrates the application of the Public Systems Evaluation Model to solid waste collection efforts. The measurement and assessment of the effectiveness of such effort is based on a three dimensional

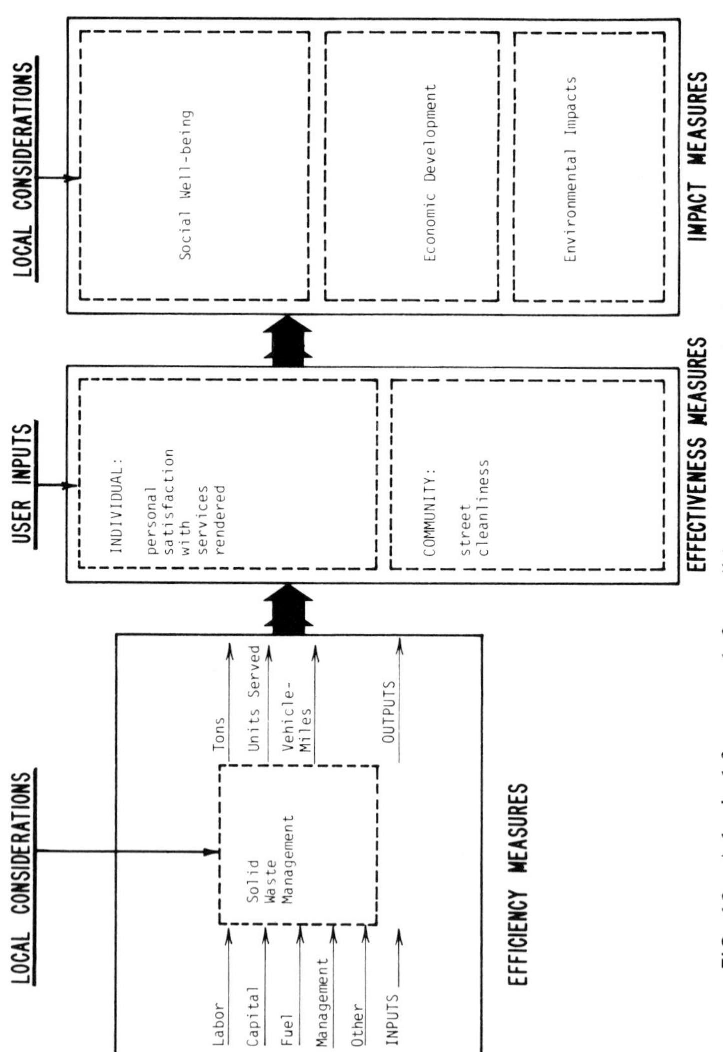

FIG. 12 A logical framework for solid waste systems evaluation. From Ref. 28.

analysis involving (a) users' satisfaction, (b) the effects of the service on the community, and (c) societal values and preferences. The users' perceptions of the level of satisfaction with the quality of service is measured by the direct questioning of a representative sample of the user population. Questions measuring the satisfaction of the citizens are analyzed by giving the various responses point values ranging from 0 to some positive value that is dependent on the importance of the question. A "Users' Satisfaction Index" (USI) is thus defined as

$$\text{USI} = \sum_{i}^{q} R_i$$

where R is the value of question i and q is the total number of questions. The USI assumes maximum value (100 in this case) when all of the responses are given maximum values, indicating complete satisfaction with all the components that constitute the overall quality of the service. The weights given to different items can be varied to reflect community preferences. An example of a questionnaire designed to measure citizen satisfaction with municipal services is given in Ref. 28.

The immediate objectives of an effective solid waste collection system are to provide a clean environment that is esthetically pleasing and free from health hazards. Attention thus should be focused on defining, measuring, and weighting the different variables that make up the community effects component of the effectiveness model. Overall street cleanliness can be measured by a visual inspection program conducted by trained inspectors. The visual inspection scheme described here is similar to the one used in Washington, D.C., and other cities. It is based on training an inspector to rate block faces on a scale of 1 to 4, with the lower end of the scale representing a clean street and the higher end representing a heavily littered street. As the inspector drives down a street, he or she records the visual cleanliness on a cassette tape. The recorded data are later transcribed and analyzed. It has been found that a trained inspector can cover as many as 250 block faces in an 8-h working day. Figure 13 provides an example of a rating scheme that might be used for this purpose. A similar scheme is described in Ref. 29. A neighborhood Community Effects Index (CEI) can be calculated by combining the aggregate effects of all block faces in the neighborhood and normalizing the values obtained so that an impeccably clean neighborhood would have a rating of 100:

$$\text{CEI} = 100 - \frac{100}{4}\left(\frac{\sum_1^b S_i}{n} - 1\right)$$

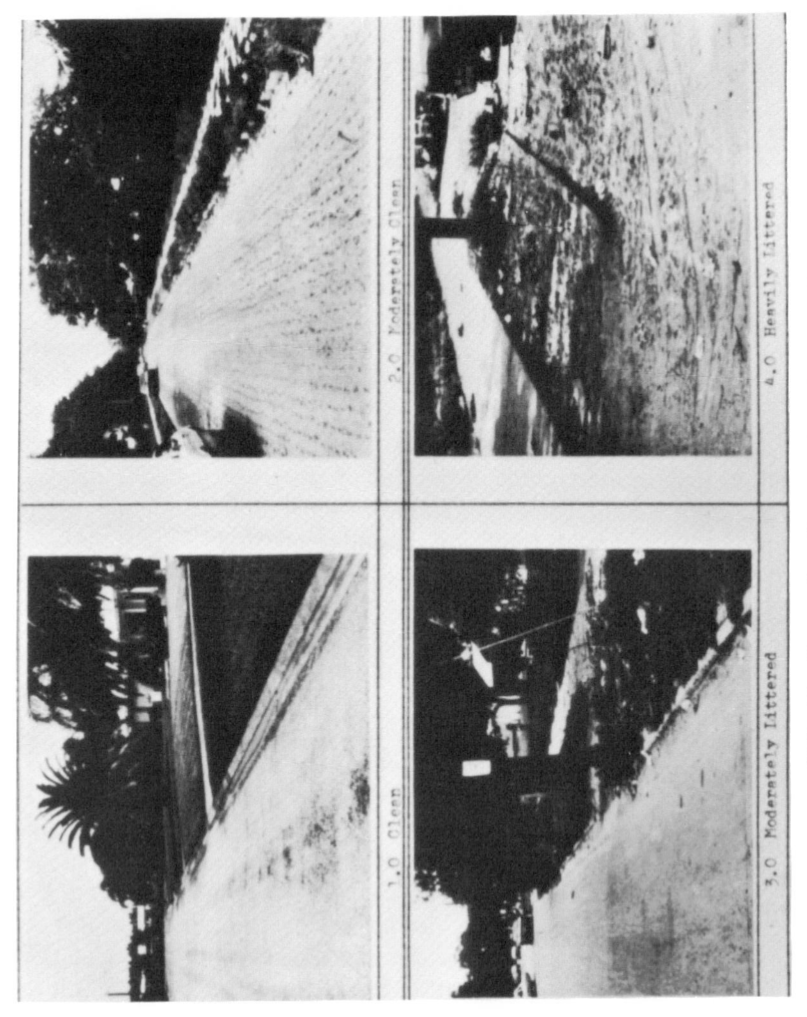

FIG. 13 Street cleanliness ratings. From Ref. 28.

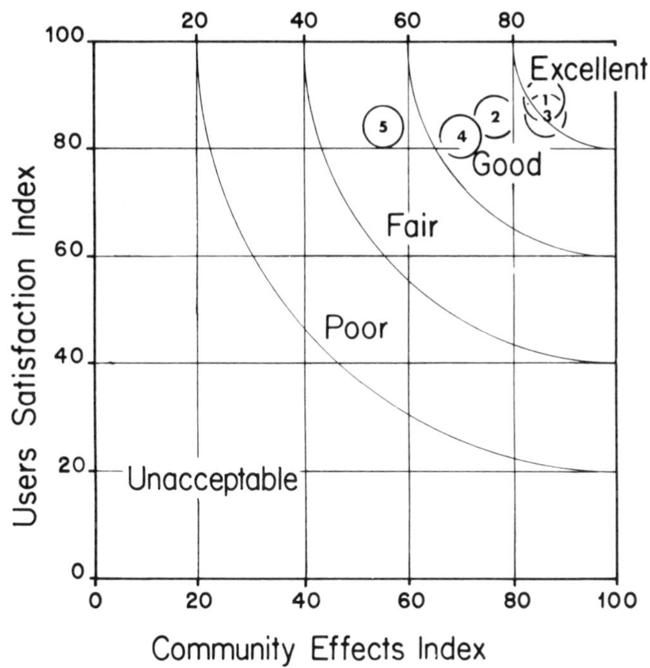

FIG. 14 The "level of service" concept. From Ref. 28.

where S_i is the cleanliness rating of the ith blockface and b is the number of block faces in the neighborhood.

Societal values are possibly the most difficult to evaluate quantitatively. Their use is suggested in the proposed model in order to provide guidelines for the development of thresholds of societal acceptability and for the formation of different combinations of the user satisfaction and community effect indexes. Societal values reflect such factors as a community's resource allocation decisions, its budgetary appropriations, priorities, and citizen participation activities. This area has not yet been fully explored, and further research is needed to translate conceptual suggestions into workable processes. Societal values clearly are interrelated and overlap in many aspects. It should be emphasized, however, that the above approach attempts independently to evaluate the performance of a system, as perceived by the users themselves, within a general framework of accepted societal values. The model basically determines the effectiveness of a public service relative to three independent variables and combines these factors into a new measure of "level of service," which is intended to describe the relative performance of alternative proposals, solutions, and management strategies. Figure

14 is a graphic representation of the level-of-service concept. Each of the numbered circles on the graph represents the combination of user satisfaction and community effects indexes prevailing in a given neighborhood. Neighborhoods (1) and (3) are thus characterized to have a combination yielding a level of service perceived to be "excellent," whereas neighborhood (5) has a combination of indexes warranting a "good" level of service. The partitioning of the two-dimensional user–community space is dictated by the priorities, values, and perceptions of a given community. It also reflects the level of physical improvement and public investment needed in each subdivision of the city. When properly applied, the level-of-service concept can be used by city management to help determine community development priorities.

The National Commission on Productivity, in a recent study of the productivity of solid waste collection [30], has described the level of service as

> . . . a matter of community choice, with citizens of any jurisdiction having the option of deciding what they want and how much they are willing to pay for it. Sometimes citizens are adamant in their preference for one level of service over another. . . In other cases, citizens have simply grown accustomed to a given level of service, never stopping to question whether it could be changed, how much a change would cost, or how much would be saved by making the change. [p. 5.]

In order to use the above model, data must be obtained for the calculation of both the level of user satisfaction and the level of community effects. These two levels then are plotted two-dimensionally and an appropriate level of service for the area in question at a given time is identified. Once this is accomplished for one town or area, the same procedure can be used for establishing the level of service in other areas of the city and in other cities, or for evaluating the change in the level of service that might result from community or governmental actions. The main value of the model lies in its capability for providing comparative, quantitative evaluations of the effectiveness of community services, which, in turn, provide useful inputs into the decision-making processes within the community.

REFERENCES

1. R. O. Toftner, *Developing a Local and Regional Solid Waste Management Plan*, SW-101ts.1, USEPA,† 1973.

† U.S. Environmental Protection Agency, Office of Solid Waste Management Programs, Washington, USEPA-OSWMP.

2. T. H. Roberts, "The Planning Process," in *Planning for Solid Waste Management*, USEPA, 1971.
3. R. de Neufville and J. H. Stafford, *Systems Analysis for Engineers and Managers*, McGraw-Hill, New York, 1971.
4. E. R. Zausner, *Financing Solid Waste Management in Small Communities*, SW-57ts, USEPA-OSWMP,† 1971.
5. "Financing," in *Resource Recovery Plant Implementation: Guides for Municipal Officials*, SW-157.4, USEPA-OSWMP, 1975.
6. R. A. Colonna and C. McLaren, *Decision-Makers Guide in Solid Waste Management*, SW-127, USEPA, 1974.
7. *The Solid Waste Disposal Act*, Title II of Public Law 89-272, October 20, 1975, as amended by Public Law 91-512, October 26, 1970, as amended by Public Law 93–14, April 9, 1973, and as amended by Public Law 93–611, January 2, 1975.
8. "Solid Waste Recovery and Management," *Environmental Quality Bond Act*, Title 5, Chap. 659, New York State Laws of 1972.
9. "Waste Disposal Bonds," *Referendum Bill 26*, Chap. 127, Washington State Laws of 1972.
10. *Second Report to Congress: Resource Recovery and Source Reduction*, SW-122, USEPA-OSWMP, 1974.
11. *Third Report to Congress: Resource Recovery and Waste Reduction*, SW-161, USEPA-OSWMP, 1975.
12. B. P. Helms and R. Clark, "Locational Models of Solid Waste Management," *Am. Soc. Civil Eng. J. Urban Plan. Develop. Div.* **97**, No. UP1, 1–14 (April 1971).
13. J. F. Hudson, D. S. Grossman, and D. H. Marks, *Analysis Models for Solid Waste Collections*, Department of Civil Engineering, MIT, Cambridge, Mass., 1973.
14. D. H. Marks and J. C. Liebman, "Locational Models: Solid Waste Collection Example," *Am. Soc. Civil Eng. J. Urban Plan. Develop. Div.* **97**, No. UP1, 15–30 (April 1971).
15. *Solid Waste Management Planning: Snohomish County, Washington*, Systems Control, Inc., Pala Alto, Calif., 1971.
16. M. M. Truitt, J. C. Liebman, and C. W. Kruse, *Mathematical Modeling of Solid Waste Collection Policies*, Public Health Service Publication (PHS) No. 2030, USDHEW,† 1970.
17. D. Marks and J. Liebman, *Mathematical Analysis of Solid Waste Collection*, PHS No. 2104, USDHEW, 1970.
18. C. G. Golueke and P. H. McGauhey, *Comprehensive Studies of Solid Waste Management*, PHS No. 2039, USDHEW, 1970.
19. *WRAP: A Model for Regional Solid Waste Management Planning* (Draft of User's Manual), USEPA, Washington, 1976.
20. *Solid Waste Collection Practice*, American Public Works Association, Chicago, 1975.
21. K. A. Shuster, *A Five-Stage Improvement Process for Solid Waste Collection Systems*, SW-131, USEPA, 1974.
22. D. A. Schur and K. A. Shuster, *Heuristic Routing for Solid Waste Collection Vehicles*, SW-113, USEPA, 1974.

† U.S. Department of Health, Education, and Welfare, Washington.

23. D. H. Marks, "Modeling in Solid Waste Management: A State-of-the-Art Review," in *Proceedings of the EPA Conference on Environmental Modeling and Simulation* (Wayne Ott, ed.), USEPA, 1976.
24. J. C. Liebman, "Model in Solid Waste Management," in *A Guide to Models in Governmental Planning and Operations*, USEPA, 1974.
25. D. H. Marks, "Routing for Public Service Vehicles," *Am. Soc. Civil Eng. J. Urban Plan. Develop. Div.* **97**, No. UP2, 165–178 (December 1971).
26. D. Warner and J. S. Dajani, *The Impact of Water and Sewer Development in Rural America*, Heath, Lexington, Mass., 1975.
27. G. Gilbert and J. S. Dajani, *Measuring the Performance of Transit Service*, Institute of Policy Sciences and Public Affairs, Duke University, Durham, N.C., 1975.
28. J. S. Dajani, P. A. Vesilind, and G. Hartman, *Measuring the Effectiveness of Solid Waste Collection*, Department of Civil Engineering and the Institute of Policy Sciences and Public Affairs, Duke University, Durham, N.C., March, 1976.
29. L. H. Blair and A. I. Schwartz, *How Clean Is Our City?*, The Urban Institute, Washington, D.C., 1972.
30. *Report of the Solid Waste Management Advisory Group on Opportunities for Improving Productivity of Solid Waste Collection*, National Council on Productivity, Washington, 1973.
31. J. H. Flandreau, "Resources Recovery: State Level Planning," *Resource Recov. Energ. Rev.* **4** (1), 16 (1977).
32. D. R. Price, "Considerations in Planning the Management of Hazardous Waste Materials," *Pollut. Eng.* **9** (2), 31 (1977).
33. Texas Water Quality Board, "Industrial Solid (Hazardous) Waste Guidelines," *Waste Age* **8** (4), 10 (1977).

INDEX

D

E

F

G

H

M

N

O

P

R